12th International Conference on Intelligent Technologies in Logistics and Mechatronics Systems (ITELMS'2018)

AA004881

Panavezys, Lithuania
26-27 April 2018

Editors:

László T. Kóczy
Daiva Žostautienė
Olga Strikulienė
Elvyra Zacharovienė

ISBN: 978-1-7138-3363-5

Printed from e-media with permission by:

Curran Associates, Inc.
57 Morehouse Lane
Red Hook, NY 12571

Some format issues inherent in the e-media version may also appear in this print version.

Copyright© (2018) by Editografica s.r.l.
All rights reserved.

Printed with permission by Curran Associates, Inc. (2021)

For permission requests, please contact Kaunas University of Technologies
at the address below.

Kaunas University of Technology
K. Donelaičio St. 73
44249 Kaunas, Lithuania

Phone: +370 (37) 300 000
Fax: +370 (37) 324 144
Email: ktu@ktu.it

https://itms.ktu.edu

Additional copies of this publication are available from:

Curran Associates, Inc.
57 Morehouse Lane
Red Hook, NY 12571 USA
Phone: 845-758-0400
Fax: 845-758-2633
Email: curran@proceedings.com
Web: www.proceedings.com

12th International Conference on Intelligent Technologies in Logistics and Mechatronics Systems

ITELMS'2018

April 26-27, 2018
Panevėžys, Lithuania

Editors
László T. Kóczy
Daiva Žostautienė
Olga Strikulienė
Elvyra Zacharovienė

EDlearning

International Scientific Committee

- **Lütfi ATAY** (Çanakale Onsekiz Mart University, TURKEY)
- **Paulo Tavares de CASTRO** (Porto University, PORTUGAL)
- **Krishna DHIR** (University of Hawaii at Hilo, USA)
- **Jan FURCH** (Brno University of Defence, CZECH REPUBLIC)
- **Anda GHEORGHIU** (National University of Political and Administrative Studies & Hyperion University, ROMANIA)
- **Ausma GOLUBEVA** (University of Latvia, LATVIA)
- **Simona GRAMA** (Groupe ESC Troyes, FRANCE)
- **Volodymyr HUTSAYLYUK** (Military University of Technology, POLAND)
- **Isak KARABEGOVIC** (BOSNIA AND HERZEGOVINA)
- **Marios KASINOPOULOS** (University of Central Lancashire, CYPRUS)
- **Askin KESER** (Uludag University, TURKEY)
- **Hilal Yildirir KESER** (Uludag University, TURKEY)
- **László T. KÓCZY** (**Chair**, Széchenyi István University and Budapest University of Technology and Economics, HUNGARY)
- **Demetris KOURSAROS** (Cyprus University of Technology, Cyprus)
- **Piotr KULCZYCKI** (AGH University of Science and Technology, POLAND)
- **Nida KVEDARAITĖ** (Kaunas University of Technology, LITHUANIA)
- **Anatoly LEVCHENKOV** (Technical University, LATVIA)
- **Liudvika LEISYTĖ** (Technical University of Dortmund, GERMANY)
- **Diana LIPINSKIENĖ** (Kaunas University of Technology, LITHUANIA)
- **Zygmunt MIERCZYK** (Military University of Technology, POLAND)
- **Vytautas OSTAŠEVIČIUS** (Kaunas University of Technology, LITHUANIA)
- **Juan Jesus PEREZ** (Technical University of Catalonia, SPAIN)
- **Claudiu POZNA** (Transilvania University of Brasov, ROMANIA)
- **Aušra REPEČKIENĖ** (Kaunas University of Technology, LITHUANIA)
- **Leonīds RIBICKIS** (Riga Technical University, LATVIA)
- **Marek SITARZ** (Silesian Technical University, TRANSMEC, POLAND)
- **Aleksandr SLADKOWSKY** (Silesian Technical University, POLAND)
- **Lucjan SNIEZEK** (Military University of Technology, POLAND)
- **Daniel SPRECIC** (BOSNIA AND HERZEGOVINA)
- **Brigita STANIKŪNIENĖ** (Kaunas University of Technology, LITHUANIA)
- **Dalia SUSNIENĖ** (**Co-chair**, Kaunas University of Technology, LITHUANIA)
- **Arūnas TAUTKUS** (Kaunas University of Technology, LITHUANIA)
- **Dainius VAIČIULIS** (Kaunas University of Technology, LITHUANIA)
- **Dmitri VINNIKOV** (Tallinn University of Technology, ESTONIA)
- **Darius VIRŽONIS** (Kaunas University of Technology, LITHUANIA)
- **Katarzyna WASOWSKA** (Siedlce University of Natural Sciences and Humanities, POLAND)
- **Violetta WEREDA** (Co-chair, Military University of Technology, WARSAW)
- **Marzena WOJCEK-AUGUSTYNIAK** (University of Natural Sciences and Humanities, POLAND)
- **Daiva ŽOSTAUTIENĖ** (**Co-chair**, Kaunas University of Technology, LITHUANIA)

All papers were reviewed.

© Copyright 2018 by EDITOGRAFICA s.r.l.

www.edlearning.it • congressi@editografica.com

All rights reserved. No part of this publication may be reproduced, stored in a retrieval system, or transmitted, in any form, or by any means, electronic, mechanical, photocopying, recording or otherwise, without the prior permission, in writing, from the publisher.

Printed in September 2018 by Editografica • Bologna (Italy)

Foreword

The 12th International Conference "Intelligent Technologies in Logistics and Mechatronics Systems" ITELMS'2018 was organized as part of the 2nd Forum on Innovative Technologies and Management for Sustainability that hosted two conferences: Changes in Social and Business Environment CISABE'2018 and ITELMS'2018. The main idea of this joint event was to stimulate interdisciplinary approaches on sustainability issues since the most advanced, innovative solutions are created by close cooperation between technological and social sciences. Such is definitely Logistics that has a clearly engineering aspect but at the same time it is closely connected with business and management, especially marketing.

Sustainable development focuses on balancing the fine line between competing needs - the need to always move forward technologically and economically, and the need to protect the environment where we live. However, sustainability is not only about the environment, it is also about human health and about the health of the society. The lead conference topic encouraged researchers to offer deeper insights on the issue, since in the context of the industrial revolution 4.0 it is clearly understood that technological solutions without interwoven social ones would never work.

The conference participants were encouraged to discuss issues about how technological sciences can help overcome challenges, such as the ones identified in the new Horizon 2020 Global Research Program; such as safe energy and transport, the effects of climate change and the conservation of natural resources, the demographic change, and the development of innovative and safe societies. The conference provided a platform for new forms of cooperation, for the identification of emerging obstacles, and for the provision of solutions to overcome them.

The topics of the 12th International Conference "Intelligent Technologies in Logistics and Mechatronics Systems" ITELMS'2018 included, among others:

- Intelligent Logistics Systems;
- Multicriteria Decision Making;
- Composites in Infrastructures;
- Intelligent Mechatronics Systems;
- Sensors and sensing phenomena;
- Mechanisms of Transport;
- Transport Systems;
- Modern Building Technologies;

- Military technologies;
- Production Engineering.

This conference emerged as the result of collective efforts and here, we take the opportunity to express our recognition of the efforts and hard work of all the people who have made ITELMS' 18 possible, who have done the hard work of preparation and organisation. Through these lines of welcome we would like to especially pay tribute to the Scientific Committee which has assured the quality of the pool of accepted papers, and to the members of the Organising Committee for their strong motivation and volunteer work, and eventually, to all the people who have directly or indirectly influenced the smooth progress towards the conference. Finally, we would like to express our gratitude to all the authors for contributing their papers to this volume.

László T. KÓCZY
Daiva ŽOSTAUTIENĖ
Olga STRIKULIENĖ
Elvyra ZACHAROVIENĖ

Contents

ITMS'2018

panevėžys faculty of technologies and business

The 12th International Scientific Conference Intelligent Technologies in Logistics and Mechatronics Systems (ITELMS'2018), 26–27 April 2018, Panevėžys, Lithuania

Synchronization of Heat Pump Operation with Positive Daily Gains of Ambient Air Temperature

Vytautas Adomavičius[a], Jonas Valickas[b*], Gytis Petrauskas[a]

[a]*Kaunas University of Technology, Studentų 48–205, LT–51367 Kaunas, Lithuania*
[b]*Kaunas University of Technology, Nemuno g. 33, LT–37164 Panevėžys, Lithuania*

Abstract

Goal of this paper is to present concept of air source heat pump control dependently on the predictable daily gains of ambient air temperature in order to achieve more efficient operation of the pump and savings of electric energy consumed by the pump. Therefore variations of ambient air were measured and recorded in the site of our university campus during the space heating season, and regularities of the variations were analysed. It was determined that differences of air temperature measured in daytime and night, as the rule, are positive and usually can make up from 2–3 °C to 15–16 °C. This fact allows presuming that that it would be useful to change currently practiced concept of heat pump control. The pump control system should be set on the positive daily gains of air temperature to produce more heat energy for the less electric energy consumed. Heat energy produced by the pump should be stored in hot water tank. System of heating premises have to operate independently from the system mentioned above. Efficiency of the proposed heat pump control system can be significantly enhanced by means of solar collectors for air heating specially installed in the building for this purpose. Theoretical concept given in this paper is intended to check experimentally.

© 2018 V. Adomavičius, J. Valickas, G. Petrauskas
Peer-review under responsibility of the Kaunas University of Technology, Panevėžys Faculty of Technologies and Business

Keywords: renewable energy, heat energy, air source heat pump, energetically self-sufficient buildings, energy efficiency

* Corresponding author. Tel.: +370 656 24494
E-mail address: jonas.valickas@ktu.lt

© 2018 V. Adomavičius, J. Valickas, G. Petrauskas
Peer-review under responsibility of the Kaunas University of Technology, Panevėžys Faculty of Technologies and Business

1. Introduction

Pollution of environment in Earth planet by all sectors of global economy is rapidly coming to its natural and very dangerous limits. Trespassing of these limits will induce more and more strong natural disasters and anomalies. Scientists of the world already published and keep publishing many scientific facts and proofs confirming this trend. Concern of leading scientists regarding this trend is expressed in their paper *World Scientists' Warning to Humanity: A Second Notice*, which was published in December 2017 in journal *BioScience* [1]. The paper is already supported and signed by more than 20 000 scientist-signatories from around 200 countries. Mostly unfair and not educated well enough people are denying climate change. Some of them, as businessmen working in areas of fossil fuels, are doing this because of the profit reasons. However, as it was proved many cases, majority of population in many countries support measures taken to reduce environmental pollution and to stop climate change. It is evident for the majority of older people without any researches and substantiations because they can see what kind of climate is now and remember what climate was before decades.

Quantification of the climate change process carried out by scientists evaluates these phenomena much more exactly. Harmful impact of fossil fuels combustion in transport vehicles, power plants, industry, boiler houses, buildings, agricultural machinery on atmosphere, forests, entire biosphere, climate change, health of people, and rates of mass extinction of fauna and flora species is proved and disclosed in many thousands of publications including [2–5]. The information presented in these scientific publications is helpful for prediction of climate change further development, its possible threats and consequences, and for working out a strategy how to stop the dangerous climate change. This information also induces motivation for many people to take adequate measures in order to solve this tremendous problem as soon as possible.

According to the presently prevailing opinion, shifting from fossil fuels to the renewable energy sources now is the most appropriate way to mitigate processes of environmental pollution and climate change. Expansion of renewables is already happening in many sectors of economy. Revolutionary changes are especially significant in power production and transport sectors. Revolution in the transport sector comes mostly through electricity, which is rapidly going green. Buildings in many countries are going green too. Rapidly increasing number of buildings of various types already have installed autonomous RES-based heat and power plants, mostly solar collectors, heat pumps and rooftop solar power plants and small-scale wind turbines in windy regions. This trend must be supported and strengthened.

Heat pumps are used in buildings for space heating and domestic hot water production. There are some proposals in scientific publications to enhance efficiency of air source heat pumps operation at low ambient temperatures in winter by assisting the heat pump by means of solar collectors for water heating. This concept is described in a number of publications, including [6–8]. According to this concept, heat energy of higher potential produced by the solar collector is brought to the plate heat exchanger. They both makes up the water cycle loop. After this, the heat is passed from the heat exchanger into the refrigerant cycle loop of the heat pump what allows increasing coefficient of performance (COP) of the pump. The proposed system has better efficiency of operation, however, in this case additional equipment have to be included in the system – the solar collector for water heating and the plate heat exchanger. But additional equipment requires more costs.

In this paper, enhancement of air source heat pump performance efficiency is proposed to realize by means of the synchronization of heat pump operation with positive daily gains of ambient air temperature. Apart from the control system, significant supplementation of the air source heat pump system is not necessary in this case.

Nomenclature

E_{di} the daily heat energy needs per month "i", kWh;
E_{dj} the daily heat energy output by the heat pump per day "j", kWh;
E_{nj} the electric energy consumed by the heat pump per day "j", kWh;
e_a the relative annual thermal energy needs for space heating of building, kWh/m^2;
k_i the relative annual thermal energy needs per month "i", %;
k_{opj} the coefficient of performance of the heat pump per day "j";
n_i the number of days per month "i";

P_e	the rated electric power of heat pump;
S	the heated area of building, m^2;
Δt_j	the available daily time of the heat pump operation for the day "j" at higher temperature, h;
Δt_{sj}	the sufficient daily time of the heat pump operation for the day "j" at higher temperature, h;
τ_r	the reference value of the ambient air temperature allowing to start heat pump, °C;
τ_{av}	the average value of the ambient air temperature of heat pump operation according this concept, °C.

2. Analysis of ambient air temperature fluctuations

Variations of outdoor ambient air temperature during the heating season 2017/2018 were measured in campus of Kaunas University of Technology (KTU). Curve of ambient air temperature alteration covering almost entire period of heating season in the campus was recorded and presented in Figure 1. Analysis of ambient air temperature instability over days and nights shows more or less significant gains of the air temperature during the days in the vast majority of cases. The bright daytime has a lower air temperature than during the night only in extremely rare cases.

Fig. 1. Curve of ambient air temperature alteration in campus of Kaunas university of Technology from November 5, 2017 till March 8, 2018

Investigation of the ambient air temperature alteration curve allowed to formulate the following conclusions, which could be important for consideration of heat pumps operation only synchronously with positive daily gains of ambient air temperature:

– differences of air temperature measured in the night and in the next daytime usually are positive and can make up from +(2–3) °C to +(15–16) °C;

- number of considerable differences (about +6 °C and more) between air temperatures in the night and in the next daytime makes up about 50 %;
- duration of the coldest period in Kaunas city in 2017/2018 heating season, when air temperature at nights reached about -(5–15) °C, made up about two weeks (very similar air temperatures in winters were during the last four years).

3. Concept of the method

More efficient operation of air source heat pumps at higher ambient air temperatures is well known and it would be useful to exploit this peculiarity as much as possible. It can be evidently explained by means of the curve of air source heat pump coefficient of performance versus temperature of ambient air (Fig. 2). However, concept of heat pump control, when system of space heating control is set to keep the desirable temperature in the rooms of building, predominantly is practiced up to present days. We propose to set the pump control system on the positive daily gains of air temperature in order to produce more heat energy for the less electric energy consumed. The produced heat energy can be stored in well insulated tank for hot water and used as the domestic hot water and for the heating of premises. System of heating premises have to operate by the exploiting heat energy stored in the hot water tank independently from the system of heat energy generation mentioned above.

The most problematic period for space heating in buildings in high latitudes is winter months when required amounts of heat energy are the highest. The relative heat energy demand ratio shows what monthly percentage of the annual heat energy consumed per heating season can be expected in average. The distribution of this ratio for local region is presented in Figure 3. Information given in Figure 2 and 3 is used in further simplified quantification of air source heat pump operation possibility based on the positive gains of ambient air temperature.

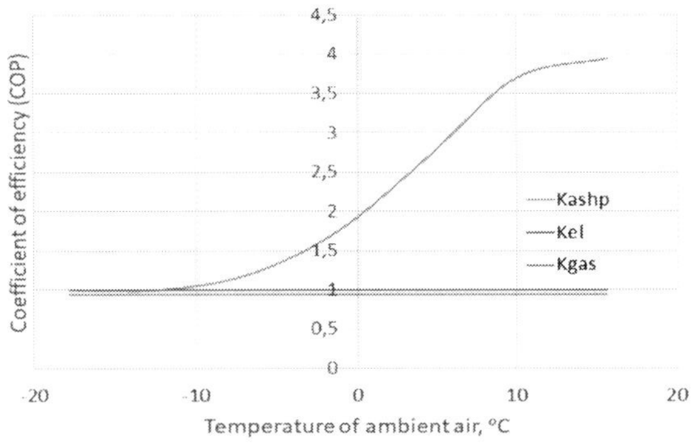

Fig. 2. Comparison of efficiency of heat generation technologies based on the air source heat pump, electrical and natural gas heaters

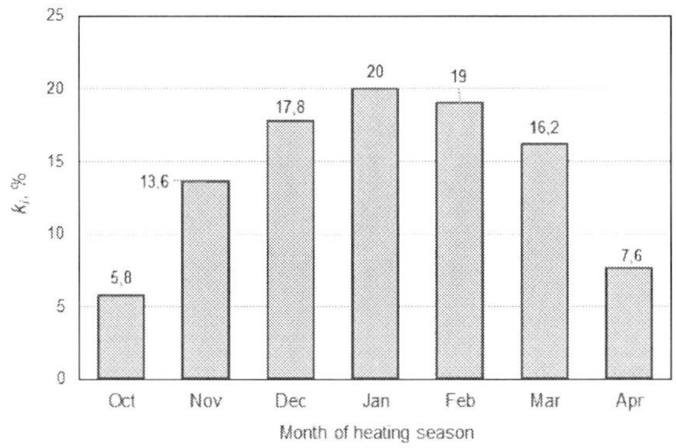

Fig. 3. Perennial average distribution of the relative heat energy demand ratio for months per heating season in Kaunas region

Amount of heat energy per day necessary for space heating in any building approximately can be calculated on basis of the relative annual thermal needs per square meter of the heated area:

$$E_{di} = \frac{e_a \cdot k_i \cdot S}{100 \cdot n_i}; \qquad (1)$$

where E_{di} is the daily heat energy needs for space heating per month "i", kWh;

e_a – the relative annual thermal energy needs for space heating, kWh/m^2;

k_i – the relative heat energy demand ratio for month "i", %;

S – the building's heated area, m^2;

n_i – the number of days per month "i".

Daily heat energy output by the heat pump during the ambient air temperature increase period can be approximately calculated by means of the following formula:

$$E_{\mathrm{dp}} \approx P_{\mathrm{e}} \cdot k_{\mathrm{op}}(\tau) \cdot \Delta t_j; \tag{2}$$

where E_{dp} is the daily heat energy output by heat pump, kWh;

P_{e} – the rated electric power of heat pump, kW;

k_{op} – the coefficient of performance of heat pump;

τ – the variable temperature of the ambient air, ℃;

Δt_j – the available daily time of heat pump operation for the day "j" at higher temperature, h.

Justification of air source heat pump operation possibility when positive gains of ambient air temperature are exploited was carried out by means of quantification of the case described by experimental curve of ambient air temperature alteration in KTU campus. The curve was recorded in March 9–14, 2018 and presented below in Figure 4. Intervals of time Δt_j available for generation of heat energy in this figure are indicated as Δt_{09}–Δt_{14}.

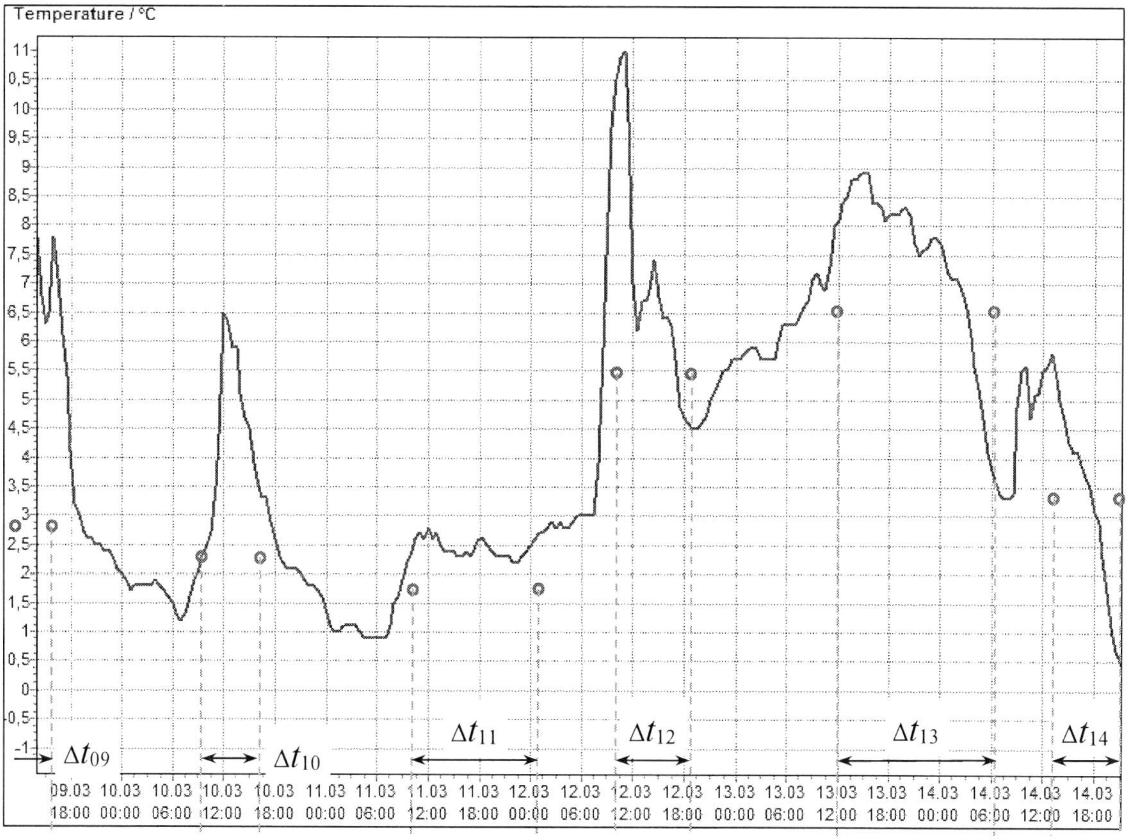

Fig. 4. Curve of ambient air temperature alteration in campus of Kaunas university of Technology from March 9 till 14, 2018

Results of approximate calculations performed in order to justify air source heat pump operation on the positive gains of ambient air temperature are presented in Table 1. The relative annual thermal energy needs for space heating in this table correspond energy class of building A++. This is preliminary information which can be helpful for taking decision regarding expediency of this concept practical implementation in a pilot project. Operation of the proposed space heating system modification can be consider as acceptable at the following conditions:

- if the daily heat energy output by the air source heat pump exceeds the daily heat energy needs for month "i" ($E_{dp} > E_{di}$);
- if the consumption of electricity for space heating decreases, when control system of the heat pump is set on the exploitation of positive daily gains of air temperature.

Table 1. Quantification of air source heat pump operation possibility on the positive gains of ambient air temperature

Initial data:							
March 9–14, 2018. e_a = 15 kWh/m^2 (energy class A++); k_i = 16.2 %; n_i = 31; P_e = 3 kW; S = 200 m^2 (Fig. 2; Fig. 3, Fig. 4)							
Parameter	Day of month	09	10	11	12	13	14
Δt_{dj}	h	4.5	6.5	14.2	8.0	17.2	7.5
τ_r	°C	3.5	3.0	2.4	6.0	7.0	4.0
τ_{av}	°C	6.5	4.5	2.4	8.2	7.7	4.9
E_{di}	kWh	16.53	16.53	16.53	16.53	16.53	16.53
k_{opj}	–	3.1	2.7	2.35	3.43	3.3	2.8
E_{dj}	kWh	43.20	53.82	100.10	82.32	170.28	63.00
Δt_{sj}	h	1.78	2.04	2.34	1.61	1.67	1.97
E_{nj}	kWh	5.34	16.12	7.02	4.83	5.01	5.91

Figure 5 summarizes the main findings of our investigation. As it is shown in the figure, the estimated daily heat energy output by the air source heat pump significantly exceeds the daily average heat energy needs for March for the building under our analysis (the red line 16.53 kWh). Consumption of electricity for space heating are low (the blue bars) because heat pump is operating with significantly higher COP, which is available at higher daily temperatures of ambient air.

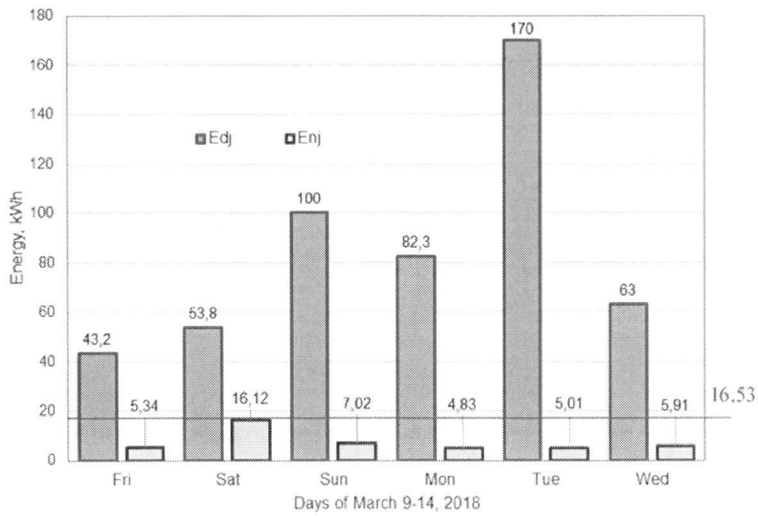

Fig. 5. Estimated daily heat energy output by the air source heat pump (E_{dj}), daily average heat energy needs for month "i" (16,53 kWh) and electricity consumption by the air source heat pump (E_{nj}) for space heating

Promising results reflected in the Figure 5 allow us proceeding this research in higher level. The concept, assumptions and quantifications described in this paper would be expedient to check by means of natural experiments, which are more reliable and accurate.

The concept described in this paper is not the only way to increase efficiency of the air source heat pump performance. Temperature of air, which circulates in open area, is not high enough as it could be in a spaces closed by means of transparent materials, where dissipation of heat is restricted. Well known glassed-in loggias installed in multi-storey buildings or greenhouses added on to the one family houses from the south side already could be used as solar collectors for preheating of air. This air preheating installations could enhance efficiency of heat pumps operation significantly, especially when sun is shining. Similar space heating systems based on air source heat pumps with ambient air preheating are described in references [9, 10]. Apart from this, it would be useful to install massive heat energy storage facilities inside the solar collectors for air heating. It could be made from rocks or concrete.

Conclusions

1. Pollution of environment in the world is rapidly coming to its natural and very dangerous limits and trespassing of these limits will result in stronger natural disasters. The most appropriate way to mitigate processes of environment pollution and climate change is shifting from dirty fossil fuels to the RES-based economy as soon as possible.
2. Rapid expansion of renewables currently is seen in many sectors of world economy. Buildings in many countries are going green too. Increasing number of buildings already have installed autonomous RES-based heat and power plants. This trend must be supported and strengthened.
3. Our experimental analysis of ambient air temperature instability over days and nights disclosed rather significant gains of the air temperature during the days: differences of air temperature in the night and in the next day as the rule are positive and make up from +(2–3) °C to +(15–16) °C; number of considerable differences (about +6 °C and more) make up about 50 %, duration of the coldest period in Kaunas city in 2017/18 heating season, when air temperature at nights was about -(5–15) °C, made up about two weeks.
4. Presently concept of heat pump control, when system of space heating control is set to keep the desirable temperature in building, is predominant. We propose to set the pump control system on the positive daily gains of air temperature in order to produce more heat energy for the less electric energy consumed.
5. Justification of air source heat pump operation expediency when positive gains of ambient air temperature are exploited was carried out by means of quantification of the case, which was conditioned by the experimental curve of ambient air temperature alteration.
6. The case study carried out for the local meteorological conditions disclosed that estimated daily heat energy output by the air source heat pump significantly exceeds the daily average heat energy needs for March for the building under our analysis. Consumption of electricity for space heating are low because heat pump is operating with significantly higher efficiency, which is available due to a higher daily temperatures.
7. Efficiency of the air source heat pump performance also can be increased by preheating the air in the glassed-in loggias installed in multi-storey buildings or greenhouses added on to the one family houses from south.

References

[1] Ripple WJ. Wolf Ch., Newsome TM., Galetti M., Alamgir M., Christ E., Machmoud MI., Laurance WF. and 15,364 scientist signatories from 184 countries. World Scientists' Warning to Humanity: A Second Notice. *BioScience*, Volume 67, Issue 12, 1 December 2017, pp. 1026–1028. Access via internet: https://doi.org/10.1093/biosci/bix125.

[2] Hooftman N., Oliveira L., Messagie M., Coosemans T., and Van Mierlo J. Environmental Analysis of Petrol, Diesel and Electric Passenger Cars in a Belgian Urban Setting. *Energies,* Vol. 9 (2016), 84, pp. 1–22. Access via internet: http://www.mdpi.com/1996-1073/9/2/84/htm.

[3] Guttikunda SA., Jawakar P. Atmospheric emissions and pollution from the coal-fired thermal power plants in India. *Atmospheric Environment*, 92 (2014), pp. 449–460. Access via internet: http://www.indiaairquality.info/wp-content/uploads/docs/2014-08-AE-Emissions-Health-Coal-PPs-India.pdf.

[4] Aber JD., Nadelhoffer KJ., Steudler P., Melillo JM. Nitrogen Saturation in Northern Forest Ecosystems: Excess nitrogen from fossil fuel

combustion may stress the biosphere. *BioScience*, Volume 39, Issue 6, 1 June 1989, pp. 378–386. Access via internet: https://doi.org/10.2307/1311067.

[5] Keenan RJ., Reams GA., Achard F., de Freitas JV. Grainger A., Lindquist E. Dynamics of global forest area: Results from the FAO Global Forest Resources Assessment 2015. *Forest Ecology and Management*, 352: 9–20.

[6] Liu Y., Zhou GH., Ma J. Experimental Study of a Solar Assisted Air Source Heat Pump in Heating Condition. May 2011. Access via internet: https://doi.10.4028/www.scientific.net/AMR.243-249.4909.

[7] Wu J., Chen Ch., Pan S., Wei J., Pan T., Wei Y., Wang Y., Wang X. and Su J. Experimental Study of the Performance of Air Source Heat Pump Systems Assisted by Low-Temperature Solar-Heated Water. *Advances in Mechanical Engineering* Volume 2013, 8 p. Access via internet: http://dx.doi.org/10.1155/2013/843013.

[8] Su S., Huang Q., Chen J. Analysis of air source heat-pump system assisted by solar energy. *Advanced Materials Research*, Vols. 516–517 (2012), pp. 257–260. Access via internet: www.scientific.net. DOI:10.4028/www.scientific.net/AMR.516-517.257.

[9] Adomavičius V., Valickas J., Petrauskas G., Pušinaitis L. Potential of village house for sustainable energy production. Proceedings of 6th International Conference on Trends in Agriculture Engineering 2016 – Part I. Prague, Czech Republic, September 7–9, 2016. pp. 17–25.

[10] V. Adomavičius, Valickas J., Petrauskas G., Pušinaitis L. Concept of RES-based microgrid for apartment building. The 11th International Conference on Intelligent Technologies in Logistics and Mechatronics Systems (ITMS'2016). Panevėžys, Lithuania; 28–29 April 2016. Medimond Publishing Company, 2016, pp. 1–6.

ITMS'2018

panevėžys faculty of technologies and business

The 12th International Scientific Conference Intelligent Technologies in Logistics and Mechatronics Systems (ITELMS'2018), 26–27 April 2018, Panevėžys, Lithuania

Multi-Objective Assessment of the Effectiveness of Attaining a First-Cycle Degree in Construction Engineering in Lithuania

Donatas Aviža[a*], Alfredas Bartulis[a], Danguolė Striukienė[a], Elvyra Zacharovienė[a]

[a]*Kaunas University of Technology, Nemuno g. 33, LT–37164 Panevėžys, Lithuania*

Abstract

In the course of the empirical research, first-cycle full-time studies awarding the Bachelor's degree of Construction Engineering in Lithuanian universities and colleges were assessed. SAW method was selected for the multi-objective effectiveness assessment of attaining the degree. Findings of the performed empirical research suggest that it is the most effective to complete the first-cycle Construction Engineering studies in Vilnius College of Technologies and Design (0.86 points from 1), and least effective – in Šiauliai University (0.74 points of 1). The research findings are relevant for the employers and prospective students who choose between higher education institutions for their Construction Engineering studies.

© 2018 D. Aviža, A. Bartulis, D. Striukienė, E. Zacharovienė
Peer-review under responsibility of the Kaunas University of Technology, Panevėžys Faculty of Technologies and Business

Keywords: SAW method, bachelor of construction engineering, college, university

1. Introduction. Higher education system in Lithuania

Until 2000, there was a unitary system functioning in the country (university sector organised studies only, while the majority of research institutes were separated from universities). After 2000, binary structure of higher education was introduced and non-university higher education institutions – colleges – joined the system of higher education institutions [1]. The principle of binarity is expressed by treating university and non-university studies equally. Such higher education system enables parallel functioning of university and non-university studies, which meet the market needs by preparing the people of different professions. According to Statistics Lithuania, in academic years

* Corresponding author. Tel.: +370-623-75003
E-mail address: donatas.aviza@ktu.lt

© 2018 D. Aviža, A. Bartulis, D. Striukienė, E. Zacharovienė
Peer-review under responsibility of the Kaunas University of Technology, Panevėžys Faculty of Technologies and Business

B427H0002 . EdLearning.it

2015–2016, there were 39,772 students studying in colleges (82 % in state and 18 % in non-state colleges), whereas the universities had 93,524 students (95 % were in state and 5 % in non-state universities). In the recent years, a general trend is observed: the colleges lose about 2,000 students, and universities – 7,000 students annually. This demographic decline will continue until 2020; then, a slight growth is expected.

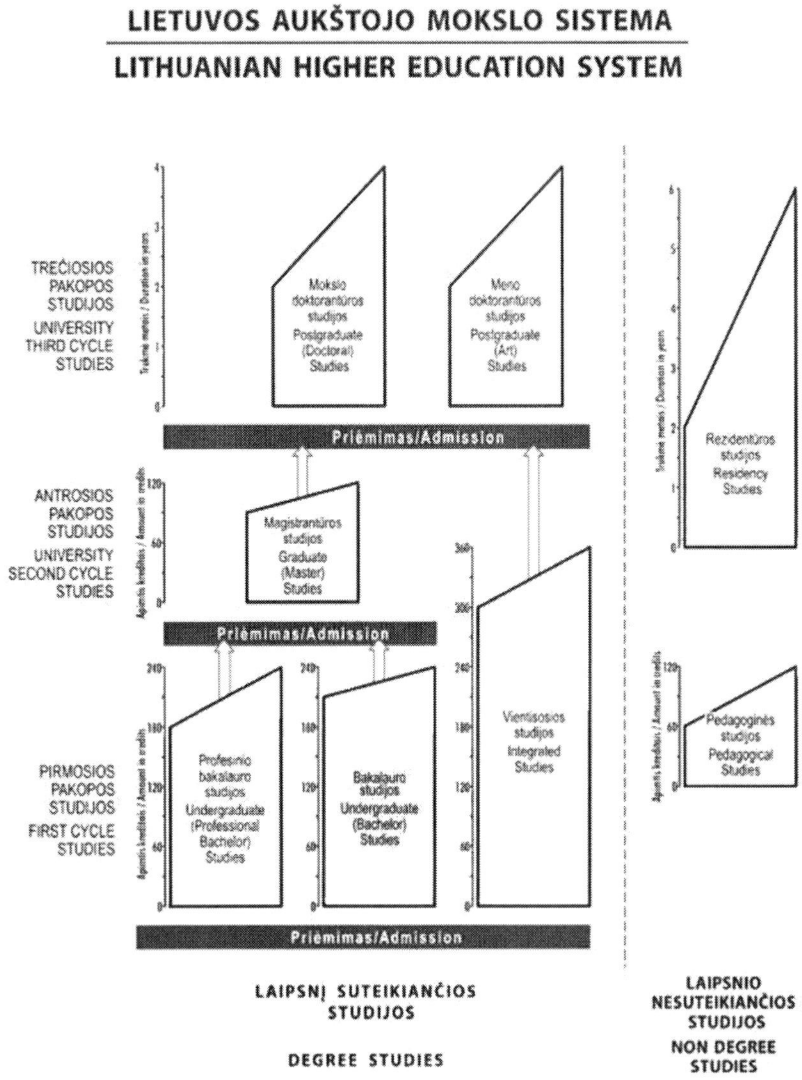

Fig. 1 Higher education system in Lithuania [1]

The Law on Higher Education and Research clearly identifies the following two sectors:

- two types of higher education institutions are distinguished: universities (22; 14 of them are state-owned) and colleges (22; 12 of them state-owned);
- study programmes of two types are identified: university (1,440) and college (451) programmes;
- universities and colleges confer the qualification of different profile: colleges award the Professional Bachelor‘s degree, whereas the universities may award a Bachelor‘s, Master‘s and scientific degrees (see Fig. 1);
- college studies are more oriented to preparation for vocational activities and enable the individuals to acquire the professional qualification based on applied scientific research and/or applied scientific activity;

university studies are more oriented to the universal general education, theoretical preparation and highest-level professional skills;

– qualifications conferred in Lithuania are divided into 8 levels by defining the learning outcomes in terms of complexity, autonomy and range of activities the individual acquiring the qualification is able to perform.

The Lithuanian Qualifications Framework is in line with the Recommendation of the European Parliament and of the Council of 23 April 2008 on the establishment of the European Qualifications Framework for lifelong learning (2008/C111/01) and the 8 levels of qualifications framework established therein. The Government of the Republic of Lithuania approved the Description of the Lithuanian Qualifications Framework in 2010 [2].

The levels of qualifications are arranged hierarchically according to the competence substantiating the qualifications, ways of acquiring the qualifications and criteria defining the qualification levels [3]:

1) complexity of activities – a qualification criterion used to describe the character of activities, variety of tasks and degree of responsibility;
2) autonomy of activities – a qualification criterion used to describe the changes in the activity organisation and nature of subordination;
3) range of activities – a qualification criterion used to describe the changes in technological and organisational environment of activities.

Level VI qualifications are acquired by completing the first cycle of university or college studies, and in the cases and in accordance with the procedure established by the laws – by completing the study or requalification programmes that do not lead to a degree and/or through professional experience and independent learning [4].

2. Research methodology

Multi-objective decision-making methods are applied in a multitude of fields for the solving of engineering as well as economic problems [5–7]: options are assessed according to several indicators, which may be measured in different units; assessment criteria may be contradictory. The purpose of multi-objective assessment is the ranking of all options of the phenomenon under consideration [8].

A decision-making process comprises three stages:

1) compiling the option list;
2) defining the indicators according to which the options are to be assessed;
3) ranking the options [9].

Based on classifications of multi-criteria decision-making techniques described by various researchers [6, 10–12;] presented a classification system of multi-objective methods (Table 1).

Table 1. A taxonomy of methods for the multiple-attribute decision-making [12]

Class of methods	Information about the indicators obtained from the decision-maker	Method titles
Methods based on multi-attribute utility theory	Quantitative measurements	SAW, TOPSIS, TOPSIS-G, COPRAS, COPRAS-G, ARAS, MOORA, VIKOR, MULTIMOORA
Class of methods	Information about the indicators obtained from the decision-maker	Method titles
Methods of analytical hierarchy and fuzzy sets	Qualitative measurements are expressed in qualitative terms	AHP, FUZZY
Verbal decision analysis methods	Qualitative measurements, quantitative variables are not involved	ZAPROS, PARK, ORKLASS, CLARA, DIFLASS, CIKL
Comparative preference methods	Quantitative and qualitative measurements	ELECTRE, PROMETHEE, MELCHIOR, UTA, MAUT, TACTIC

In consideration of the complexity of the problems in question and after applying the Saaty's method suitability screening set [13], in this paper, the SAW multi-objective quantitative decision-making method will be applied.

SAW refers to the simple additive weighing method. The rules of this method were outlined by MacCrimmon [14]. The method is well-known from traditional multi-criteria methods and is frequently used in research. The baseline data for this method include the decision matrix, which may not have non-numerical values and significance of each indicator. Its result is the sum of constant indicators' significances for each option [15].

Steps of the SAW method are as follows:

1) a decision matrix is compiled;
2) the decision matrix is normalised;
3) each element of the same option of the normalised matrix is multiplied by its significance and added up with other elements of the option (row).

Initial data when using this method for decision-making are the solution matrix P and the significance values of the effectiveness indicator, for instance, values of integrated significance (q_j^*) that fulfil the equation:

$$\sum_{j=1}^{n} q_j^* = 1. \tag{1}$$

All elements of the solution matrix P, which have to be maximised, are normalised according to the formula below:

$$\bar{x}_{ij} = \frac{x_{ij}}{x_j^{max}}. \tag{2}$$

whereas those, which have to be minimised, are normalised according to the other formula:

$$\bar{x}_{ij} = \frac{x_j^{min}}{x_{ij}}. \tag{3}$$

When determining the rationality of an option, the appropriate elements of the normalised matrix are multiplied by the significance value of the effectiveness indicators and the products received are added up. The sum of the rational option products will be the maximum:

$$A = \left\{A_i \middle| \max_i \sum_{i=1}^{n} q_j^* \bar{x}_{ij}\right\}; \tag{4}$$

where $\bar{x}_{ij}$ – normalised decision-making matrix;

q_j^* – indicator significance values.

Seeing that the scientific problems under consideration are of complex nature, when solving the tasks of this paper it is proposed to apply the multi-objective decision-making method – SAW. This method is one of the sufficiently easy to understand and the most broadly applied multi-objective assessment methods in the world.

3. Research

The research assessed the first-cycle full-time studies of Construction Engineering leading to a Bachelor's degree in Lithuanian universities and colleges. SAW method was chosen for the multi-objective assessment of the degree attainment effectiveness.

In the first stage, a decision-making matrix (Table 2) is compiled and 9 higher education institutions are assessed according to 5 indicators: annual tuition fee; duration of studies; overall institution evaluation by MOSTA [16]; possibility to continue studies in graduate study programmes for the Master's degree; average insured income of the alumni.

In the second stage, all elements of the matrix are minimised or maximised, i. e. normalised, by turning them into dimension-less units of measurement (Table 3).

Table 2. Decision-making matrix (for 2016-2017 data)

Options	Indicators				
	Annual tuition fee, €	Duration of studies in years	Overall institution evaluation (by MOSTA), in points	Possibility to continue studies in graduate programmes, in points	Average monthly insured income of alumni, €
1. Kaunas University of Technology [17]	2242	4,00	59,00	1,0	822,20
2. Panevėžys University of Applied Sciences [18]	1386	3,00	64,00	0,5	619,60
3. Šiauliai University [19]	2242	3,50	19,00	1,0	801,40
4. Šiauliai State College [20]	1520	3,00	60,00	0,5	589,20
5. Vilnius Gediminas Technical University [21]	2242	4,00	58,00	1,0	882,60
6. Vilnius College of Technologies and Design [22]	1386	3,00	62,00	0,5	724,10
7. Kaunas University of Applied Engineering Sciences 23]	1390	3,00	60,00	0,5	690,90
8. Klaipėda University [24]	2160	4,00	41,00	1,0	888,20
9. Klaipėda State University of Applied Sciences [25]	1386	3,00	53,00	0,5	692,80
Min. or max. indicator	min.	min.	max.	max.	max.
Best value of the indicator	1386,000	3,000	64,000	1,000	888,200

Table 3. Normalised matrix

Options	Indicators				
	Annual tuition fee, €	Duration of studies in years	Overall institution evaluation (by MOSTA), in points	Possibility to continue studies in graduate programmes, in points	Average monthly insured income of alumni, €
1. Kaunas University of Technology	0,618	0,750	0,922	1,000	0,926
2. Panevėžys University of Applied Sciences	1,000	1,000	1,000	0,500	0,698
3. Šiauliai University	0,618	0,857	0,297	1,000	0,902
4. Šiauliai State College	0,912	1,000	0,938	0,500	0,663
5. Vilnius Gediminas Technical University	0,618	0,750	0,906	1,000	0,994
6. Vilnius College of Technologies and Design	1,000	1,000	0,969	0,500	0,815
7. Kaunas University of Applied Engineering Sciences	0,997	1,000	0,938	0,500	0,778
8. Klaipėda University	0,642	0,750	0,641	1,000	1,000
9. Klaipėda State University of Applied Sciences	1,000	1,000	0,828	0,500	0,780
Min. or max. indicator	min.	min.	max.	max.	max.
Significance of indicators	0,20	0,20	0,20	0,20	0,20

In the third stage, each element of the same option is multiplied by its significance and added up to other elements of the option (row) (Table 4). In this study, the weight of the indicators' significance is assumed to be equal and proportionate for all indicators seeking to avoid subjective assessment.

Table 4. Matrix of values (row of priorities)

Options	Indicators					Row of priorities
	Annual tuition fee, €	Duration of studies in years	Overall institution evaluation (by MOSTA), in points	Possibility to continue studies in graduate programmes, in points	Average monthly insured income of alumni, €	
1. Kaunas University of Technology	0,618	0,750	0,922	1,000	0,926	3
2. Panevėžys University of Applied Sciences	1,000	1,000	1,000	0,500	0,698	5
3. Šiauliai University	0,618	0,857	0,297	1,000	0,902	9
4. Šiauliai State College	0,912	1,000	0,938	0,500	0,663	8
5. Vilnius Gediminas Technical University	0,618	0,750	0,906	1,000	0,994	2
6. Vilnius College of Technologies and Design	1,000	1,000	0,969	0,500	0,815	1
7. Kaunas University of Applied Engineering Sciences	0,997	1,000	0,938	0,500	0,778	4
8. Klaipėda University	0,642	0,750	0,641	1,000	1,000	7
9. Klaipėda State University of Applied Sciences	1,000	1,000	0,828	0,500	0,780	6

Upon completion of the multi-objective assessment, effectiveness points were calculated for each higher education institution (Fig. 2).

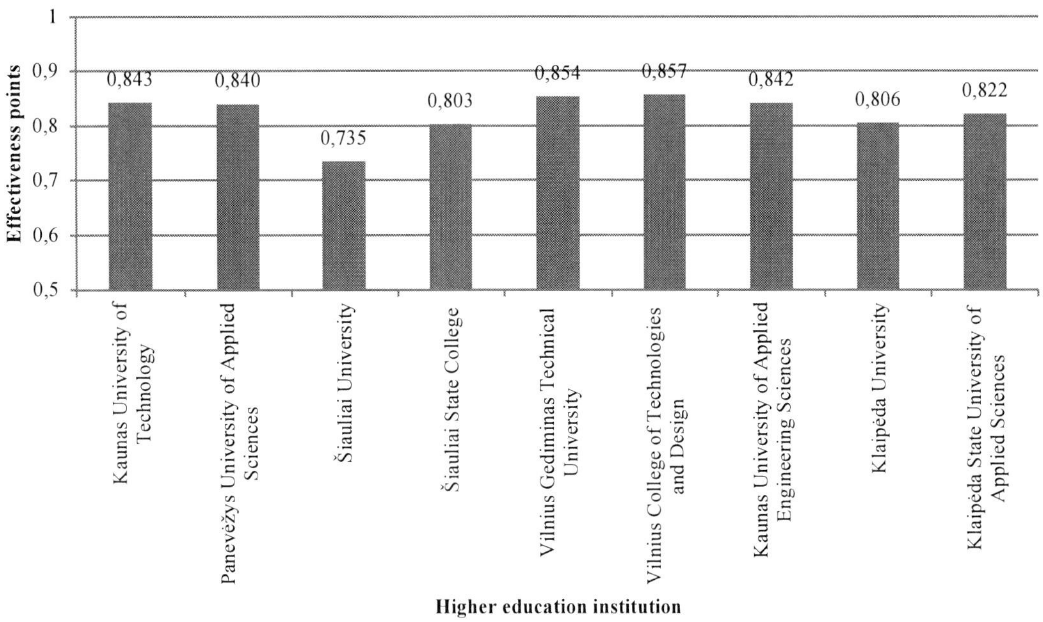

Fig. 2. Assessment of effectiveness of Construction Engineering studies in higher education institutions

Thus, the conducted empirical research suggests that it is the most effective to complete the first-cycle Construction Engineering studies in Vilnius College of Technologies and Design (0.86 points of 1), and the least effective – in Šiauliai University (0.74 points of 1). When assessing the studies in Panevėžio city, Panevėžys University of Applied Sciences and Kaunas University of Technology collected a similar number of points (0.84 of 1), and this represents an effectiveness that is lower than the first place rating by 2.3 % only.

Conclusions

1. Upon completion of the first-cycle Construction Engineering studies at the university and college, the same level VI qualification is acquired.
2. The most effective studies are in Vilnius College of Technologies and Design and the least effective – in Šiauliai University.
3. The Bachelor's degree is earned in colleges within a period that is 25 % shorter, whereas the average tuition fee there is 2.2 times lower than at the university.
4. It is better to study at the university than at the college if the student has free of charge studies (because it is more expensive).

References

[1] Aukštojo mokslo sistemos aprašas ir schema. Access via internet: https://www.smk.lt/uploads/files/studiju_sistema/AUK%C5%A0TOJO%20MOKSLO%20SISTEMOS%20APRA%C5%A0AS_2016.pdf [Accessed: 2018-01-24].

[2] Lietuvos Respublikos vyriausybė, Nutarimas dėl Lietuvos kvalifikacijų sandaros aprašo patvirtinimo, 2010 m. gegužės 4 d. Nr. 535, Vilnius. Access via internet: https://www.e-tar.lt/portal/lt/legalAct/TAR.BC967702800C/TAIS_405326 [Accessed: 2018-01-24].

[3] STUDIJŲ KOKYBĖS VERTINIMO CENTRAS. Aukštasis mokslas. Access via internet: http://www.skvc.lt/default/lt/lietuvos-svietimo-sistema/aukstasis-mokslas [Accessed: 2018-01-24].

[4] Švietimo ir mokslo ministerija. Išsilavinimo prilyginimas ir kvalifikacijų pripažinimas. Access via internet: https://www.smm.lt/web/lt/pedagogams/kvalifikacija/isilav_palyginimas [Accessed: 2018-01-24].

[5] Zavadskas E. K., Turskis Z. Multiple criteria decision making (MCDM) methods in Economics: an overview, Technological and Economic Development of Economy, 2011, 17(2): 397–427. Access via internet: http://dx.doi.org/10.3846/20294913.2011.593291 [Accessed: 2018-01-24].

[6] Hwang C. L., Yoon K. 1981. Multiple attribute decision making – methods and applications. A State of the Art Surwey. Springer Verlag, Berlin, Heidelberg, New York, 250 p.

[7] Zavadskas E., Kaklauskas A. The new method of multicriteria evaluation of projects in Deutsch-Litauisch-Polnisches Kolloquium zum Baubetriebswesen. HTWK: Beitrage Zu Lehre und Forschung, 1996, J. 3, pp. 3–13.

[8] Ginevicius R., Podvezko V. Daugiakriterinio vertinimo taikymo galimybės kiekybiniam socialinių reiškinių vertinimui, Verslas: teorija ir praktika [Business: theory and practice], 2008, 9(2): 81–87.

[9] Mitkus S. Investicinių projektų efektyvumo tyrimas kompiuterinės grafikos metodais, Technological and Economic Development of Economy [Ūkio technologinis ir ekonominis vystymas], 1996, 13: 142–147.

[10] Ustinovicius L., Zavadskas E. K. Statybos investicijų efektyvumo sistemotechninis ivertinimas [Assessment of investment profitability in construction from technological perspectives]. Vilnius: Technika, 2004, 220 p. ISBN 9986-05-806-6.

[11] Turskis Z. 2008. Multi-Attribute Contractors Ranking Method by Applying Ordering of Feasible Alternatives of Solutions in Terms of Preferability Technique. echnological and Economic Development of Economy, 14(2): 224 – 239.

[12] Simanavičienė R. Kiekybinių daugiatikslių sprendimo priėmimo metodų jautrumo analizė. Daktaro disertacija. Vilnius, Technika, 2011.

[13] Saaty T. L., Ergu D. When is a Decision-Making Method Trustworthy? Criteria for Evaluating Multi-Criteria Decision-Making Methods, International Journal of Information Technology & Decision Making, 2015, 14(06): 1171–1187.

[14] MacCrimmon K. R. Decision making among multiple–attribute alternatives: A Survey and Consolidated Approach. RAND Memorandum, RM-4823-ARPA, 1968.

[15] Kareivaitė R. Daugiakriterių metodų panaudojimas vertinant darnų vystymąsi. Journal of Management, 2012, 1 (20) 121–127.

[16] MOSTA. Interaktyvūs duomenys. Access via internet: http://www.mosta.lt/lt/interaktyvus-duomenys [Accessed: 2018-01-24].

[17] KTU. Bachelor programmes. Access via internet: https://stojantiesiems.ktu.edu/bakalauras/#ba-kaina [Accessed: 2018-01-24].

[18] Panevėžio kolegija. Metinės studijų kainos. Access via internet: http://panko.lt/metines-studiju-kainos/ [Accessed: 2018-01-24].

[19] Access via internet: http://www.su.lt/index.php?option=com_content&view=article&layout=edit&Itemid=1390&id=4433&lang=lt [Accessed: 2018-01-24].

[20] Šiaulių University. Bachelor programmes. Access via internet: https://www.svako.lt/lt/stojantiesiems/studiju-programos-formos-kaina [Accessed: 2018-01-24].

[21] Vilnius Gediminas Technical University. Access via internet: http://www.vgtu.lt/norintiems-studijuoti/bakalauro-ir-vientisosios-studijos/studiju-programos/pagal-krypciu-grupes-ir-kryptis/334?pid=118501&y=2017 [Accessed: 2018-01-24].
[22] Vilnius College of Technologies and Design. Access via internet: http://www.vtdko.lt/index.php/component/content/article/177-stojantiesiems/studiju-kainos/497-metine-studiju-kaina [Accessed: 2018-01-24].
[23] Kaunas University of Applied Engineering Sciences. Access via internet: http://www.ktk.lt/busimiems-studentams/finansai-ir-mokesciai/ [Accessed: 2018-01-24].
[24] Klaipėda University. Access via internet: https://www.ku.lt/priemimas2016/priemimas-i-ku-studijas/bakalauro-studijos/juros-technologiju-ir-gamtos-mokslu-fakultetas/statybos-inzinerija/ [Accessed: 2018-01-24].
[25] Klaipėda State University of Applied Sciences. Access via internet: http://www.kvk.lt/lt/studiju-programu-kainos/ [Accessed: 2018-01-24].

ITMS'2018

panevėžys faculty of technologies and business

The 12th International Scientific Conference Intelligent Technologies in Logistics and Mechatronics Systems (ITELMS'2018), 26–27 April 2018, Panevėžys, Lithuania

Principles of Determining the Basic Functions of Railway Hubs

Yuriy Barash[a], Volodymyr Bobyl[a], Tanya Charkina[a], Natalya Bozhok[a], Olena Chornovil[a]

[a]*Dnipropetrovsk National University of Railway Transport named after Academician, V. Lazaryan, 2 Lazarian St., 49010 Dnipro, Ukraine*

Abstract

Goal. During the past two years, the question about creation of passenger transportation hubs has been raised at the conferences held by Ukrzalisnitsya (Ukrainian Railways). And it became evident that every specialist has a different understanding of the concept. At the moment, several railway junctions for the passenger hubs have been chosen. These locations are Kiyiv, Kharkiv, Dnepr, Lviv and Odesa. The developers' goal is to set functions of these hubs and to improve them due to optimization of train schedule, shortening of trip time and providing passengers with comfortable conditions.
Results. The work analyzes the existing functions of railway stations and provides a list of some new functions that can be introduced to transform existing stations into railway transport hubs.
Practical significance. The implementation of the new passenger transportation junction functions listed in the work will allow the management of the Ukrainian Railways to provide additional facilities for passengers and improve the quality of services due to train schedule optimization, reducing travel time, providing additional transfer services for passengers and baggage as well as shopping and entertainment services.

© 2018 Yu. Barash, V. Bobyl, T. Charkina, N. Bozhok, O. Chornovil
Peer-review under responsibility of the Kaunas University of Technology, Panevėžys Faculty of Technologies and Business

Keywords: passenger stations, transport interchange nodes, HUB functions, train schedule, transfer of passengers and baggage

1. Introduction

There are several definitions of the term HUB on the Internet and in the literature. Hub (literally – wheel hub, centre) is in the general sense, a junction point of a network.

- In aviation (in English, often a *gateway*) is a hub airport, a sufficiently large interchange and reloading transport hub with the necessary set of services, having a suitable geographic location.
- In transport – a transfer, reloading point.
- On off-road vehicles, it is a freewheel clutch for the front, connectable, axle. There are manual and automatic hubs.

© 2018 Yu. Barash, V. Bobyl, T. Charkina, N. Bozhok, O. Chornovil
Peer-review under responsibility of the Kaunas University of Technology, Panevėžys Faculty of Technologies and Business
B427H0003 . Ed*Learning*.it

- In computer technology and networks, it is a network hub (for example, a USB hub).
- In the Direct Connect file-sharing network, hubs are network servers.
- In Fidonet, the hub is the main node through which the mail is sent.
- In power engineering, a hub is a specialized distribution centre that divides one route into two or more.
- Habs is the nickname of NHL team Montreal Canadiens.

But we are interested in the transport concept of HUB. The Wikipedia says it is a transport interchange, where traffic flows converge. Initially, this term was used for large airports, where the passengers transferred from one direction of travel to another. Subsequently, the term HUB began to be used for any transport interchange points. Taking into account the above, it is possible to organize railway HUBs for convenient passenger travel in Ukraine.

Publications on the organization of HUBs are mainly in foreign literature [4-10]. In Ukraine, this issue is investigated only at scientific conferences and in the office of PJSC "Ukrainian Railways" [1–3]. But this issue will require further research, since large cities have long been formed transport infrastructure, which is squeezed by urban buildings.

2. Purpose of the study

For two consecutive years at the conferences of Ukrzaliznytsia the question arises about the organization of transport passenger HUBs in Ukraine, but each specialist understands this issue in different ways. Now even there are defined a few railway junctions, where the passenger HUBs will be organized: Kiev, Kharkiv, Dnipro, Lviv and Odessa. The authors of the article set out to define the functions that HUBs will perform and to improve them by optimizing the train schedule, reducing the travel period and providing maximum comfort for passengers.

The transport hubs methodology has been improved along with approaches as for the hubs construction which differs from the present ones in the complex vision of passenger transportation by several transport means simultaneously and corresponding transfers. The latter allows to shorten time term of trips and make them more comfortable for a single ticket.

3. Methodology

At present, there exist no scientific approach in Ukraine as for the construction of hubs and definition of their main functions. Now to solve this problem general methodology of passenger transportation is in practice, which does not take into account additional functions of railway stations connected with improvement of schedules of several transport means taken simultaneously along with transfers. As the problem under research concerns interaction of railways with other transport means and transfer transportation, the authors use the systematic approach that makes it possible to determine all the necessary functions, which railway hubs must perform.

4. Findings

Currently, railway stations in Ukraine – future HUBs perform the following functions (Fig. 1):

- The transfer of passengers from one direction of travel to another is the main function. Now it is executed by Ukrzaliznytsya, but it is not completely optimized and needs to be improved;
- Transfer of passengers from long-distance passenger transport modes to high-speed and commuter trains and in opposite direction. This function is now also being implemented, but an additional linking of the schedules of the specified trains is needed, which requires a reduction in the total passenger travel time;
- Transfer of passengers from railway trains to all possible types of local transport (metro, tram, trolleybus, bus) for transportation to their homes or other local stations. This function is now performed by local authorities.
- If you study foreign experience, then you can determine the additional functions of railway HUBs (Fig. 1):
- Providing shopping, entertainment and information services for convenient accommodation of passengers whilst waiting for the following modes of transport;
- Organization of transfers of passengers for their further travel by other means of transport (buses, airplanes, river and sea vessels, etc.);

- Organization of transfers of baggage of the passengers for their further travel by other means of transport (buses, planes, river and sea vessels, etc.);
- Providing passengers with single ticket travel services for different types of transport.

It is possible to further increase the list of functions and services that can be provided by railway HUBs, but first they need to be economically substantiated

The most important function of HUBs is the optimization of the traffic schedule between the passenger trains of various types of communication to shorten the length of stay of passengers at railway stations whilst waiting for other trains and transfers.

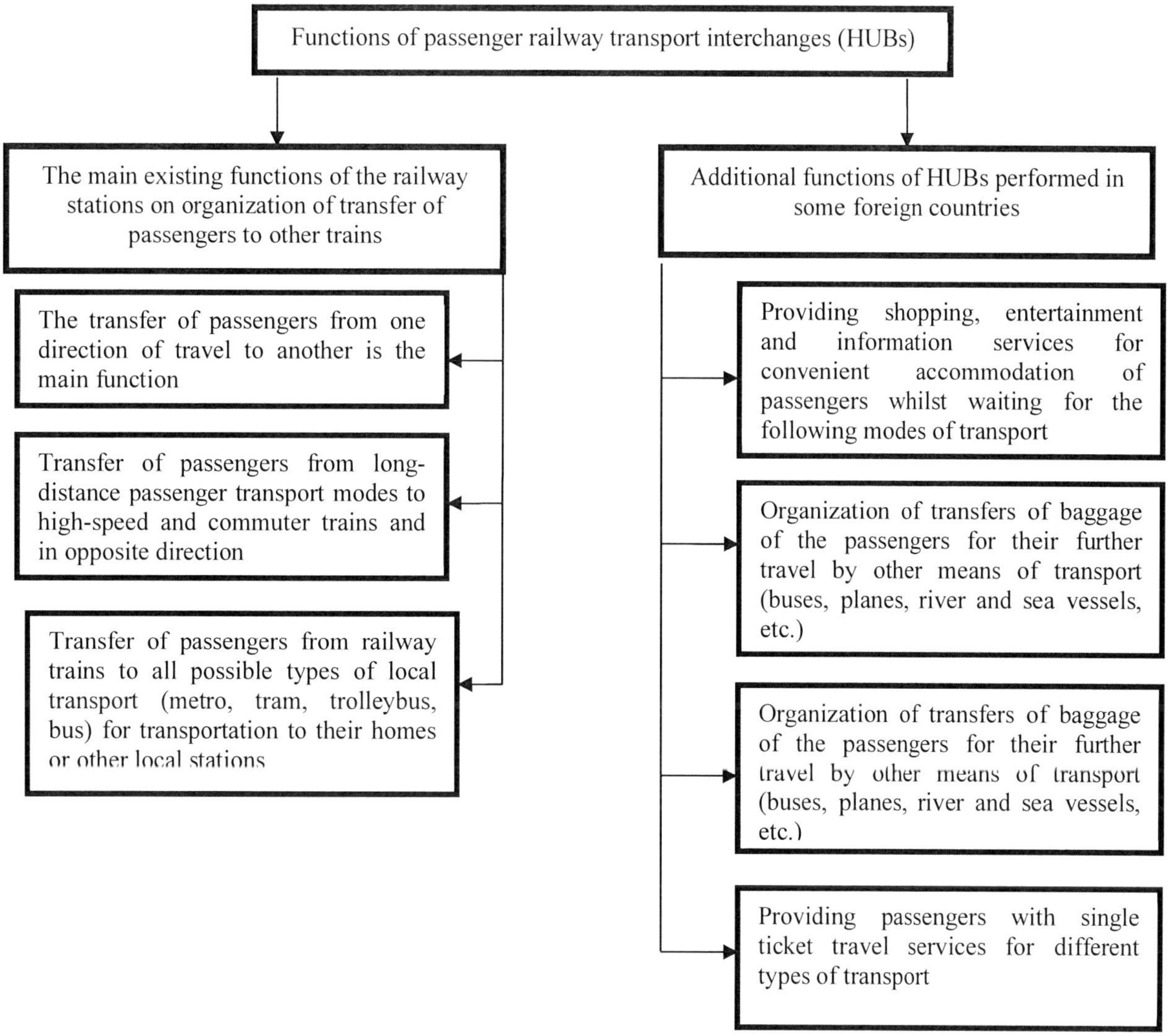

Fig. 1 Possible functions of passenger rail transport interchanges (HUBs)

How to understand the optimization of the schedule of passenger trains? Suburban trains must arrive at the train station (HUB) in advance of the arrival of a passenger or high-speed train, taking into account their delay. In this case, it is possible to introduce on the railways for the convenience of passengers a single ticket, which will allow them to travel by two types of trains. All these measures will allow passengers to reduce the total travel time only by rail.

The most important function of HUBs is the optimization of the traffic schedule between the passenger trains of various types of communication to shorten the length of stay of passengers at railway stations whilst waiting for other trains and transfers. How to understand the optimization of the schedule of passenger trains? Suburban trains must arrive at the train station (HUB) in advance of the arrival of a passenger or high-speed train, taking into account their delay. In this case, it is possible to introduce on the railways for the convenience of passengers a single ticket, which will allow them to travel by two types of trains. All these measures will allow passengers to reduce the total travel time only by rail.

There are many cases when a passenger travels a complex route, for example, from a small town in the Dnipropetrovsk region by a suburban train to Dnipro, and then by a high-speed train to Kiev, then by bus to Boryspil airport, and from there on an airplane to spend vacation in any country. This complex journey requires a lot of effort from the passenger, while the presence of children and suitcase creates many problems.

If we introduce additional functions at the Dnipro station, namely, link the schedule of the suburban and high-speed trains and organize a single ticket trip, which will also include the cost of a shuttle to Boryspil Airport, it will significantly reduce the passengers' travel time and improve the quality of the services provided to them.

By the way, now in Ukraine, some travel companies use transfer services to transport tourists from some cities to major transport points (airports, railways, river and seaports). The price of such a transfer is included in the cost of the tour. In view of the above, it is possible to add the organization of transfers of passengers for their further travel by other modes of transport (buses, airplanes, river and sea vessels, etc.) to the main functions of the railway HUB.

Railway HUBs can also use an additional service that is currently used by Ukrainian airports for arranging transfers of baggage for their further travel by other modes of transport (buses, airplanes, river and sea vessels, etc.). This will allow passengers to have only the necessary things with them, and spend free time before traveling at their pleasure. For this purpose, it is advisable to provide shopping and entertainment services at railway stations.

One can give an example, as in Hungary, the functions of HUBs are used at the Kispest (Budapest) railway junction. The station serves the transportation of passengers in the long-distance, high-speed and suburban traffic. Near across the land crossing there is a bus station, a subway, a tram, a taxi and a large shopping and entertainment centre. All this together is a compact and very comfortable HUB.

5. Practical value

The implementation of the new functions of the transport interchange points will allow the management of PJSC "Ukrainian Railways" to create additional amenities for passengers and improve the quality of their services by optimizing the train schedule, reducing the travel period, providing additional transfers for passengers and baggage, as well as shopping and entertainment services.

Conclusions

On the basis of the above studies, the following conclusions can be drawn:

1. Organization of HUBs on the basis of large railway stations of Ukraine is expedient and necessary, as it will reduce the total travel time of passengers and significantly improve the quality of their services. At the same time, this will increase the efficiency of the use of rail transport and generate additional revenue from the provision of new services.
2. Each city should take into account the existing features of the construction of the transport infrastructure, to customize the organization of HUBs and use only some of the proposed functions.

References

[1] Materials of the V International Scientific and Practical Conference "Marketing and Logistics in the Rail Transport Management System Mil-2015". Kamyanets-Podilskyi city LLC "Sprint-Service", 2015.

[2] Materials of the X Anniversary International Scientific and Practical Conference "Problems of Economics and Management in the Railroad Transportation of the CCEU-CCEU-2016". Kyiv, 2016.

[3] Materials of the IV International scientific and practical conference "Marketing and logistics in the system of management on the railway transport". Zaporozhye, 2016.

[4] Kreutzberger E., Konings R. The challenge of appropriate hub terminal and hub-and-spoke network development for seaports and intermodal rail transport in Europe. *Research in Transportation Business & Management* 2016; 19: 83–96. DOI: https://doi.org/10.1016/j.rtbm.2016.05.003.

[5] Bartosik M., Wiak S. Multi-annual Program "By Railway to the 21st Century" as Key Factor in the Development of Rail Transport in Poland. *Transportation Research Procedia* 2016; 14: 518–527. DOI: https://doi.org/10.1016/j.trpro.2016.05.107

[6] Rau H., Hynes M., Heisserer B. Transport policy and governance in turbulent times: Evidence from Ireland. *Case Studies on Transport Policy* 2016; 4: 45–56. DOI: https://doi.org/10.1016/j.cstp.2015.11.006

[7] Yatskiv I., Budilovich E. A comprehensive analysis of the planned multimodal public transportation HUB. *Transportation Research Procedia* 2017; 24: 50–57. DOI: https://doi.org/10.1016/j.trpro.2017.05.067.

[8] Zhu Y., Hu C., Xu D., Tang J. Research on Optimization for Passenger Streamline of Hub. *Procedia – Social and Behavioral Sciences* 2014; 138: 776–782. DOI: https://doi.org/10.1016/j.sbspro.2014.07.258.

[9] Motraghi A. Rail research projects: Case studies. *Research in Transportation Economics* 2013; 41: 76–83. DOI: https://doi.org/10.1016/j.retrec.2012.10.001

[10] Hisham E. Sustainable development criteria set for the transportation hubs of the national association of provinces planning. *Procedia Engineering* 2011; 21: 1042–1055. DOI: https://doi.org/10.1016/j.proeng.2011.11.2110.

ITMS'2018

The 12th International Scientific Conference Intelligent Technologies in Logistics and Mechatronics Systems (ITELMS'2018), 26–27 April 2018, Panevėžys, Lithuania

Vehicle Exhaust Manifold Thermal Fatigue Test

Zilvinas Bazaras[a], Algirdas Laskys[a], Vaidas Lukosevicius[a*]

[a]*Kaunas University of Technology, Studentų str. 53, LT-43124, Kaunas, Lithuania*

Abstract

This article covers one of the main exhaust systems components fatigue analysis. Three different design types of the exhaust manifold were designed to find optimal one.
In this study "SolidWorks Simulation" and "SolidWorks Flow" programs were used to get the results. First of all, flow analysis was done, after that the results were taken to static analysis and after that to fatigue analysis.

© 2018 Z. Bazaras, A. Laskys, V. Lukosevicius
Peer-review under responsibility of the Kaunas University of Technology, Panevėžys Faculty of Technologies and Business

Keywords: exhaust system, exams manifold, flow analysis, static analysis, fatigue analysis

1. Introduction

The purpose of the car's exhaust system is to safely remove any accumulated gas from the engine during operation and to reduce exhaust noise. The exhaust system not only reduces noise but also prevents toxic gases from entering the car interior.

The aim of this work is to investigate the influence of temperature on the car's exhaust manifold. The SolidWorks application was used to draw the manifold. Analysis performed in FLOEFD environment.

2. Temperatures in exhaust system

Depending on the types of internal combustion engine (petrol or diesel), the temperature of the exhaust system also varies. For a diesel engine, the temperature of the exhaust system depends on the type of catalyst and the level

* Corresponding author. Tel.: +370 6730 3836
E-mail address: vaidas.lukosevicius@ktu.lt

© 2018 Z. Bazaras, A. Laskys, V. Lukosevicius **B427H0004** . Ed*Learning*.it
Peer-review under responsibility of the Kaunas University of Technology, Panevėžys Faculty of Technologies and Business

of particulate filter application. With a high content of soot in the catalyst or particulate filter (soot amount > 60 %), the temperature of the system increases as the exhaust gas flow in the system drops. With lower flow, the gas is released too slowly, resulting in an overheating of the exhaust system, which affects the longevity of the system [1].

3. Modeling manifolds

Three types of the manifolds for a four-cylinder internal combustion engine were chosen. Type A manifold is of Type 4–1, where the direction of the exhaust pipe coincides with the direction of flowing gas. The gas emitted from each cylinder flows into the common tube and flows through it to the next part of the exhaust system. Type B manifold is of Type 4–2 (Fig. 1). The gases are flowing separately to each manifold pipe, which moves from the four tubes to the two configurations. Type C manifold is a Type 4–1, where the direction of flow of gas does not coincide with the source of the manifold outlet [2].

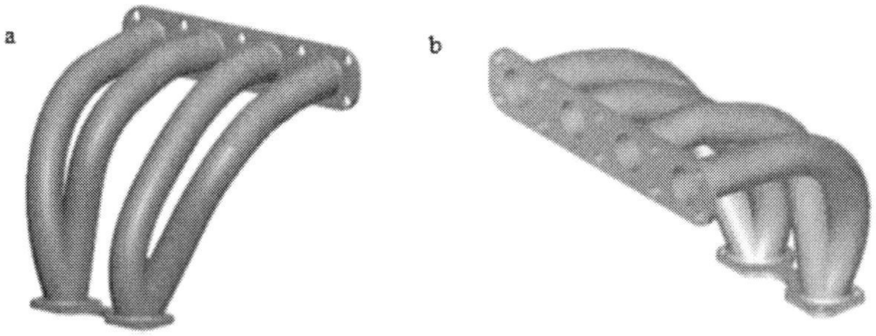

Fig. 1. a) B type exhaust manifolds front view; b) back view [3]

4. Results of analysis on engine load

During the analysis, the volume of flowing gas is set to 100 liters [4]. In Table 1 we can see the minimum number of cycles that causes fatigue. At 850 rpm, the engine load, most likely, after 120300 cycles, changes in the Type C manifold. The maximum number of cycles from the analyzed manifolds until the fatigue emerges is the type B manifold, which is 166900 cycles. Increasing the engine load up to 6000 rpm the maximum number of cycles – 31120, is offset by a Type B manifold too.

Table 1. Flow analysis of manifolds

Manifold	Number of cycles	
	850 rpm	6000 rpm
A	132000	23360
B	166900	31120
C	120300	22500

In Figure 2 we can see that the maximum stresses in the exhaust manifold are $1.163 \cdot 10^8$ Pa. Minimum stresses in the manifold $1.175 \cdot 10^6$ Pa. In Figure 3, we see the number of floating cycles, which indicates the changes in the different model locations in the case of loads. The maximum number of cycles is 1000000. From the number of cycles in the graph, we see that the initial changes in the model occur after 166900 cycles. In Figure 4 we see the probability of a fatigue break up. Exceeding the 100 % confidence threshold, we see the resulting changes in the model.

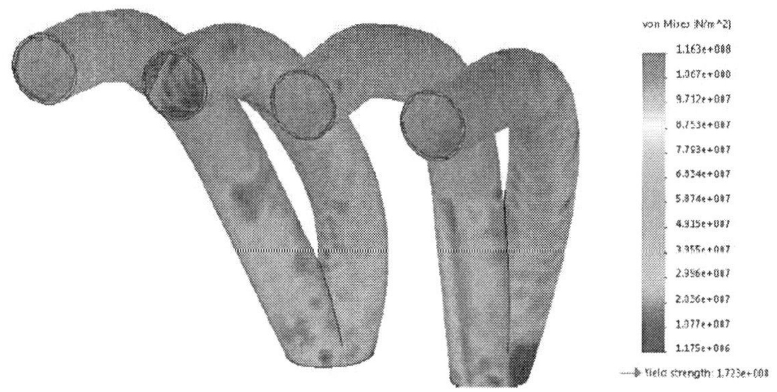

Fig. 2. Stress distribution in B type manifold

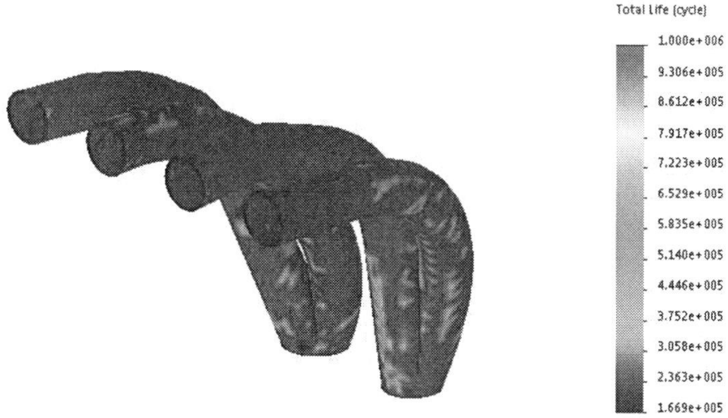

Fig. 3. Number of cycles in manifold

Table 2 shows the stresses and shifts that occur during the analysis. Minimum stresses $1.102 \cdot 10^8$ N / m^2, obtained in a Type C manifold. By increasing the engine load to 6000 rpm, the minimum stresses are $1.308 \cdot 10^8$ N / m^2, obtained in a Type C manifold. Minimum shifts of $7.611 \cdot 10^{-3}$ mm, at 850 rpm for engine pumping, are obtained in type A manifold. Maximum displacement, $4.736 \cdot 10^{-2}$ mm, is obtained in Type A manifold. Increasing the engine load to 6000rpm, the maximum displacement, $5.196 \cdot 10^{-2}$ mm, is obtained in Type B manifold.

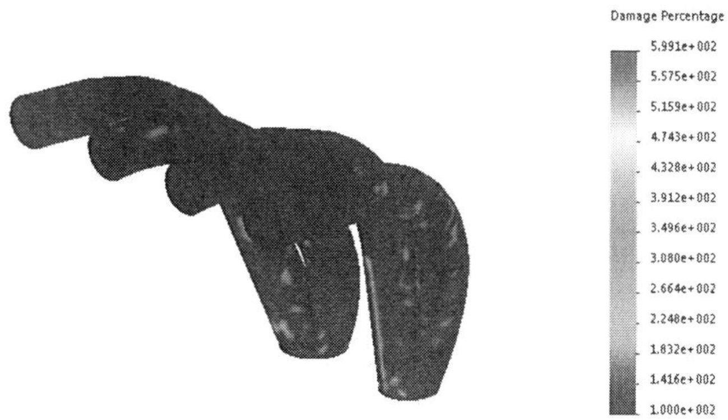

Fig. 4. Probability of fatigue

Table 2. Stress and deformation in manifolds

Manifold	Stress, N/m^2		Deformation, mm	
	850 rpm	6000 rpm	850 rpm	6000 rpm
A	$1.170 \cdot 10^8$	$1.492 \cdot 10^8$	$7{,}611 \cdot 10^{-3}$	$1{,}118 \cdot 10^{-2}$
B	$1.163 \cdot 10^8$	$1.41 \cdot 10^8$	$4{,}736 \cdot 10^{-2}$	$5{,}196 \cdot 10^{-2}$
C	$1.102 \cdot 10^8$	$1.308 \cdot 10^8$	$2{,}389 \cdot 10\text{-}^2$	$2{,}542 \cdot 10^{-2}$

Conclusions

1. The minimum static pressure of 82635 Pa at 850 rpm for engine load is in Type B manifold by increasing the engine load to 6000 rpm, the minimum static pressure of 101325 Pa remains in Type B manifold.
2. At an engine speed of 850 rpm, it is likely that this is through 200 cycles of gas flowing through the Type B manifold. By increasing the engine load to 6000 rpm, it is likely that in 97 cycles, gas flows through the Type B manifold.
3. The maximum stresses of $1.170 \cdot 10^8$ N / m^2 at 850 rpm for motor loading are obtained in type A manifold. Minimum stresses $1.102 \cdot 10^8$ N / m^2, obtained in a Type C manifold. Increasing the engine load to 6000 rpm, with the lowest stresses of $1.308 \cdot 10^8$ N / m^2, obtained in the Type C manifold. Maximum stresses $1.492 \cdot 10^8$ N / m^2, obtained in type A manifold.
4. The maximum number of cycles from the manifolds to be analyzed until the fatigue is sustained by the Type B manifold, which is 166900 cycles.

References

[1] Merker G. P., Schwarz Ch., Teichmann R. Combustion Engines Development, 2009.
[2] Smith Ph. H., Morrison J. C. Scientific Design of Exhaust and Intake Systems, Third Edition.
[3] Haegeman A., Baelman M. Experimental Flow Dynamics in Automotive Exhaust Systems with a Close-Coupled Catalyst.
[4] Kanazaki M., Morikawa M., Obayashi Sh., Nakahashi K. Exhaust Manifold Design for a Car Engine Based on Engine Cycle Simulation, Parallel Computational Fluid Dynamics, International Conference Parallel CFD 2002.

ITMS'2018

ktu 1922 panevėžys faculty of technologies and business

The 12th International Scientific Conference Intelligent Technologies in Logistics and Mechatronics Systems (ITELMS'2018), 26–27 April 2018, Panevėžys, Lithuania

Modelling and Optimization of Engine Mounting Bracket

Zilvinas Bazaras[a], Rolandas Makaras[a], Vaidas Lukosevicius[a*]

[a]*Kaunas University of Technology, Studentų str. 53, LT-43124, Kaunas, Lithuania*

Abstract

The paper presents the original model of the engine mounting bracket and overview the defects of the existing design. The main objective of the work was to prepare the topological optimization of the engine mounting bracket and to model new more stiff optimal design engine mounting bracket. The design, mass and stiffness of original mounting bracket was changed. The final model of the engine mounting bracket has been compared with the preliminary model according to the obtained data, as well as the conclusions formulated.

© 2018 Z. Bazaras, R. Makaras, V. Lukosevicius
Peer-review under responsibility of the Kaunas University of Technology, Panevėžys Faculty of Technologies and Business

Keywords: engine mounting bracket, finite element method, topology optimization

Introduction

The purpose of the engine mounting bracket is to reduce engine vibrations, prevent the engine from turning or wasting when the torque is suddenly increased or reduced, keep the engine in turning and attach the engine to the car body (Fig. 1). Engine mounting brackets must meet two basic criteria. First, they must be stiff enough and strongly suppressed by the vibration of the engine idling at 5 ± 30 *Hz* frequencies, must control the motor vibration resonance and ensure that these vibrations are isolated. Secondly, the mounting bracket must be in the form of shock absorbers when semi-static load conditions occur, such as traveling on uneven roads, sudden increase or decrease in speed, braking or turning over. Engine mounting brackets are made of stainless steel and rubber, only traction-shaped brackets are made of aluminum alloy and rubber, as they operate under lower load. Engine mounting brackets rubber is very important parts because it isolates the vehicle from vibration of the engine. Mounting brackets have

* Corresponding author. Tel.: +370 6730 3836
E-mail address: vaidas.lukosevicius@ktu.lt

© 2018 Z. Bazaras, R. Makaras, V. Lukosevicius **B427H0005** . Ed*Learning*.it
Peer-review under responsibility of the Kaunas University of Technology, Panevėžys Faculty of Technologies and Business

been designed to be rigid and resistant to fatigue and have a maximum stiffness under maximum load conditions. High stiffness rubber has a good insulator at low engine speeds, i. e. good dynamic characteristics with low frequency oscillations, but weakly isolates high-frequency oscillations, and low stiffness rubber has good isolation to high frequency oscillations. So, it has been trying to reconcile these two characteristics in light of the prevailing load and fatigue lifecycle. The most suitable is a natural rubber. Tensile strength, elongation and abrasion resistance are excellent throughout the stiffness range. With its strength, fatigue resistance and excellent resistance to deformation, this rubber is ideally suited for use in areas with high dynamic loads at low and natural temperatures. Natural rubber can be used in areas exposed to water, dilute acid, alkaline or other chemicals. Its compounds are not adequately protected against the effects of oil, oil or gasoline. Low resistance to high temperatures and sensitive to atmospherically conditions, unless there is a special blend of natural rubber with special accessories.

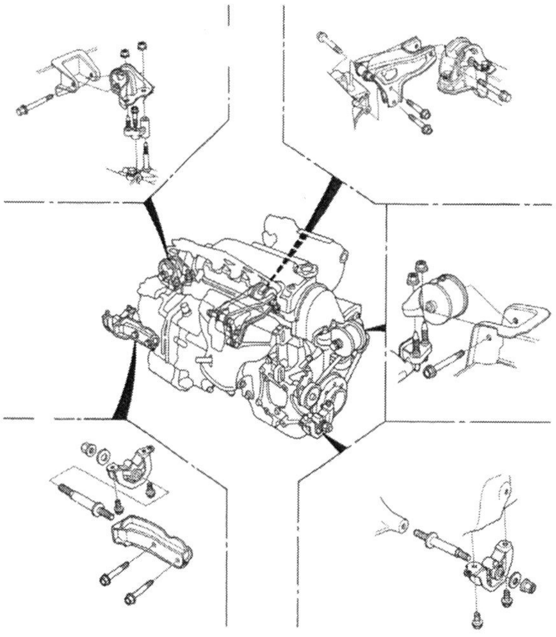

Fig. 1. The arrangements of engine mounting brackets [1, 2]

The type and use of the engine mounting bracket depends on the engine type of the car and how it is mounted in the car (Fig. 2).

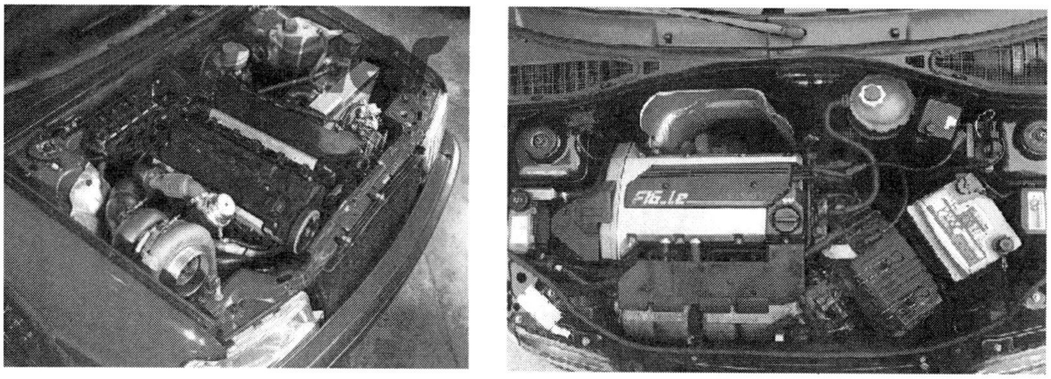

Fig. 2. Engine mounting brackets schemes in the vehicles

Lubricating engine mounting brackets (Fig. 3) are commonly used in diesel engines, because diesel engines have higher vibrations than petrol engines. The lubricating engine mounting brackets provide a noisy stiffness and damping properties that are different from the frequency and excitation amplitude. They are the conceptually the best passive engine fastening element known today.

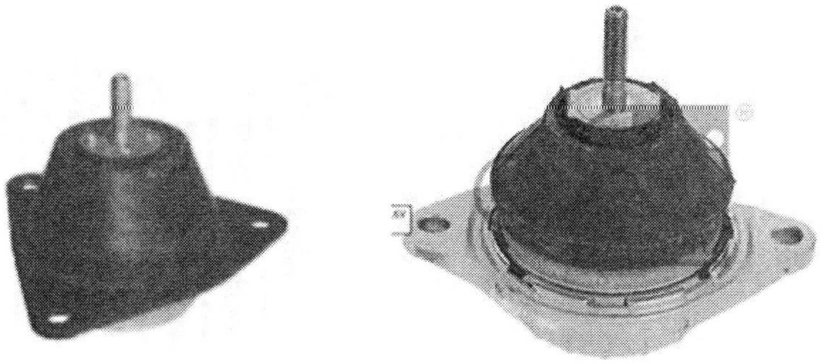

Fig. 3. The lubricating engine mounting brackets [3]

2. Engine mounting bracket optimization studies

An important problem in the automotive industry is achievement a better conceptual model combining performance and production costs in the early stages of product development. The purpose of topology design and optimization is to provide the initial model of the concept, which is achieved through the use of computer technology. The purpose of this research is to demonstrate the design of an engine mounting pillow, using a topology optimization method, optimal structure of the components of the pillow under static loading conditions. The task of this research is to create the concept of an initial project, which has an optimal stroke model. Effectiveness is proven by experimental results.

The structure of the final product is highly dependent on the concept of the original design for which the car's price depends. In order to meet today's vehicle design requirements and increase cost and fuel efficiency, more and more attention is being paid to the weight of the car parts and what materials they are made of. Therefore, an important problem in the automotive industry is how to achieve a better design of the product by combining the product's usefulness and production costs at an early stage of their design and product development.

The biggest drawbacks are due to:

- there is no basic knowledge of the product layout;
- insufficient experience of the designer, lack of creativity;
- there are contradictory factors during the design.

Optimization methods in the automotive industry are used to create products that, by their low weight, improve the cost and fuel efficiency, without sacrificing strength and performance. During optimization, the shape and size of the structure can be changed, but the structure of the topology remains unchanged. Therefore, the optimization method must be combined with a conceptual project to create an optimal layout of the original design.

The new component topology is determined during the concession project. The designer is considering a lot of topology methods; from which he chooses the final component layout. This is a test and error method that greatly depends on the designer's experience and creativity. After this method, the final component layout is possible, which is usually not optimal. However, in terms of optimization of the topology, the designer is not able to choose the right topology from all possible ones, since he is not obliged to know everything about alternative topologies. Another problem is the conflicting factors that arise during the design process. Typically, there are several factors that can reduce weight and increase stiffness. The essential principle of topological optimization is that with the initial shape of the aggregate or detail, and knowing the details of the factors - forces, fortifications, environmental effects, we can choose the optimal form of material and form filling. The method is based on the fact that the initial model is loaded with the already known or calculated steps of the operating loads, and the program chooses and displays the concentration points of the elements by deciding and applying the optimization algorithm. Using these data, we can graphically see which model space requires more stiffness, thus defining the preliminary form of an enhanced model. The goal of the topology optimization method is usually to increase stiffness (slack reduction) or maximize the self-clocking of the structure. In this study, several practical examples are being carried out to determine the effect of specific frequencies on the layout of the product, to reduce mass and to increase the rigidity of the product. Over the past few years, various research has been done using topology design and form optimization techniques. Although a lot of work has been done on topology design, only two methods are widely used by scientists, namely homogenization and density determination. So most topology optimization applications in literature are based on either homogenization or density detection methods. In the homogenization method, the density of the material is considered as a variable and the main objective is to minimize internal stresses that are equal to the maximum specific structure of the structure [7, 8]. Using this method, the density of the material is selected as a variable size, and the purpose is the same as the homogenization method. The density of a single finite element is selected as a variable, and its relation is expressed in the empirical formula of the Jung's module. This connection is between density and material properties irrespective of their microstructure. They use the density method to determine the structure of the topology. The main objectives would be to increase the strength of the structure and increase stiffness. The change of engine position did not fit the motor mounting pillow to the engine. It was necessary to redesign the existing engine mounting pillow. The engine shifted 30 mm in a horizontal direction. The engine cushion began to crack.

The original model (Fig. 4) was prepared with SolidWorks software. It was modelled in accordance with the constraints imposed, i. e. to geometric dimensions. Detailing the old engine mounting bracket and using a new dimension, i. e. displacement 30mm. The new model met the geometric requirements, but it was not known whether this form was optimal. Therefore, the original engine mounting bracket model was primitive and very simplified (Fig. 5). Topology optimization was done using Ansys software package. Since the material of the engine mounting bracket elements are known, the same material properties were assigned to this model. The finite element mesh is chosen as finer as possible to make the calculations as accurate as possible [5, 6]. The smaller the grid, the more complex the calculation, and the longer this calculation takes, and it requires a lot of computer resources.

Fig. 4. Original engine mounting bracket

The design of the new bracket is sufficiently stiff, but it is not tested or has an optimal geometric shape. For this purpose, a more technologically acceptable model was designed and an optimal form was calculated and an optimized final model was modelled.

The preliminary design model was loaded with a force of 1000 *N* and mounted as already mentioned in the previous section (Optimization of the original engine mounting pillow model), like the original model. After calculating the static case, we obtained that the vertical displacement is 7.8 *mm* and the first natural frequency is 125 *Hz*. Model mass 520 *g*. The final model is designed optimally. After performing the static analysis and adding the same force, the obtained results are also presented in Table 1 and Figure 6–7.

Fig. 5. Finite element engine mounting bracket model (left) and topology optimization results. Iterations number – 100; mass reduction – 70%

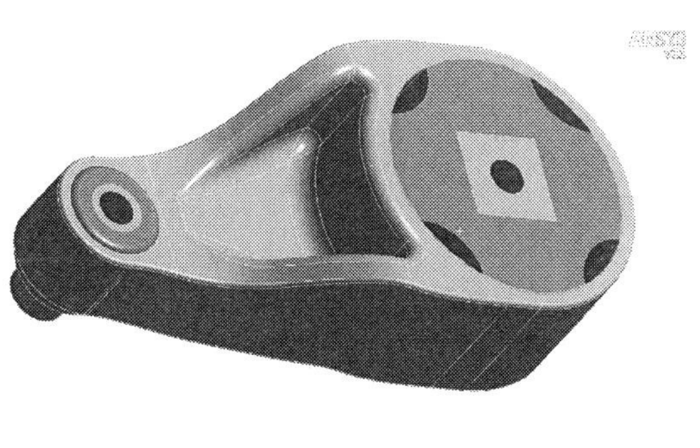

Fig. 6. The final design of engine mounting bracket

Table 1. Comparison of initial and optimized engine mounting brackets calculations

Parameter	Initial variant	Optimized variant
Mass, *g*	520	512
Compliance, *mm/kN*	7.8	3
Stiffness, *kN/mm*	0.128	0.33
First natural frequency, *Hz*	125	273

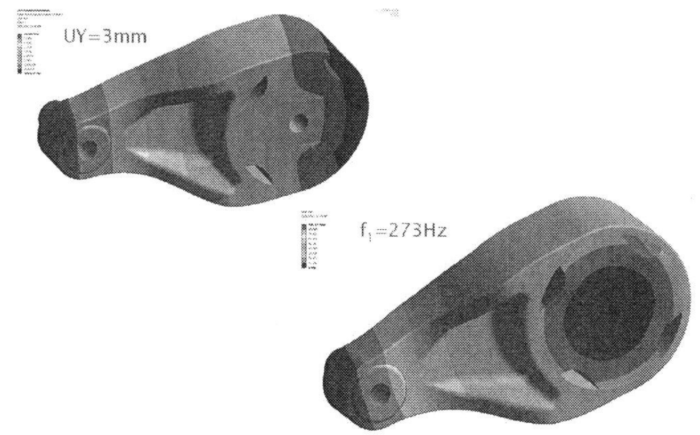

Non è stata trovata alcuna voce dell'indice delle figure.Conclusions

The work was done to optimize the topology of the car engine mounting bracket using Ansys software package. After topological optimization analysis results, changes to the vehicle engine mounting bracket design has been performed. Newly estimated engine mounting bracket shape have positive results as subsequently determined by numerical static and dynamic tests. It is statistically and dynamically computed that the stiffness of the final model increased by 2.6 times and the first natural frequency increased by 2.2 times. The optimization method has proven to be more robust and somewhat lighter than initial structure.

References

[1] Engine mounting brackets. Access via internet: http://www.mscsoftware.co.kr/upfile/conference_pdf/3-8.pdf [Accessed: 2017-01-20].
[2] Engine mounting brackets. Access via internet: http://pid.sagepub.com/content/221/7/823.full.pdf [Accessed: 2017-01-20].
[3] Lubricating engine mounting brackets. Access via internet: www.ias.ac.in/sadhana/Pdf2000June/Pe879.pdf [Accessed: 2017-01-20].
[4] Engine mounting brackets rubber. Access via internet: http://www.elastomer.co.uk/natural_rubber [Accessed: 2017-01-20].
[5] Barauskas R. Baigtinių elementų metodo pagrindai. Kaunas: Technologija, 1998. 375 p.
[6] Baranauskas R, Kačinauskas R, Belevičius R. Baigtinių elementų metodo pagrindai. Vilnius: Technika, 2004. 612 p.
[7] Finite element method. Access via internet: http://e-stud.vgtu.lt/files/dest/7342/3_baigtiniu%20elementu%20metodas.pdf [Accessed: 2017-02-20].
[8] Topology and shape optimization. Access via internet: http://fe04.fe-design.de/fileadmin/publikationen/publikationen2001/2001-06-06_CIRP-final.pdf [Accessed: 2017-02-20].

ITMS'2018

panevėžys faculty of technologies and business
ktu 1922

The 12th International Scientific Conference Intelligent Technologies in Logistics and Mechatronics Systems (ITELMS'2018), 26–27 April 2018, Panevėžys, Lithuania

Control Algorithm of Multiple Unmanned Electrical Aerial Vehicles for Their Collision Prevention

Anna Beinarovica[a*], Mikhail Gorobetz[a], Anatoly Levchenkov[a]

[a]Riga Technical University, Institute of Industrial Electronics, 12 Azenes street, Riga, LV-1048, Latvia

Abstract

Authors propose the algorithm to prevent collisions between various unmanned electrical aerial vehicles (UAV) performing the common goal or moving at the same area. The developed algorithm is proposed for use in case of some unmanned vehicles are moving towards or through the same coordinates of the target point or the trajectory. All the vehicles are communicating with each other. The algorithm is implemented in the developed embedded device of each UAV that corrects the flight height and sets necessary motor rotation speed to avoid the collision. The computer models of the quadcopters, as an example of UAV system, are presented in the paper and prove the workability of the algorithm.

© 2018 A. Beinarovica, M. Gorobetz, A. Levchenkov
Peer-review under responsibility of the Kaunas University of Technology, Panevėžys Faculty of Technologies and Business

Keywords: unmanned electrical vehicle, anti-collision system, aerial vehicle, electric drive

1. Introduction

The use of aerial vehicles for industrial purposes began relatively recently. Photographers all over the world offer to use modern technologies to gain a better result of taking photos of celebrations, such as wedding. In Australia drones are equipped with artificial intelligence powered software that can distinguish sharks from dolphins, wales, boats, and other marine life in real-time with 90% accuracy. Drones could help detect potential terrorists in public spaces, merely by measuring anomalies in their heart rates, according to Chahl, a Professor of Sensor Systems in

* Corresponding author. Tel.: +371 25998958
E-mail address: anja19892@inbox.lv

© 2018 A. Beinarovica, M. Gorobetz, A. Levchenkov
Peer-review under responsibility of the Kaunas University of Technology, Panevėžys Faculty of Technologies and Business

UniSA's School of Engineering. Amazon legally delivered its first Prime order in the United Kingdom in December 2016. All these possibilities of use aerial vehicles prove the relevance of the topic.

In most cases, man-operator, who sets the speed and the trajectory of motion by using the remote control, controls the aerial vehicle. However, it is not productive use of human time and force in the automatization and optimization century. Moreover, man, who is driving aerial vehicle, has ability to make mistakes, caused by human factor. For example, a woman was seriously injured by a falling drone in 2016 in Quebec, a drone crashed through a Manhattan woman's 27th-floor window, and a New Jersey man was arrested for accidentally crashing a drone into the Empire State Building.

All over the world, companies are trying to embed artificial intellect (AI) in the vehicles and to provide the autonomous drive, and unmanned aerial vehicles, such as drones or quadcopters, not an exception. This may help to minimize amount of aerial vehicles collisions caused by human factor. In 2016. S. Roelofsen, A. Martinoli and D.Gillet proposed a collision avoidance algorithm for unmanned aerial vehicles with limited field of view constraints [1]. Authors presented a safe collision avoidance algorithm based on potential fields for fixed-wing unmanned aerial vehicles (UAVs) with constrained field of view sensors such as cameras. They demonstrated the effectiveness of the proposed method with several simulations, including one with randomized trajectories covering a large set of possible configurations. In difference with a mentioned research, this paper deals with a collision avoidance of rotorcraft vehicles. In [2] conflict resolution was achieved with obstacle trajectory data taken from a simulated camera and range-finder in the presence of their respective measurement uncertainties. M. Hehn and R. D'Andrea proposed an algorithm for the real-time trajectory generation for quadcopters [3]. The ability to plan trajectories from nonrest conditions was used in conjunction with waypoints i1n order to guide vehicles around obstacles without stopping. In this study, the main focus is still on the trajectory generation for the multiply UAV collision prevention between themselves.

2. Problem formulation

Usually unmanned aerial vehicles are operated by the human. Therefore, providing totally unmanned control system for the aerial vehicle needs the assessment of all the possible risks. The most common risks are: collision with other objects, fast or undelayed battery discharge, wrong route planning etc. In cases of solving tasks with several UAV, one more risk appears – risk of collision between UAV. Let's suppose, that three UAV are looking for the same object, and each of them is programmed to reach his goal. It means that each of them will try to reach approximately the same point coordinates. So, how can each of them reach an aim without interfering each other? The algorithm described in this paper solves this problem and prevents collisions between UAV in tasks of several UAV working in one area.

In 2014 anti-collision system for navigation inside an UAV, using fuzzy controllers and range sensors [4] was proposed. In that research authors were working to provide a system that will help to prevent UAV collisions with obstacles, but nothing about collisions with other UAV was said. In 2013 a comparative study of collision avoidance techniques for unmanned aerial vehicles was presents and published by A. Alexopoulos and others [5]. The first collision avoidance method in that study was based on a geometric approach which computes a direction of avoidance from the flight direction and simple geometric equations. The second technique used virtual repulsive force fields causing the UAV to be repelled by obstacles. The last method was a grid-based online path re-planning algorithm with A* search that finds a collision free path during flight. Various flight scenarios were defined including static and dynamic obstacles. In difference of that research, we provide a system, were dynamic obstacles are UAV, and all these dynamic obstacles are communicating with each other. This proves the novel of the proposed research.

3. Anti-collision system structure

The proposed anti-collision system is not centralized and is distributed among the vehicles. All calculations and decision-making are made by the anti-collision system embedded in each UAV separately. Figure 1 shows the structure of the proposed system.

There are communication components ensuring data transmission, such as satellite positioning system – GNSS (such as GALILEO, GPS etc.) and radio frequency modules – RF.

Each UAV has embedded electronic device D_{TR} – control components to obtain the position, to calculate the motion parameters, to communicate with other devices and to control the electric drive of UAV.

Authors assume that D_{TR} is embedded into each mechanical vehicle for the most effective proposed control system operation.

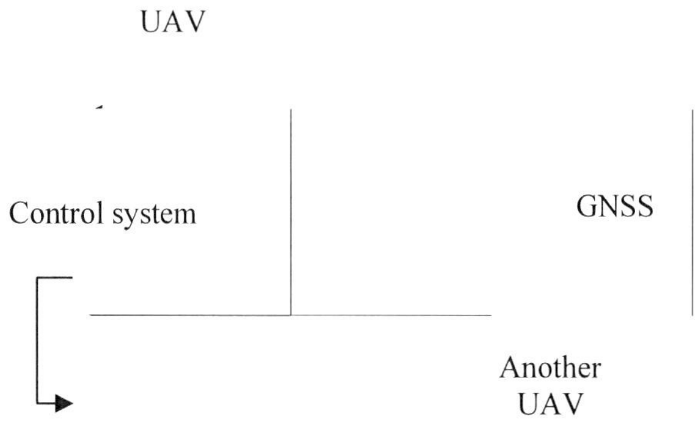

Fig. 1. Structure of the anti-collision system of UAVs

D_{TR} device of the UAV receives the information about UAV location by GNSS and information about other vehicles location, speed and movement direction by RF. Risk assessment module (RA) calculates the distance between UAV. If the distance is smaller than specified, the necessary safe height of flight is found out by target function (TF) and decision about height change is made. In decision, making module (DM) the necessary engine speed is calculated also and is send to the D_{TR} device. D_{TR} controls the UEVs electric drive (electronic speed controller and electric motor M) and affects the engine speed.

4. Mathematical model of the proposed system of UAV

The team of the UAVs is a group of autonomous unmanned aerial vehicles whose actions agree with certain rules and with only common interests.

The set of UAVs is given:

$$UAVS = (UAV1, \ldots UAVn). \quad (1)$$

The **utility** function as a common interest function is:

$$U = w(x, a1, a2, \ldots, an) \rightarrow opt; \quad (2)$$

where u – **utility** function – common interest function;
x – **state** of the environment;
a_i – **action** of the i-th UAV.

The **information** of each UAV depends on the state of environment:

$$yi = \alpha i(x); \quad (3)$$

where y_i – **information** of i-th UAV.

The **decision rule** of i-th UAV results an action of i-th quadcopter and depends on the information:

$$ai = \beta i(yi)l; \tag{4}$$

where β_i – **decision rule** of i-th UAV.

Interaction between i-th and j-th UAV:

$$q_{ij} = \frac{\partial w}{\partial a_i \partial a_j}. \tag{5}$$

A set of decision rules is optimal if:

$$E(S) = E(\, w(\, x, (\, \beta 1(y1), \ldots, \beta n(yn)\,)\,)\,) \rightarrow \text{max for a given probability distribution on x.} \tag{6}$$

The location L^{UAVS} of UAVs is represented by three subsets $<\chi_c^{UAVS}, \psi_c^{UAVS}, \eta_c^{UAVS}>$, that are latitutde χ, longitude ψ and altitude η:

$$\chi_c^{UAVS} = \{\chi_c^{UAV_1}, \chi_c^{UAV_2}, \ldots, \chi_c^{UAV_n}\}, \psi_c^{UAVS} = \{\psi_c^{UAV_1}, \psi_c^{UAV_2}, \ldots, \psi_c^{UAV_n}\}, \eta_c^{UAVS} = \{\eta_c^{UAV_1}, \eta_c^{UAV_2}, \ldots, \eta_c^{UAV_n}\}; \tag{7}$$

where χ_c^{UAV} – Latitude of the current point;
ψ_c^{UAV} – Longitude of the current point;
η_c^{UAV} – Altitude of the current point.

The common goal defined in this paper is to reach the same target point TP = $<\chi_{tp}, \psi_{tp}, \eta_{tp}>$.
Geographical coordinates of the target point are defined by following set:
χ_{tp} – Latitude of the target point;
ψ_{tp} – Longitude of the target point;
η_{tp} – Altitude of the target point.

But the common safety, i. t. collision prevention, criterion is:

$$D = |UAV_i UAV_j| = \sqrt{(\chi_c^j - \chi_c^i)^2 + (\psi_c^j - \psi_c^i)^2 + (\eta_c^j - \eta_c^i)^2} > S; \tag{8}$$

where S is safety distance limit for each pair of $< UAV_i, UAV_j >$, $i = 1..n$, $j = 1..n$, $i \neq j$.

It is obvious that in case if $\chi_c^{UAV_i} = \chi_c^{UAV_j} = \chi_{tp}$ AND $\psi_c^{UAV_i} = \psi_c^{UAV_j} = \psi_{tp}$ AND $\eta_c^{UAV_i} = \eta_c^{UAV_j} = \eta_{tp}$ the safety criteria cannot be satisfied, because $D = 0$.

So, the common target function with anti-collision criteria is following:

$$\begin{cases} \eta_c^{UAV_i} = \eta_c^{UAV_j} \rightarrow \eta_{tp} \\ \chi_c^{UAV_i} = \chi_c^{UAV_j} \rightarrow \chi_{tp} \\ \psi_c^{UAV_i} = \psi_c^{UAV_j} \rightarrow \psi_{tp} \\ D = |UAV_i UAV_j| > S \\ i = 1..n, \\ j = 1..n, \\ i \neq j \end{cases}. \tag{9}$$

Each UAV is tending to reach the target point, but if it is not possible than the altitude becomes a subject to be changed first. If it is impossible to change the altitude, then latitude and longitude might be changed.

If the distance D till other UAV is less or equal to S, the collision possibility is high and the decision about the altitude η_c^p change is made.

5. Control algorithm of multiple UAV for their collision prevention

Control algorithm of multiple electrical UAV for their collision prevention was developed:

STEP 0. Initialization: Let's assume that all UAVs have the common goal TP. The UAV_1 is the vehicle, where the following algorithm is working. And it is assumed that the same algorithm is working at any other UAVS.

STEP 1. Determination of the UAV_1 own coordinates: χ_c^{UAV1}, ψ_c^{UAV1}, η_c^{UAV1}.

STEP 2. Selecting the UAV_i and requesting of its coordinates.

STEP 3. Receiving coordinates of other UAV_i: χ_c^{UAV2}, ψ_c^{UAV2}, η_c^{UAV2}.

STEP 4. Calculating the distance D between the UAVs using (8).

STEP 5. Decision making about speed change if D<=S.

If $\eta_c^{UAV1} \geq \eta_c^{UAVi}$ then UAV_1 sets the $\eta_{tp}^{UAV1}=\eta_{tp}^{UAV1} * 1.5$.

If $\eta_c^{UAV1} < \eta_c^{UAVi}$ then UAV_1 sets the $\eta_{tp}^{UAV1}=\eta_{tp}^{UAV1} * 0.5$.

If distance D is bigger or equal than allowed, then nobody is changing the height and $\eta_{tp}^{UAV1} = \eta_{tp}$.

STEP 6. Calculating target rotation speed (tw) for each motor.

STEP 7. Target Euler angles of motion (rotP – pitch, rotR – roll, rotY – yaw) calculation.

STEP 8. Traction forces (F_Z, F_{XY}) calculation.

STEP 9. Energy calculation.

STEP 10. Battery capacity calculation.

STEP 11. Processing the decision to control motors.

STEP 12. i = i+1, if i>n then i=2.

STEP 13. Starting from the STEP 1 and checking the collision risk with another UAV_i.

6. Computer models and experiment

Any maneuvres of the quadcopter requires to know the exact rotation speed of the propeller engines and necessary Euler angles. For this purpose, a Simulink model for the quadcopter UAV for angles calculations was developed (Fig. 2).

For the experiment target, point (300; 300; 100) was chosen. Additionally, program shows parameters for each UAV:

- time (t) of UAV's engine work;
- current coordinates (X, Y, Z);
- Euler angles of motion (rotP – pitch, rotR – roll, rotY – yaw);
- engines speeds (w1, w2, w3, w4);
- force vector on z axis (Fz);
- force vector on xy axis (Fxy);
- traction force (F);
- energy;
- battery capacity.

Start of the experiment is shown in the Figure 3. Three UAVs have different start point and each of them starts to move towards coordinates of the target point. After they are close enough to each other, they start to change height in case of prevention collision.

In Figure 4 the result of the experiment is shown. Three UAV reached the target point coordinates without collide and each of them has changed their height according to the developed algorithm. The first UAV took up a position at the height of 200, the second one is almost on the target height, but the third one is descended to the height 50. As the result, collision possibility was reduced to zero by changing only one target coordinate Z – height.

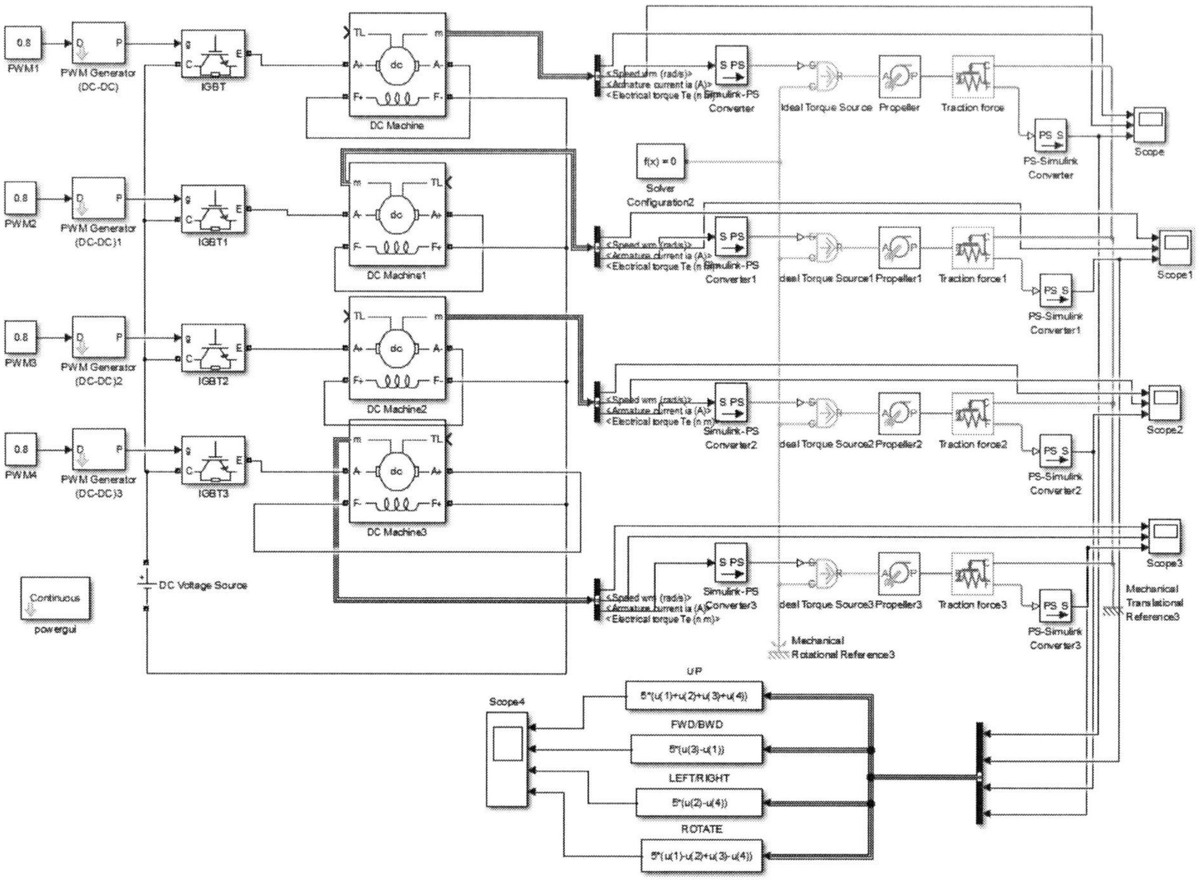

Fig. 2. Simullink model for the quadcopter UAV

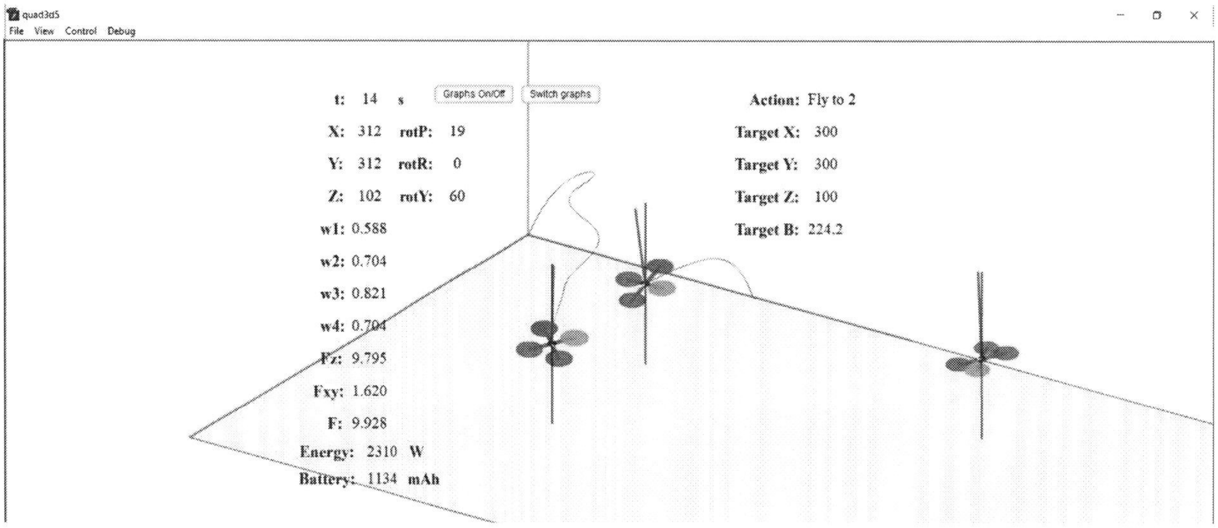

Fig. 3. Start of the experiment of multiply UAV collision prevention algorithm

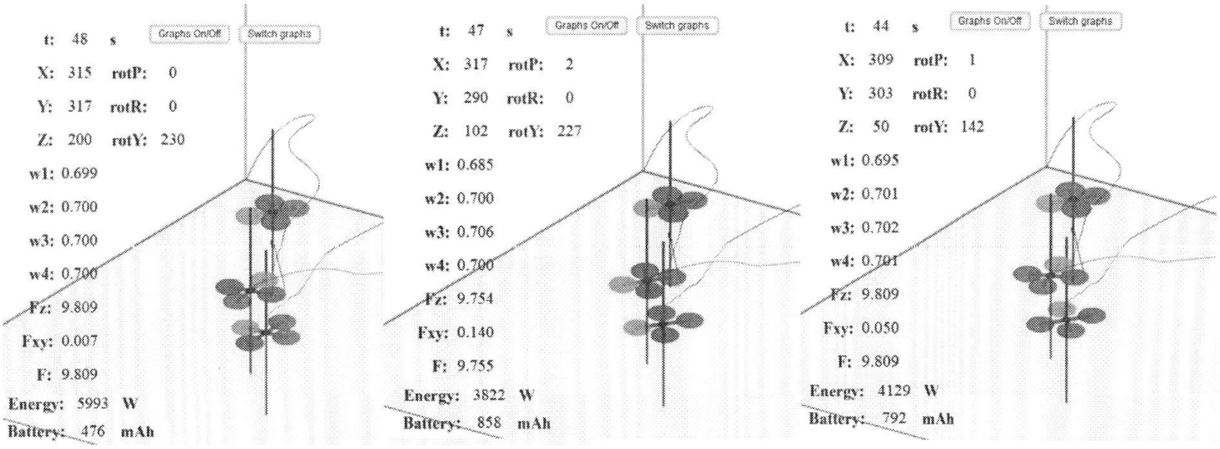

Fig. 4. UAV reaching the target point coordinates without collide

Conclusions

The proposed in this paper algorithm is working correctly. Experiment shows, that all three unmanned aerial vehicles reached their target point without any collisions by changing only one target coordinate – height of the flight.

As the proposed devices are embedded into the UAVs there is no necessity to involve infrastructure and devices can work regardless of location, also in forest or over reservoirs.

References

[1] Roelofsen S., Martinoli A. and D. Gillet. 3D Collision Avoidance Algorithm for Unmanned Aerial Vehicles with Limited Field of View Constraints// In Proceedings of 55th Conference on Decision and Control (CDC), Las Vegas, USA, December 12–14, 2016.

[2] Douthwaite J. A., De Freitas A., Myhailova L. S. An Interval Approach to Multiple Unmanned Aerial Vehicle Collision Avoidance// In Proceedings of Sensor Data Fusion: Trends, Solutions, Applications (SDF), Germany, 2017.

[3] Hehn M. and R. D'Andrea. Real-Time Trajectory Generation for Quadrocopters // Published in IEEE Transactions on Robotics, 2015.

[4] Darío Fonnegra Tarazona R., Ríos Lopera F., Góez Sánchez G.-D. Anti-collision system for navigation inside an UAV using fuzzy controllers and range sensors// In Proceedings of Signal Processing and Artificial Vision (STSIVA), Armenia, Colombia, 2014.

[5] Alexopoulos A., Kandil A., Orzechowski P. and E. Badreddin. A Comparative Study of Collision Avoidance Techniques for Unmanned Aerial Vehicles// In Proceedings of IEEE International Conference on Systems, Man, and Cybernetics, 2013.

ITMS'2018

ktu 1922 panevėžys faculty of technologies and business

The 12th International Scientific Conference Intelligent Technologies in Logistics and Mechatronics Systems (ITELMS'2018), 26–27 April 2018, Panevėžys, Lithuania

Determination of Integrated Indicator for Analysis of the Traffic Safety Condition for Traction Rolling Stock

Boris Bodnar[a], Yaroslav Bolzhelarskyi[a], Oleksandr Ochkasov[a*], Tatyana Hryshechkina[a], Laura Černiauskaitė[b]

[a]*Dnipropetrovsk National University of Railway Transport named after Academician V. Lazaryan, Lazaryan St., 2, Dnipro, Ukraine, 49010*
[b]*Vilnius College of Technologies and Design, Petras Vileisis Railway Transport Faculty, K. Kalinausko str. 7, Vilnius, Lithuania*

Abstract

Traffic safety is a major priority in railway transport operation. Locomotive facility is one of the responsible units in general system of railways. A complex and cumbersome system of indicators is used in locomotive facilities to analyze the operation. The existing system makes it difficult to analyze the general level of work organization in the locomotive facilities. Purpose of the study is to determine the methodology of forming a certain dimensionless indicator (or group of indicators) that will reflect the general level of safety in the locomotive facilities. As the research methodology it was chosen principal component analysis as the corresponding mathematical apparatus, which makes it possible to analyze the existing indicators characterizing the performed work and the traffic safety condition with the necessary degree of informativity. As a result, the main components and the degree of their influence on the general level of traffic safety in the locomotive facilities are set. The indicators that have the most influence on the technical and safety components of the integrated indicator of traffic safety condition are determined. Originality of the work is that it for the first time proposes the concept of index of traffic safety condition and the method of its determination using the principal component analysis. Practical value of the work lies in the fact that the ranking of technical and safety components according to the degree of their influence on the general traffic safety index has been performed. Also, the locomotive units which have the greatest influence on the traffic safety condition and reliability were determined.

© 2018 B. Bodnar, Y. Bolzhelarskyi, O. Ochkasov, T. Hryshechkina, L. Černiauskaitė
Peer-review under responsibility of the Kaunas University of Technology, Panevėžys Faculty of Technologies and Business

Keywords: railway transport, traffic safety, locomotive, reliability, principal component analysis, traffic safety condition

* Corresponding author. Tel.: +38 (056) 733 19 61.
E-mail address: abochkasov@gmail.com, ORCID 0000-0002-7719-7214

© 2018 B. Bodnar, Y. Bolzhelarskyi, O. Ochkasov, T. Hryshechkina, L. Černiauskaitė
Peer-review under responsibility of the Kaunas University of Technology, Panevėžys Faculty of Technologies and Business

B427H0007 . EdLearning.it

1. Introduction

Operational work at railway transport should be organized according to the requirements of the Rules of Technical Operation of Ukrainian Railways (RTO). In addition to analyzing quantitative and qualitative indicators at the railway network, constant monitoring and analysis of the traffic safety condition are carried out. The purpose of analysis is to determine the ways to improve the operation efficiency of locomotives and ensure the transportations safety.

Locomotive facilities are part of a complex organizational and technical system of the railway. The tasks of locomotive facilities include: operational work, rolling stock repair, ensuring traffic safety, planning technical and organizational support, and a number of others. Obviously, estimating the traffic safety of locomotive fleet using the existing system of indicators is a difficult task. Indicators analysis requires the use of a systematic approach that will make it possible to conduct fuller and deeper analysis.

With the existing system of analysis of the locomotive facilities operation, the available statistical information is a huge amount of information that is difficult to analyze. The conducted analysis makes it possible to assert that the existing system of indicators of traffic safety analysis is periodically changed from year to year, which also complicates the analysis. It should be realized that attempts to improve the process of analyzing the locomotive facilities performance by introducing new additional indicators will lead to the creation of even larger data arrays, the analysis of which using conventional methods will be complicated. For example, the work [1] presents 70 concepts and indicators of locomotive use, which are grouped into 4 groups. They include maintenance support, runtime and service life, availability and locomotive usage.

According to the European Classification, to analyze the level of safety and reliability of technical objects the RAMS (Reliability, Availability, Maintenance, and Safety) groups, which are provided by the standards, as well as the concept of life cycle cost LCC are used. Each of the RAMS groups includes a number of indicators. For example, the reliability group includes such indicators as failure rate, durability, failure probability, runtime and run-to-failure, average time between failures. Availability takes into account a number of coefficients (intrinsic, operational and achievable availability), rolling stock utilization rate, utilization degree of service operation schedules [2].

Specialists in the field of railway transport reliability carry out investigations of the interconnection degree between the RAMS parameters.

The work [3] presents the studied influence of the railway transport reliability on its availability. The article considered the main RAMS-characteristics, showed the connection between the reliability and availability of transport modes according to the European standard EN 50126 [4, 5]. The factors influencing the RAMS-characteristics are discussed in detail, and the hierarchical dependency schemes are given. Also, the work determines the availability of railway transport modes as a function of reliability and average downtime. The obtained dependence can be useful when making decisions on the rolling stock modernization. But this function does not provide complete information on the reliability and traffic safety at the railway. To achieve this, additional factors should be taken into account. This necessitates identification of one indicator that contains all the necessary traffic safety information.

Such models of locomotive failures as the distribution models of the engine failure number, depending on the locomotive revolutions or on the fuel consumed are analyzed in the work [6]. The authors propose to consider the density indicator of the engine failure probability in accordance with the fuel consumption as one of the indicators of engine reliability.

The work [7] notes that the growing complexity of equipment often leads to failures. In this regard, the aspects of reliability, maintainability and availability are now on the forefront. The paper considers the engine as an important component of the traction rolling stock. A failure rate analysis is carried out using the Weibull distribution.

In the paper [8] the authors analyzed the rolling stock reliability using the methods of expert research in order to determine the influence degree of electric locomotive units on the possibility of its further operation. The influence of locomotive equipment reliability on the choice of the maintenance system and the parameters of the diagnostic systems is explained in the work [9].

In the work [2] it is noted that reliability functions can be investigated using covariance. Example of the use of structural modeling to study the operation of railway transport enterprises is given in the work [10]. The authors identified the most significant indicators of the locomotive operation.

The selected information may be useful for the following reasons:

- to determine the essential factors affecting the service life of an element;
- to remove variables that distort analysis;
- to improve understanding of failure modes;
- to provide more efficient prediction of the failure rate.

2. Methodology

In the conducted research the authors propose to analyze existing indicators characterizing the traffic safety condition in the locomotive facilities [11] using the principal component analysis [12]. The study is aimed to determine the methodology of forming a certain dimensionless indicator (or group of indicators) that will reflect the general level of safety condition in the locomotive facilities. It is clear that the creation of a single "general indicator" to evaluate the safety condition in the locomotive facilities is a difficult task, but the use of modern methods of mathematical analysis and the fuzzy sets theory makes it possible to formalize the analysis of indicators. The use example of the principal component analysis for analyzing the quality of the maintenance system for electric rolling stock is given in the work [13].

The main preconditions for the transition from a large number of initial indicators of the locomotive facilities operation to a smaller number of the most informative variables, as noted above, are as follows:

- duplication of information derived from the closely interrelated signs (according to analysis of the traffic safety condition at Ukrzaliznytsia). For example, the indicators "wrong actions of employees" and "wrong actions of locomotive crews"; occurrence of faults of the traction rolling stock, faults distribution according to the equipment types and repeat of "wrong actions of locomotive crews" complicate analysis and may lead to data distortion;
- non-informative indicators, small variability, when the indicator is practically unchanged during transition from one object to another;
- possibility of aggregation according to some characteristics.

Among the variants of methods for reducing the dimension (principal component analysis, factor analysis, extreme grouping of parameters, selection of the most informative indicators in discriminant analysis models and regression models, etc.) the authors selected the principal component analysis as the corresponding mathematical apparatus, which makes it possible to analyze the existing indicators characterizing performed work and the traffic safety condition with the necessary degree of informativity.

Advantages of this method for analysis and forecasting, unlike, for example, the classical regression analysis, lie in the fact that at the latter, one attempts to include in the model the maximum possible number of factors that are often characterized by significant correlation (multilinearity). The forecast for such variables, as a rule, is not accurate. Therefore, the problem arises to replace the initial interrelated variables by a set of uncorrelated parameters. This problem is solved precisely using the principal component analysis.

This method approximates the n-dimensional set of observations to an n-dimensional ellipsoid. Reducing the dimension occurs by projection on the semi-axis of this ellipsoid. Thus, the largest amount of information is stored. And the semi-axes of the obtained ellipsoid will be the main components – new, uncorrelated indicators.

Application of this method when solving the problem of analyzing the traffic safety condition in the locomotive facilities is as follows.

We assign k-dimensional vector of output variables (indicators related to traffic safety):

$$X = \left(x_1, \ldots, x_k\right). \tag{1}$$

We find a linear combination of output data:

$$G = AX\ . \tag{2}$$

Matrix A has the following form:

$$A = \begin{pmatrix} a_{11} & \dots & a_{1k} \\ \dots & \dots & \dots \\ a_{i1} & \dots & a_{ik} \\ \dots & \dots & \dots \\ a_{k'1} & \dots & a_{k'k} \end{pmatrix} \tag{3}$$

where a_{ik} – is the k-th element of the i-th eigenvector of the matrix A.

For each matrix raw, a condition must be fulfilled:

$$\bar{a}_i \bar{a}_i^T = 1. \tag{4}$$

In order to make each subsequent linear combination identify new factors that are not related to the previous ones, we introduce an additional condition of pairwise correlation of the principal components:

$$corr(g_i, g_j) = 0 \quad i = \overline{1,k}\ ,\ \ j = \overline{1,k}\ . \tag{5}$$

To calculate the first main component, let us solve the optimization problem:

$$\begin{cases} D(g_1) \to \max\limits_{a_1} \\ \bar{a}_1 \bar{a}_1^T = 1 \end{cases}; \tag{6}$$

where $\bar{a}_1$ – is the first raw of the matrix A, eigenvector corresponding to the largest own value of the matrix A.

Similarly, all the principal components of the vector G are found. As a result, a vector of new variables – the principal components $G = (g_1, ..., g_k)$ is obtained.

To determine significant quantity of principal components, the following condition is used:

$$\sum_{i=1}^{p} D(g_i) \le \alpha\ ,\quad p \le k\ . \tag{7}$$

where α – is the required percentage of saving initial information.

In order to determine a single integrated indicator of traffic safety condition the calculations using the hierarchy analysis method [14] are performed. The initial data for the hierarchy analysis method are the results of work of the principal components analysis (namely, the number of principal components and their dispersion). Calculation using the method of hierarchies is as follows.

1. A matrix of pairwise comparisons is constructed. The number of matrix columns is the number of principal components obtained at the previous step. Elements of the first line:

$$a_{11} = 1\ ;$$

$$a_{12} = \frac{s_2}{s_1}; \tag{8}$$

$$a_{13} = \frac{s_3}{s_1};$$

where s_i – is the dispersion of *i*-th principal component.

2. The component weights are determined:

$$w_i = \frac{a_{1n}}{a_{1i}}. \tag{9}$$

3. The relative values of the weights are determined:

$$w_i^* = \frac{w_i}{\sum_{k=1}^{n} w_k}. \tag{10}$$

4. The index of traffic safety condition is determined:

$$I_{ts} = \sum_{i=1}^{n} w_i^* g_i; \tag{11}$$

where g_i – *i*-th principal component.

Information from statistical sources of Ukrzaliznytsia (UZ) [11] is taken as initial data. During calculations the possibility of creating new indicators of locomotive reliability was not considered, only the UZ classifiers adopted in the locomotive facilities were used.

Statistical data grouped in the sections were used as the initial data. Statistical data are shown in the Table 1.

Table 1. Statistical data on traffic incidents

Section	Indictor		Values by years							
			2008	2009	2010	2011	2012	2013	2014	2015
Traffic incidents	red light running (drastic accident)	x_1	3	1	0	7	3	4	3	5
	derailments and collisions of traction rolling stock during shunting (incidents)	x_2	4	7	19	21	16	0	0	0
	self-uncoupling and breakage of automatic coupling (incidents);	x_3	1	1	2	0	4	0	0	0
	wrong actions of employees (incidents);	x_4	0	0	0	6	48	47	56	40
	fault occurrence of traction rolling stock (incidents);	x_5	234	207	186	195	200	186	150	173

Table 1 (Continued)

Section	Indictor		Values by years							
			2008	2009	2010	2011	2012	2013	2014	2015
Causes of traffic incidents	insufficient roundhouse servicing	x_6	168	175	170	157	146	131	116	113
	wrong actions of locomotive crews (incidents)	x_7	87	77	78	63	72	72	68	67
	insufficient factory repair	x_8	34	19	24	12	8	10	2	0
	technical ones	x_9	29	41	36	45	29	18	18	33
	others	x_{10}	6	9	0	5	2	2	9	0
Faults of the main equipment of the traction rolling stock:	diesel	x_{11}	36	30	32	30	32	28	29	37
	main generator	x_{12}	1	3	4	1	2	1	4	2
	power transformers	x_{13}	2	4	4	2	1	3	5	4
	traction electric engines	x_{14}	27	24	35	27	27	37	20	18
	mechanical support equipment	x_{15}	9	12	2	20	11	6	8	8
	electrical support equipment	x_{16}	6	1	3	14	2	0	0	0
	elements of traction gear	x_{17}	0	0	2	2	0	0	0	0
	wheel sets	x_{18}	5	2	2	1	2	0	0	0
	axle-boxes of wheel sets	x_{19}	19	21	13	15	9	13	7	10
	wheel-motor block	x_{20}	23	11	23	5	6	5	9	3
	power circuits	x_{21}	43	43	44	23	37	32	19	24
	control circuits	x_{22}	54	57	32	21	29	15	17	15
	power electrical equipment	x_{23}	12	16	7	25	8	12	7	15
	electrical equipment of control circuits	x_{24}	1	3	2	10	1	1	0	3
	accumulator battery	x_{25}	6	9	6	3	3	4	4	3
	braking equipment	x_{26}	21	21	20	18	12	9	6	12
	security devices	x_{27}	0	1	2	0	0	0	0	0
	mechanical equipment	x_{28}	0	4	0	0	12	5	6	3
	roof equipment	x_{29}	18	20	16	10	13	15	9	12
	automatic coupling	x_{30}	2	1	0	0	0	0	0	0
	bogie, vehicle	x_{31}	2	5	8	8	0	0	0	0
	others	x_{32}	5	0	0	1	1	0	7	0
	wrong actions of locomotive crews (equipment failures)	x_{33}	32	30	51	45	49	47	56	40

If one analyzes the railway transport incidents statistical data, they are overloaded by absolute indicators that tend to decrease (Figures 1 and 2), but they do not allow making objective conclusions concerning the traffic safety condition.

For more detailed study of the traffic safety condition in the locomotive facilities, it is necessary to perform a joint analysis of qualitative and quantitative indicators and indicators that reflect the reliability, efficiency and operation quality of the traction rolling stock fleet of Ukrzaliznytsia. This article does not consider this problem.

The research sets the problem to combine information from existing indicators of traffic safety condition $x_1 \div x_{33}$ and rank (sort) the values according to their influence on the reliability, efficiency and traffic safety condition in the locomotive fleet.

As a generalization indicator, it is proposed to use the term "Index of traffic safety condition" I_{ts}, which can be found using the methodology (1)–(11).

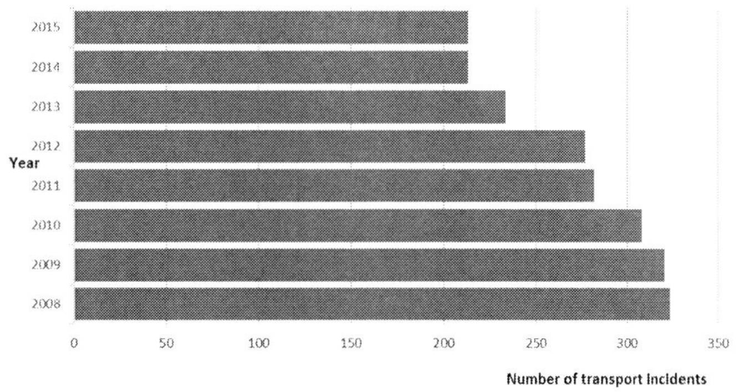

Fig. 1. Dynamics of the number of transport incidents in the locomotive facilities

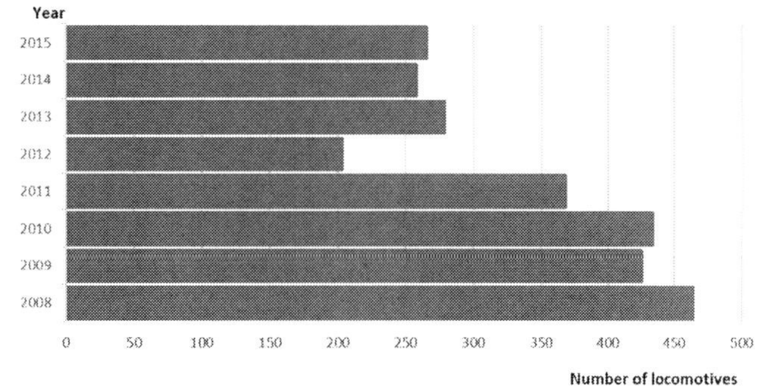

Fig. 2. Dynamics of the number of locomotives involved in transport accidents

During the calculation the required percentage of saving the initial information is taken not less than $\alpha = 85$ %. As a result of the calculation, 8 main components $g_1 \dots g_8$ are identified. In total they contain 89% of the initial information. Each of the components g_i contains a certain amount of information concerning the traffic safety condition. The structure diagram of the traffic safety condition I_{ts} is shown in the Figure 3.

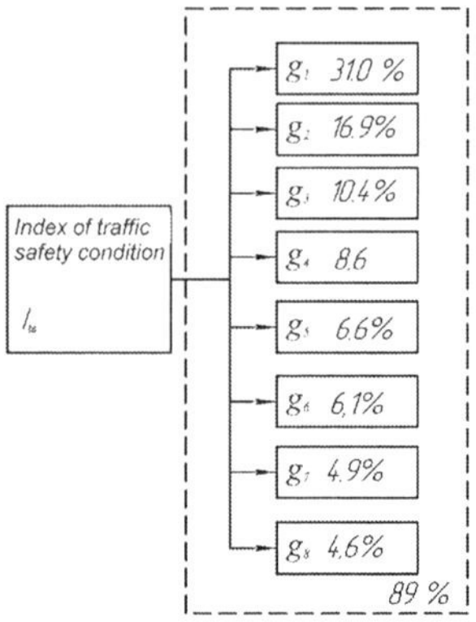

Fig. 3. Structure diagram of the Index of traffic safety condition

Each component consists of the variable values $x_1 \div x_{33}$ to different extents. Each of the variables contains a certain amount of information and influences the value of the general Index of traffic safety condition. The first component contains the largest amount of information. The diagram reflecting structure of the first principal component is shown in the Figure 4. In each of the principal components there are two components: the technical and the safety one.

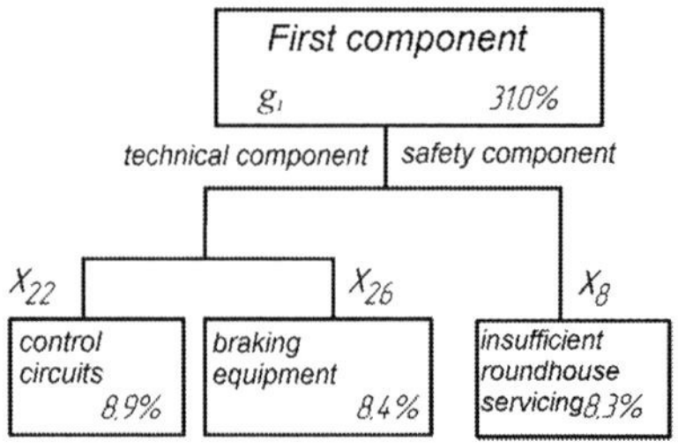

Fig. 4. Structure of the first principal component

In terms of safety component the values of such indicator as insufficient roundhouse servicing describe the traffic safety condition in the locomotive facilities most informative. The technical component includes: braking equipment and electric control circuits. Structure diagram of the second principal component is shown in the Figure 5.

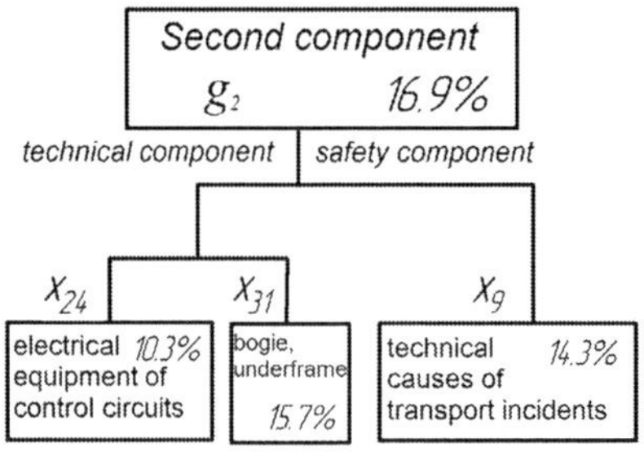

Fig. 5. Structure of the second principal component

From the standpoint of safety component, the value of technical causes of transport incidents is the most informative in the second principal component. The technical component includes: electrical control equipment, bogie and underframe.

Structure diagram of the third principal component is shown in the Figure 6.

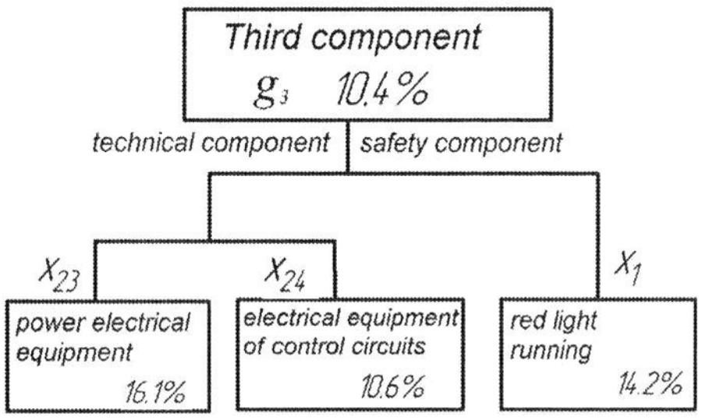

Fig. 6. Structure of the third principal component

Structural diagrams of other components are similar in composition, but others in terms of content. Taking into account that the seventh and eighth principal components have informativity less than 5 %, it is proposed to ignore the influence of these components on the general Index of traffic safety condition.

Distribution of the principal components that are the components of the Index of traffic safety condition is presented in the Table 2.

Table 2. Components distribution of the Index of traffic safety condition

Component		Technical component				Safety component	
No.	informativity	indicator	share, %	indicator	share, %	indicator	share %
1	0.310	Control circuits	8.9	Braking equipment	8.4	Insufficient roundhouse servicing	8.3
2	0.169	Electrical equipment of control circuits	10.3	Bogie, underframe	15.7	Technical causes of transport incidents	14.3
3	0.104	Power electrical equipment	16.1	Electrical equipment of control circuits	10.6	Red light running	14.2
4	0.086	Axle-boxes	7.3	Wheel-motor block	9.8	Wrong actions of locomotive crews (transport incidents)	16.6
5	0.066	Main generator	10	Accumulator battery	9.9	Wrong actions of locomotive crews (equipment failures)	19.6
6	0.061	Power transformers	15	Change gear box	10.2	Derailments and collisions during shunting (incidents)	10.8

The table ranks the technical and safety components according to the degree of their influence on the general index of traffic safety condition. Based on the analysis of the results presented in the Table 1, there were determined the locomotive units, which have the greatest influence on the traffic safety condition and reliability. Analysis of the safety component shows the contribution of each of the above-mentioned indicators to the traffic safety condition. When analyzing the influence of indicator on the general index of traffic safety, one should take into account informativity of the relevant component and the share that this indicator brings to the corresponding principal component. The proposed Index of traffic safety condition I_{ts} is an integral indicator. It takes into account the entire amount of statistical information concerning the traffic safety condition, but at the same time it makes it possible to determine the influence degree of each of the indicators on the general index of traffic safety. The value I_{ts} is a dimensionless value that reflects the general level of traffic safety. The greater the value I_{ts}, the worse is the traffic safety condition. The dynamics of I_{ts} changing by years is shown in the Figure 7.

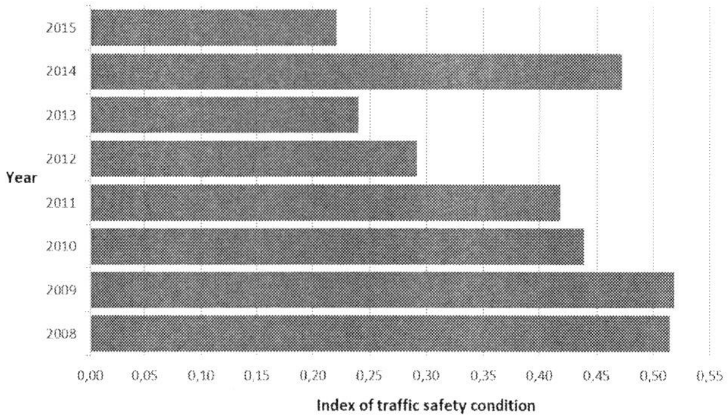

Fig. 7. Index of traffic safety condition in the locomotive facilities

The values I_{ts} are influenced by both technical and economic, social and other factors, which can explain the sharp increase of the index in 2014.

Conclusion

We can conclude that the proposed index I_{ts} can be further improved by the determination of the specific values of the index, taking into account the work performed by locomotives, the inventory locomotive fleet, the number of employees in the locomotive facilities, and others. At the initial stage, the value of the index I_{ts} allows evaluating the general traffic safety condition and can be used to compare the level of traffic safety by years.

References

[1] GOST R 56046-2014 Indicators of the use of locomotives. Terms and Definitions. "Standartinform", 2015, 28 p.

[2] Shmatchenko VV., Plekhanov PA. Standards of the CENELEC Committee as an integral part of the International Standard of the IRIS Railway Industry. Actual issues of the development of railway automation and telemechanics systems. 2013. №1. Access via internet: http://cyberleninka.ru/article/n/standarty-komiteta-cenelec-kak-sostavnaya-chast-mezhdunarodnogo-standarta-zheleznodorozhnoy-promyshlennosti-iris [Accessed: 2016-05-10].

[3] Milutinović D., Lučanin V. Relation between Reliability and Availability of Railway Vehicles. FME Transactions (2005) 33, pp. 135–139. Access via internet: http://www.mas.bg.ac.rs/_media/istrazivanje/fme/vol33/3/4._dusan_milutinovic.pdf [Accessed: 2016-05-10].

[4] EN 50126 "Railway Applications – The Specification and Demonstration of Dependability, Reliability, Availability, Maintainability and Safety (RAMS)", CENELEC.

[5] H2020-MG-8.1a-2014 RAMS data collection and failure rate analysis at component level. Revision 5, 31 p. Access via internet: http://infralert.eu/wp-content/multiverso-files/2_55780121488c4/INFRALERT-D5.1-Data-compilation-at-component-level_V5.01.pdf [Accessed: 2016-05-10].

[6] Gelumbickas G., Vaičiūnas G. Research on the influence of operational factors on the number of failures of diesel locomotives' engines Transport Problems: an International Scientific Journal. 2014, Vol. 9 Issue 1, pp. 5–12.

[7] Bose D., Ghosh G., Mandal K., Sau S. P., Kunar S. Measurement and Evaluation of Reliability, Availability and Maintainability of a Diesel Locomotive Engine, International Journal of Engineering Research and Technology. Volume 6, Number 4 (2013), pp. 515–534.

[8] Bodnar BE., Ochkasov AB. The use of the method of expert assessments in the development of diagnostic support of locomotives // Scientific works of the Kremenchug State Polytechnic University. No. 1, 2001. 10 p.

[9] Bodnar B., Ochkasov O. System Choice of the Technical Maintenance of Locomotives Equipped with on-Board Diagnostic Systems Transport Means: Proceedings of 21st International Scientific Conference, September 20–22, 2017, Kaunas University of Technology Klaipėda University [and others]. Juodkrante, Kaunas, Lithuania, 2017. Part I. pp. 43-47. A fragment of the text.

[10] Bodnar BE., Mosendz AI. Application of structural modeling for the study of the operation of the railway transport enterprise. Interuniversity collection of scientific papers. DIET. 1998. pp. 106–116.

[11] Directory of key indicators of regional branches of PJSC "Ukrainian Railways" (2005-2015). Kyiv, 2016. 60 p.

[12] Aivazyan SA., Buchstaber VM., Enyukov IS., Meshalkin LD., Aivazyan SA. Applied Statistics: Classifications and Dimension Reduction: Moscow: Finances and Statistics Statistics, 1989.607 p.

[13] Loza P. A., Grishechkina T. S. Evaluation of the quality of the maintenance of the electric rolling stock park. Electrification of Transport, No. 9/2015, pp. 87–93.

[14] Saati T. Making Decisions. The method of analyzing hierarchies. Radio and Communication: 1993, 278 p.

ITMS'2018

The 12th International Scientific Conference Intelligent Technologies in Logistics and Mechatronics Systems (ITELMS'2018), 26–27 April 2018, Panevėžys, Lithuania

Selection of the Main Controlling Parameters of Impact-Oscillatory Loading for Maximum Improvement of Plastic Properties Of Two-Phase High-Strength Titanium Alloys

Mykola Chausov[a], Volodymyr Hutsaylyuk[b*], Pavlo Maruschak[c], Andrii Pylypenko[a]

[a]*National University of Life and Environmental Sciences of Ukraine, Heroiv Oborony str.15, Kyiv, 03041, Ukraine*
[b]*Military University of Technology, Warsaw, Gen. Witolda Urbanowicza 2 str., 00-908, Poland*
[c]*Ternopil National Ivan Pul'uj Technical University, Rus'ka str. 56, Ternopil, 46001, Ukraine*

Abstract

It is now well established from experiments that by subjecting two-phase high-strength titanium alloys to a specific mode of impact-oscillatory loading at room temperature, dynamic non-equilibrium processes associated with the self-organizing structure can be realized in alloys. As a result, plastic deformation, impact strength and crack resistance of alloys vary in a broad range under subsequent loading. It is shown that the main controlling parameter of this process is the intensity of the impulse introduction of energy into the alloys, which in this study is evaluated by the jumps of deformation ε_{imp} in the process of impact-oscillatory loading. In addition, the revealed effects depend significantly on the level of the previous static deformation of alloys, in which the impact-oscillatory loading is realized. It is demonstrated on the example of tests of titanium alloys VT22, VT23 and VT23M how it is possible to maximally improve the plastic properties due to impact-oscillatory loading depending on the initial structure of alloys.

© 2018 M. Chausov, V. Hutsaylyuk, P. Maruschak, A. Pylypenko
Peer-review under responsibility of the Kaunas University of Technology, Panevėžys Faculty of Technologies and Business

Keywords: titanium alloys VT22, VT23 and VT23M, dynamic non-equilibrium processes, impulse introduction of energy, impact-oscillatory loading

* Corresponding author. Tel.: +48 22 261 839 245; fax: +48 261 837 366
E-mail address: volodymyr.hutsaylyuk@wat.edu.pl

© 2018 M. Chausov, V. Hutsaylyuk, P. Maruschak, A. Pylypenko
Peer-review under responsibility of the Kaunas University of Technology, Panevėžys Faculty of Technologies and Business

B427H0008 . EdLearning.it

1. Introduction

Two-phase, high-strength titanium alloys are widespread in advanced structures of aviation, missile, space equipment and other special equipment. Compared to high-strength titanium β-alloys, titanium (α + β)-alloys have a smaller number of expensive and scarce alloying elements. They have better mechanical and working properties of the main material and welded joints made using special technologies, as well as enhanced characteristics of crack resistance [1–2].

Titanium (α + β)-alloys are also used in parts of special equipment, in particular, protective plates of modern body armor. However, their "typical" disadvantage is a relatively low initial plastic deformation.

The analysis of the results obtained earlier by the authors, which are related to the testing of materials of various classes (steel, aluminum alloys, titanium alloys) under impact-oscillatory loading (this process is called the dynamic non-equilibrium process – DNP), suggests that under transient modes of loading with short impulses of additional force loads, when a mass transfer is taking place, but the dissipation of energy in the classical sense (the transformation of mechanical energy into heat) has not yet begun, there occur processes of the material self-organization with the formation of new forms of dissipative structures, the physical and mechanical properties of which differ significantly from the properties of the source materials [3–6]. One of the important results of these studies has been deducted from experiments: with correctly selected modes of impact-oscillatory loading, materials of all classes are plasticized significantly without a noticeable impairment of strength properties.

Traditionally, in order to improve the mechanical properties of two-phase titanium alloys, thermocycle loading at different temperatures is used followed by holding of relevant duration. They also use thermal force loading both under simple modes of monotonous loading, such as tension, shrinkage, torsion, tension with one-piece twist, and complex modes of thermal force loading – the so-called methods of intensive plastic deformation. However, the application of these methods requires the use of complex technological equipment and technological procedures of thermal force loading.

One should pay attention to another interesting feature of production of titanium alloys, which is considered in this work. The change in the production technology of two-phase (α + β)-alloys leads to a significant change in the percentage composition of α and β phases and the properties of these phases. Table 1, for example, shows the influence of the production technology of alloy VT23 on the percentage composition of α and β phases.

Table 1. The phase composition of the VT23 alloy depending on the heating temperature and the cooling rate [7]

Heating temperature, °C	Cooling environment	Phase composition	Percentage of β-phase,%	Parameters of the crystal lattice, phase		
				αβ	α0.α'	c0.α'
600	air	α+β	29	0.3220	0.293	0.467
700	air	α+β	38	0.3230	0.293	0.466
750	-«-	βм+α	70	0.3245	0.293	0.467
800	-«-	βм+α(α')	53	0.3245	0.293	0.467
900	-«-	α(α')+βм	38	0.3244	0.293	0.467
1000	-«-	α(α')+βм	21	0.3242	0.294	0.467
600	water	α+β	23	0.3200	0.293	0.467
700	water	α+β	32	0.3238	0.293	0.467
750	-«-	α+β	38	0.3245	0.293	0.467
800	-«-	βм+α	76	0.3250	0.292	0.466
900	-«-	α"(α')+((βм))	0-5	-	0.292	0.466
1000	-«-	α'	0	-	0.289	0.465

The development of an effective technological method for improving the plastic properties of two-phase high-strength titanium alloys due to impact-oscillatory loading at room temperature, which can be realized on any hydraulic test machine and which makes the procedure used in the well-known methods for improving the plastic properties of titanium alloys ten times cheaper, is relevant.

Under impact-oscillatory loading, nature itself forces the self-organization of the titanium alloy structure in such a way that a more intense mechanism of energy transfer through the material is provided. It is obvious that during the formation of dissipative structures, the energy is transferred to the existing damage of the material under impact-oscillatory loading or to the damage being formed in the material. In addition, due to the accumulation of energy, the modification of the material itself is possible on the boundaries of this damage. The excess energy is spent on the creation of the modified material with new properties and on the formation of damage in the material. The higher the hierarchy and cascade nature of dissipative structures, the more energy and mass they can accumulate, and, therefore, the greater the changes in the mechanical properties of titanium alloys, including plastic ones, in comparison with their initial state.

2. Material and experimental research

The method of impact-oscillatory loading is realized on the basis of a modified hydraulic setup for static tests ZD-100Pu and is described in detail in [8, 9]. The main idea of the proposed method is a high-speed stretching of the material with the application of the high frequency (several kilohertz) vibration process, which corresponds to the own frequency of the test machine.

Structurally this is achieved by adding the inner contour to the outer contour (load-bearing frame of the test machine) of the test machine. The inner contour represents the simplest statically undeterminable structure in the form of three parallel elements, which are loaded simultaneously – the central specimen and two satellite specimens (brittle samples) of different cross sections made of tempered steels 65G or U8...U12. During stretching, the satellite specimens are destroyed, and impulse introduction of energy into the specimen material is performed. Satellite specimens can be involved at any stage of the previous static stretching. Thus, it is possible to investigate the effect of impulse introduction of energy on the degradation of mechanical properties due to the formation of damage in the material under static stretching. By changing the initial diameter of the satellite specimens, the intensity of the impulse introduction of energy into the material can be controlled.

Mechanical tests were performed on specimens (Fig. 1) from sheet industrial titanium alloys VT22, VT23 and VT23M with a thickness of 3 mm. The strain measurement base was 16 mm.

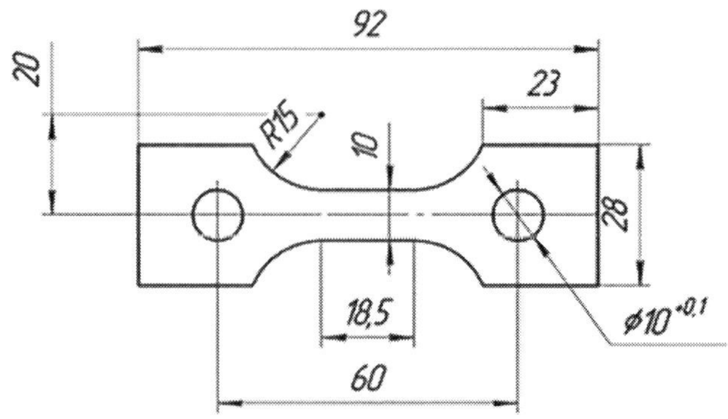

Fig. 1. Test specimen

Mechanical properties of alloys in the initial state are shown in Table 2.

Modes of mechanical loading of the specimens are summarized in Table 3.

Table 2. Mechanical properties of titanium alloys VT22, VT23 and VT23M

Titanium alloys	Mechanical properties		
	σ_{ys}, MPa	σ_{us}, MPa	δ, %
VT22	980-1180	1080-1280	6-8
VT23	980-1100	1080-1200	13-15
VT23M	1000-1050	1080-1180	19-20

Table 3. Modes of mechanical loading of specimens from titanium alloys VT22, VT23 and VT23M

Titanium alloys	Loading mode	Preliminary static deformation, ε_{stat},%	Impact loading, Fimp, kN	Ultimate tensile strength, σ_u, MPa	Failure strain, ε_f,%
VT22	Static tensile (specimen #1)	-	-	1201	8.18
	Static tensile – DNP - Static tensile (specimen #2, 7)	0.07	94.6	1186	17.9
	Static tensile – DNP - Static tensile (specimen #3, 8)	0.07	108	1169	18.86
	Static tensile – DNP - Static tensile (specimen #4, 9)	0.07	120	1131	19.01
VT23	Static tensile (specimen #19)	-	-	1129	15.7
	Static tensile – DNP - Static tensile (specimen #28)	0.267	122	1116	16.39
	Static tensile – DNP - Static tensile (specimen #31)	0.322	125	1170	19.81
	Static tensile – DNP - Static tensile (specimen #27)	0.363	125	1107	21.09
VT23M	Static tensile (specimen #33)	-	-	1086	21.14
	Static tensile – DNP - Static tensile (specimen #42)	0.025	98	1092	15.85
	Static tensile – DNP - Static tensile (specimen #37)	0.031	100	1102	16.77
	Static tensile – DNP - Static tensile (specimen #40)	0.034	103	1114	17.2
	Static tensile – DNP - Static tensile (specimen #41)	0.046	106	1095	18.33
	Static tensile – DNP - Static tensile (specimen #2)	0.093	105	1094	21.69
	Static tensile – DNP - Static tensile (specimen #7)	0.288	108	1093	22.28
	Static tensile – DNP - Static tensile (specimen #5)	0.585	97	1101	21.5
	Static tensile – DNP - Static tensile (specimen #4)	0.754	98	1104	19.31
	Static tensile – DNP - Static tensile (specimen #3)	0.852	97	1115	18.86
	Static tensile – DNP - Static tensile (specimen #39)	1.488	101	1131	12.82
	Static tensile – DNP - Static tensile (specimen #32)	3.75	99	1125	2.95

3. Results and discussion of experimental research

The results of the diffractometric research showed (Fig. 2) that β-phase is 30.3% by weight and α-phase is 69.7% by weight in the titanium alloy VT22, β-phase is 43.5% by weight and α-phase is 56.5% by weight in the titanium alloy VT23, β-phase is 22% by weight and α-phase is 78% by weight in the titanium alloy VT23M. Characteristically, both of the phase components in the specimens have a texture along the crystallographic direction (002). This may be due to the rolling of specimens or other mechanical exposure. While analyzing the data of the radiographic research (Fig. 2) and Table 1, one can see that there is no direct correlation between the percentage composition of the more plastic β-phase and the plastic deformation of alloys in the initial state. This again confirms the fact that it is difficult to develop methods for improving the plastic properties of two-phase high-strength titanium alloys.

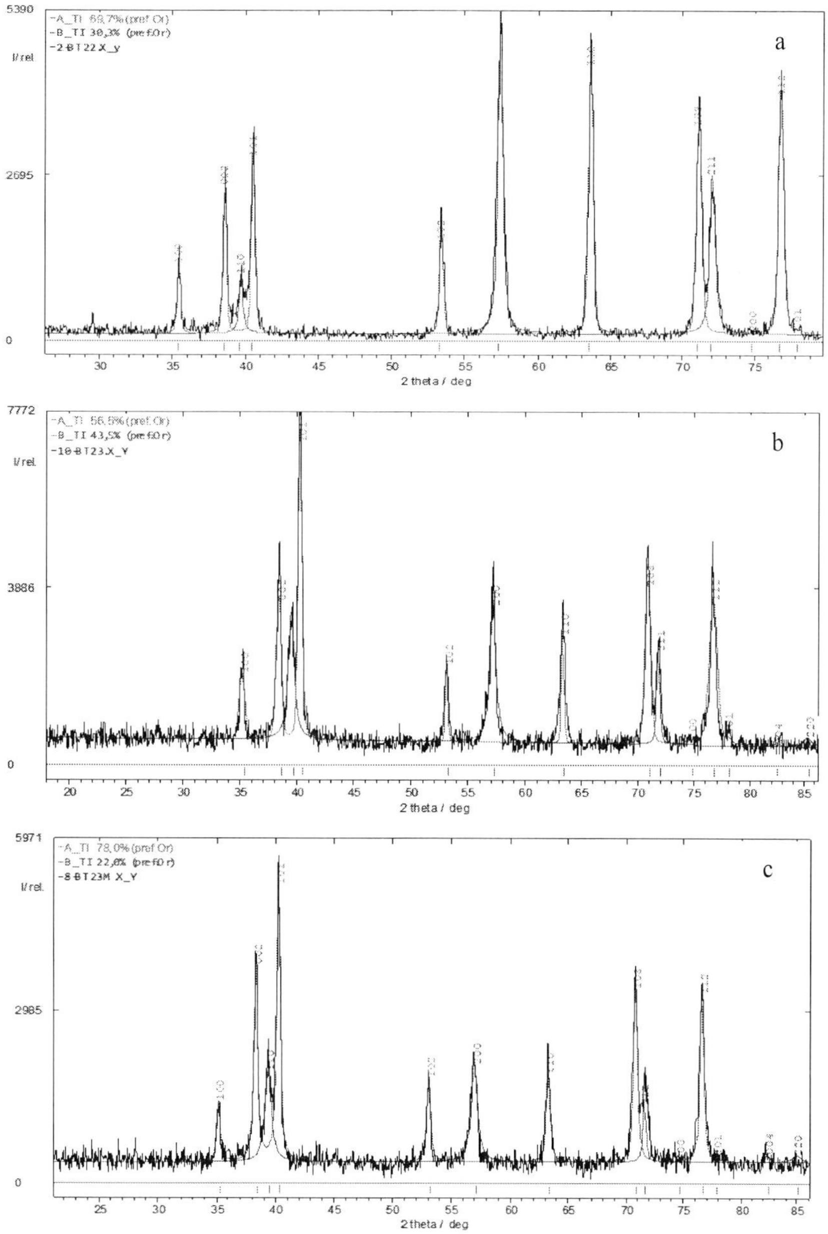

Fig. 2. Results of diffractometric research to estimate the percentage of α and β phases in the investigated titanium alloys: a) alloy VT22; b) alloy VT23; c) alloy VT23M

When evaluating the effect of impact-oscillatory loading on improving the plastic properties of the titanium alloy VT22, the first variant of controlling the impulse introduction of energy into titanium alloys was used. At first, a series of specimens was successively loaded with static stretching to achieve the identical strain level of 0.07%, and then, the intensity of impulse loading was increased in the range of 90...130 kN for each of the subsequent specimens.

3.1. Physical and mechanical peculiarities of deformation of titanium alloys

It is known that in the process of any loading, including static one, materials accumulate damage. As is shown by the authors of [11], the damage accumulation curve for static stretching has an S-shaped form with three pronounced sections: in the first section, micro-breaking of the material is associated with the formation of pores and microcracks; in the second section, micro-breaking is accompanied by an increase in the number of micro-damage, but the growth of this damage is insignificant. And, finally, in the third section, micro breaking changes due to an increase in the number and size of microdefects, in particular, due to merging of individual micro pores and microcracks. Since all spatial dissipative structures are formed due to the concentration of defects in the volume of materials, it is essentially important, in the case of the previous static stretching, to create a uniform field of damage with optimal dimensions in the structure of alloys. Then, during the impulse introduction of energy into alloys, the main part of it will be spent on creating such forms of dissipative structures that will be really effective in terms of further plasticization of alloys. This working hypothesis is fully confirmed by the experimental research conducted by the authors of [3–6].

Thus, the main controlling parameters of the impact-oscillatory loading include the degree of preliminary deformation of titanium alloys under static stretching and the intensity of the impulse introduction of energy into the alloys.

In this paper we consider two possible practical variants for controlling the impulse introduction of energy into titanium alloys under the impact-oscillatory loading:

- The first control variant is as follows: a batch of specimens was statically pre-stretched to a rather low level of stress and strain. With such strains, microdefects are practically not formed in the material or are reversed. Then, the intensity of the impulse loading was increased for each of the test specimens. Immediately after the impulse introduction of energy, the specimens were completely unloaded and statically loaded again to fracture. Random strain gains under impulse introduction of energy – ε_{imp} were chosen as a parameter characterizing the intensity of the impulse introduction of energy into titanium alloys. The choice of ε_{imp} as a parameter characterizing the intensity of impulse introduction of energy into the alloys greatly simplifies the test procedure. Therefore, the modes of impact-oscillatory loading can be created on hydraulic test machines of varying rigidity, which eliminates the need for complex calculations on the transfer of specific force directly on the specimen depending on the total impulse applied to the mechanical system. In addition, such a procedure is directly aimed at actual technological processes. By constructing the dependence curve of ε_{plast} on ε_{imp} under a repeated stretching of specimens, one can unequivocally estimate the rational modes of impulse introduction of energy into the alloys to improve their plastic properties.
- The second control variant consists in the following: each specimen from the series is loaded statically to a predetermined increasing level of strain or stress and then subjected to the same impulse.

The main difference between the two control variants is that in the first case, the dissipative structures are created in the material by means of disposing energy into those damages that are formed during impact-oscillatory loading. Moreover, in the case of the material modification, the energy is disposed on the boundaries of these damages.

In the second case, dissipative structures are created by disposing energy into the existing damages, which were formed during the previous static stretching, as well as into such damages that were formed during impact-oscillatory loading. Moreover, in the case of the material modification, the energy is disposed on the boundaries of these damages.

It is obvious that if identical strains are attained in a complex loading mode (previous static stretching – impulse introduction of energy) by two different methods of control, the dissipative structures of the materials will differ. At this stage it is impossible to predict in advance, which of these structures will be optimal for maximum improvement of the plastic properties of high-strength titanium alloys. Figure 3 shows some results of mechanical tests of alloy VT22.

The obtained results indicate that if the total stress in the specimen, including impulse loading, does not exceed the yield strength of the alloy, then the plastic deformation increases by 2.5 times at a repeated static stretching in comparison with the initial state. Also, studies have shown that in the case of DNP in the plastic region of the alloy (Fig. 3), the total plastic deformation can also increase by up to 2.5 times. At the same time, depending on the sudden increments of plastic deformation of the alloy under the DNP – from 2.5% to 6.0%, the plastic deformation of the alloy varies widely in the case of further static stretching. However, at the same time, there is a decrease in the strength properties of the alloy by up to 5.0 %. (Fig. 3).

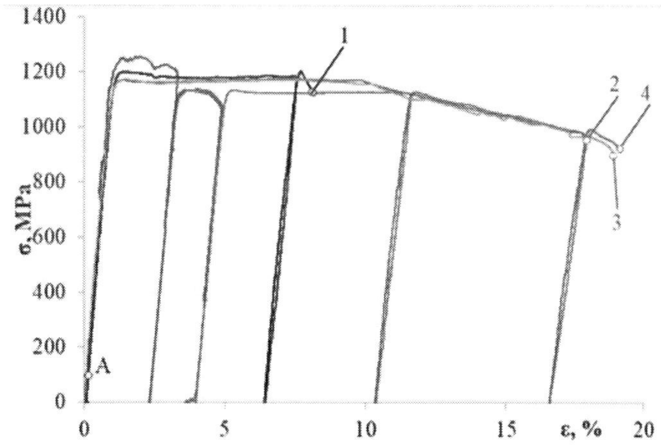

Fig. 3. Stress-strain curves of titanium alloy VT22: 1 – static stretching; 2, 3, 4 – complex loading mode (previous static stretching - realization of DNP – further static stretching). Explanations are given in the text

While evaluating the effect of the impact-oscillatory loading on the improvement of the plastic properties of titanium alloys VT23 and VT23M, the second control variant of the impulse introduction of energy into titanium alloys was used. In this case, a series of specimens was statically loaded to an increasing stress level of 0.03...0.4 σ_{ys}, and then the specimens were subjected to additional force impulse loading within the narrow range of 50.7 ± 7 kN.

The detected dependence of the degree of sudden increments of plastic deformation under the impulse introduction of energy on the plastic deformation of alloy VT23 under subsequent static stretching was used to construct an experimental dependence curve of ε_{plast} on ε_{imp} for alloy VT23 (Fig. 4). Here, the numbers indicate those specimens that were then used for metal-physical research.

As is seen from Fig. 4, the impulse introduction of energy into alloy VT23, as well as into alloy VT22, significantly increases plastic deformation of the alloy. Based on the analysis of the experimental curve given in Fig. 4, it is possible to substantiate the optimal mode of impulse introduction of energy using the second control variant for maximizing the plastic deformation of alloy VT23 compared to the initial state. In particular, for ε_{imp} = 3.0...3.5%, the plastic deformation of titanium alloy VT23 can be increased by 30...35%.

Similarly, the second control variant of the impulse introduction of energy is used to improve the plastic properties of alloy VT23M under the same additional force impulse loading within the narrow range of 50.7 ± 7 kN. The analysis of the obtained results showed that the plastic deformation of alloy VT23M may increase by up to 5% (Fig. 5) under the repeated static stretching only in the range (ε_{imp} = 0.2...1.0%). In other cases, the plastic deformation of the alloy decreases after the impulse introduction of energy. Figure 5 gives some generalized results of experiments, where figures indicate the numbers of specimens subjected to the impulse introduction of energy.

The results concerning the impulse introduction of energy into alloy VT23M proved that technological improvements to enhance the initial structure of two-phase high-strength titanium alloys reduce the effects of increasing plastic deformation after the impulse introduction of energy. However, this does not mean that DNP does not enhance other plastic properties of alloy VT23M, in particular, the impact strength and crack resistance.

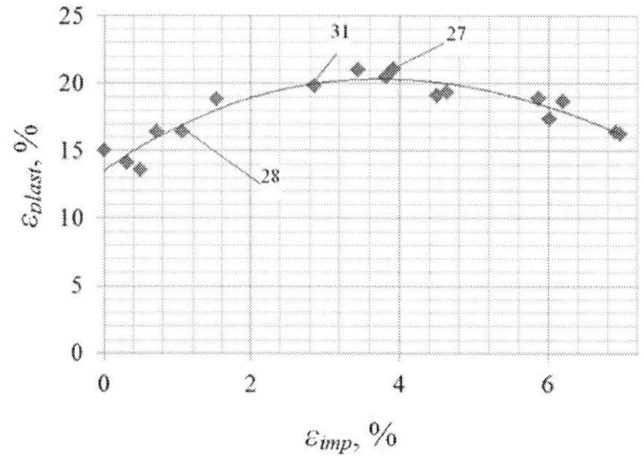

Fig. 4. Dependence curve of plastic deformation under repeated static stretching of specimens from titanium alloy VT23 on sudden increments of deformation in the alloy under impulse introduction of energy

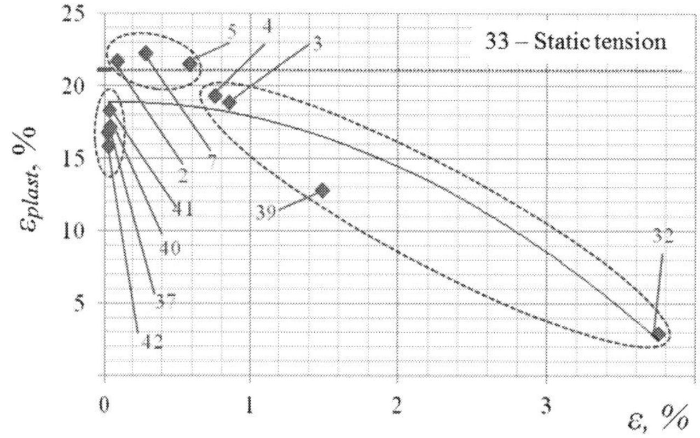

Fig. 5. Dependence of plastic deformation of specimens from titanium alloy VT23M under repeated static stretching on specific values of strain under the impulse introduction of energy into the alloy

The effect of DNP on the impact strength of the specimens after DNP (ε_{imp} = 0.2...1.0 %) was evaluated by comparison with statically deformed specimens with the same levels of plastic deformation. After applying identical notches (Fig. 6), specimens were subjected to an impact test on the PSW-30 impact testing machine.

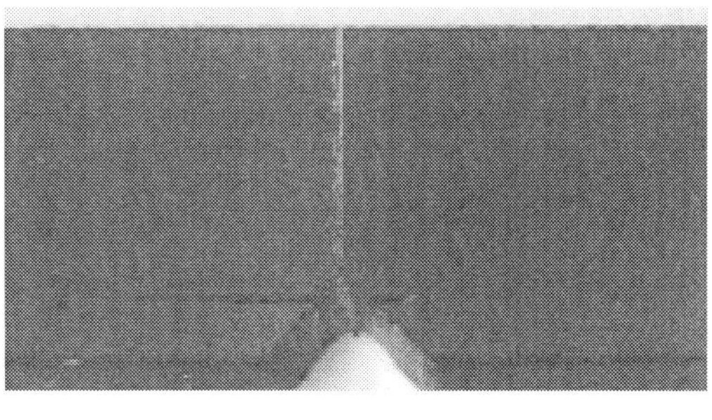

Fig. 6. Shape of the notch applied to the specimen from alloy VT23M for impact strength testing

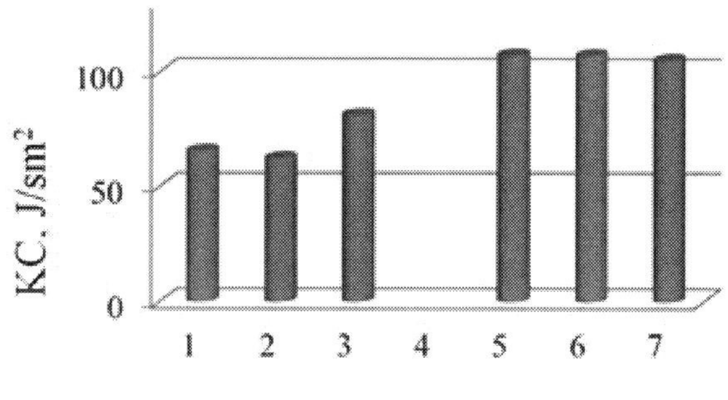

Fig. 7. Impact strength of the alloy in the initial state (1-3) and after the DNP (5-7)

The relevant test results are given in Figure 7. When comparing the test results, it can be noted that the impact strength of alloy VT23M increased by about 20 % (Fig. 7) after the DNP. The strength of the titanium alloy VT23M after different modes of loading was evaluated using the method of complete stress-strain diagrams and the method described in detail in [12, 13]. Figure 8 presents complete stress-strain diagrams of specimens from alloy VT23M (here, specimen 14u corresponds to the alloy in the initial state, and specimen 46 – to the alloy after the DNP). Evaluation of the crack resistance of the alloy in various states by the parameter K_λ [12, 13] shows that after the DNP, an increase in the crack resistance of the VT23M alloy was found to be 13–19% (the angle of inclination of the descending branch of the stress-strain diagram of the alloy after the DNP, which characterizes the energy expenditure on the growth of the macro crack by type (I + III), increases significantly).

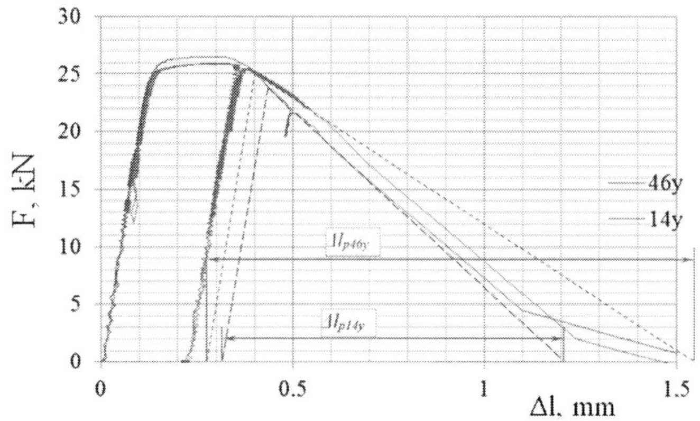

Fig. 8. Complete stress-strain diagrams of specimens from alloy VT23M in different states (explanations are given in the text)

The analysis of the obtained results has shown (Fig. 3–8) that both control variants of the impulse introduction of energy into two-phase high-strength titanium alloys lead to a significant improvement of their plastic properties in comparison with the initial state. Also, these experiments have shown that since there is no direct relationship between the percentage of the more plastic β-phase and the plastic deformation of alloys in the initial state, it is very difficult to predict the effects of impulse introduction of energy into two-phase high-strength titanium alloys to improve their plastic properties, depending only on percentage ratio of α and β phases.

Conclusions

1. Two control variants of impulse introduction of energy into two-phase high-strength titanium alloys by means of impact-oscillatory loading have been experimentally tested.
2. Complex experimental studies have been carried out, which confirmed the positive effect of the impulse introduction of energy on improving the plastic properties of sheet high-strength two-phase titanium alloys such as VT22, VT23 and VT23M. In addition, it has been established that the impact-oscillatory loading practically does not cause a decrease in the strength properties of these alloys. New data was obtained on enhancing plastic deformation of alloy VT22 by 2.5 times, alloy VT23 by 30...35 %, and alloy VT23M by up to 5% due to the impact-oscillatory loading. An increase in the impact strength of alloy VT23M by 20 %, and an increase in the crack resistance of alloy VT23M by 19 % prove that a simple technological method can be used at room temperature to significantly modify the initial mechanical properties of sheet high-strength two-phase titanium alloys.

References

[1] Khorev A. I. High-strength titanium alloy VT23 and its applications in advanced welded and brazed structures, *Welding International*, 2010, Vol. 24, Iss. 4, pp. 276–281.

[2] Khorev A. I. Alloying titanium alloys with rare-earth metals, Russian Engineering Research, 2011, Vol. 31, Iss. 11, pp. 1087–1094.

[3] Chausov M., Maruschak P., Pylypenko A., .Sergejev F. and O. Student. Effect of high-force impulse loads on the modification of mechanical properties of heat-resistant steel after service. *Estonian J. Eng.*, 2012, 18, pp. 251–258.

[4] Chausov M. G., Pylypenko A. P., Berezin V. B., Markashova L. I., Kushnariova O. S., Hutsaylyuk, V. B. Influence of Impact-Oscillatory Loading upon the Mechanical Properties of the VT-22 Titanium Alloy Sheet. *J. of Mater. Eng. And Perfom*, 2016, Vol. 25, No 8, pp. 3482–3492.

[5] Chausov M., Maruschak P., Pylypenko A. and L. Markashova. Enhancing plasticity of high-strength titanium alloys VT 22 under impact-oscillatory loading, *Philosophical Magazine*, 2017, 97, pp. 389–399.

[6] Chausov M. G., Berezin V. B., Pylypenko A. P. and V. B. Hutsaylyuk. Strain field evolution on the surface of aluminium sheet alloys exposed to specific impact with oscillation loading // *J. Strain. Anal.*, 2014, 50, pp. 61–62.

[7] Khorev A. I. Fundamentals of alloying and theory of heat treatment of welded joints in titanium β-alloys, *Welding International*, 2012, 27, 5, 2012, pp. 415–422.

[8] Chausov N. G., Voityuk D. G., Pilipenko A. P. and A. M. Kuz'menko. Setup for Testing Materials with Plotting Complete Stress–Strain Diagrams, *Strength Mater*, 2004, 36(5), pp. 532–537.

[9] Chausov N. G. and A. P. Pilipenko. Influence of dynamic overloading on fracture kinetics of metals at the final stages of deformation, *Mechanika*, 2004, 48, pp. 13–18.

[10] Chausov M., Maruschak P., Prentkovskis O., Karpets M. Risks Associated with the Use of High-Strength Titanium Alloys in Transportation Systems. In: Kabashkin I., Yatskiv I., Prentkovskis O. (eds) Reliability and Statistics in Transportation and Communication. RelStat 2017. *Lecture Notes in Networks and Systems*, Vol. 36, Springer, pp. 213–222.

[11] Lebedev A. A., Chausov N. G., Nedoseka S. A. et al. Model of damage cumulation in metallic materials under static tension, *Strength of Materials*, 1995, Vol. 27(7), pp. 379–386.

[12] Chausov N. Full deformation diagram as the source of information about accumulation of damages and materials crack resistance. *Fact Lab Mater Diagn* 2004; 70, pp. 42–49.

[13] Chausov M. G., Maruschak P. O., Hutsaylyuk V. et al. Effect of complex combined loading mode on the fracture toughness of titanium alloys / *Vacuum*, 2018, 147, pp. 51–57.

ITMS'2018

ktu 1922 panevėžys faculty of technologies and business

The 12th International Scientific Conference Intelligent Technologies in Logistics and Mechatronics Systems (ITELMS'2018), 26–27 April 2018, Panevėžys, Lithuania

Problems of Hydraulic Drive System Development in Teleoperated Anthropomorphic Manipulators

Karol Cieślik[a], Marian Janusz Łopatka[a*]

[a]*Military Univercity of Technology, Urbanowicza 2 street, 00-908 Warszaw, Poland*

Abstract

The paper presents simulation research related with possibility to carry operator's arm movement on a much larger anthropomorphic manipulator. Moreover, the article holds for the most important advantages and problems related to the widespread use of anthropomorphic manipulators with a hydrostatic drive system. The paper presents result of studies movement operator's hand. This results served as the control signal for the anthropomorphic manipulator's model. The movement operator's hand studies were carried out on a physical object with the use of stereovision camera system Were identified the maximum speed and acceleration value of the effector of the manipulator's model when it copying the characteristic movements of operator's arm. The simulation studies were made taking into account the literature and the real stiffness and dumping parameters of the hydrostatic drive system. Furthermore, checked the ability to realization of the assumed manipulators movements by the hydraulic actuators.

© 2018 K. Cieślik, M. J. Łopatka
Peer-review under responsibility of the Kaunas University of Technology, Panevėžys Faculty of Technologies and Business

Keywords: anthropomorphic manipulator, hydrostatic drive system, stifness and dumping parameters

1. Introduction

The anthropomorphic manipulator is a system similar to human limbs as to shape and actions [1, 2]. Currently, in the world market there can found the constructions of the anthropomorphic manipulators with electrical and hydrostatic drive systems (Fig. 1). The most of existing anthropomorphic manipulators use electrical drive system.

* Corresponding author. Tel.: +48-22-261-837-416
E-mail address: karol.cieslik@wat.edu.pl

© 2018 K. Cieślik, M. J. Łopatka
Peer-review under responsibility of the Kaunas University of Technology, Panevėžys Faculty of Technologies and Business

Generally, this construction is used in medicine where is important to a very precisely coping of the operator's arm movements [3].

a) b)

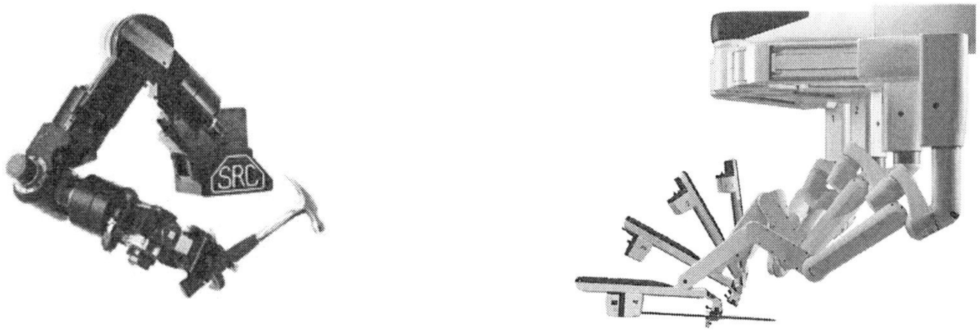

Fig. 1. Examples of constructions of the anthropomorphic manipulators: a) Sacros Slave Amr manipulator with hydrostatic drive system [4]; b) da Vinci manipulators with electrical drive system [5]

To control these structures are used the complex consoles and systems which track the movements of the operator's hand (Fig. 2).

a) b)

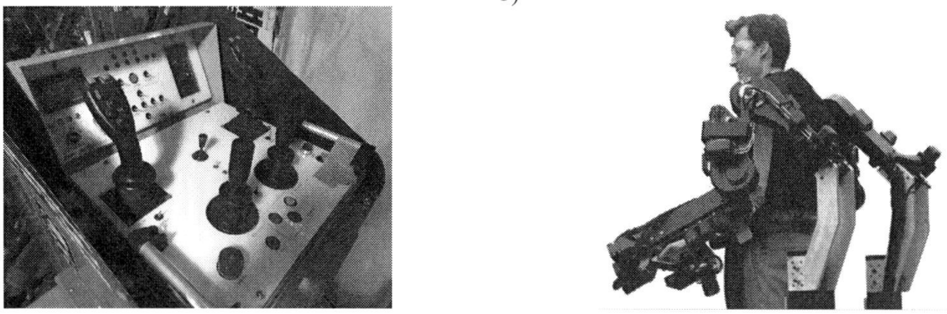

Fig. 2. The methods of control of the anthropomorphic manipulators: a) complex console of the robot Tmsuk T-52 [6]; b) tracking control system of the Raytheon Big Arm [7]

The anthropomorphic manipulators with hydrostatic drive system are only the prototypical constructions, which, despite the many advantages are not commonly used. Into their most important advantages may be included a significant increase in working possibility of human hands compared to other types of drive. In the hydrostatic drive system are susceptible components to which could include: working fluid, hoses, actuators and hydraulic motors. Their parameters of stiffness and dumping have a significant impact on works tool's dynamics of the working machines and manipulators [8–12]. This problem therefore also relates to anthropomorphic manipulators, so the values stiffness and dumping parameters of the hydraulic drive system components should be considered during of design this type of construction.

The operator's hand of the anthropomorphic manipulators used in medicine performs significantly larger movements than the effector of the manipulator. There are no problems related to impacts of the constructions a significant of the inertial forces. It is related with the relatively low velocity and acceleration with which manipulator works. This is different in manipulators where the working area is greater than work area the operator's hand [13]. The literature does not address the issue of the speed of the operator's hands during control the anthropomorphic manipulators with hydrostatic drive as well the acceleration values which impact on the construction. The article presents result of the simulation research of possibilities to copy operator's hand movement by a significantly larger anthropomorphic manipulator with hydrostatic drive system.

2. Operator's hand movements research

2.1. The program and methodology of research

The research purpose of the movement of the human hand was to identify the real values of the speed and acceleration during execution three characteristic movements of it (Fig. 3).

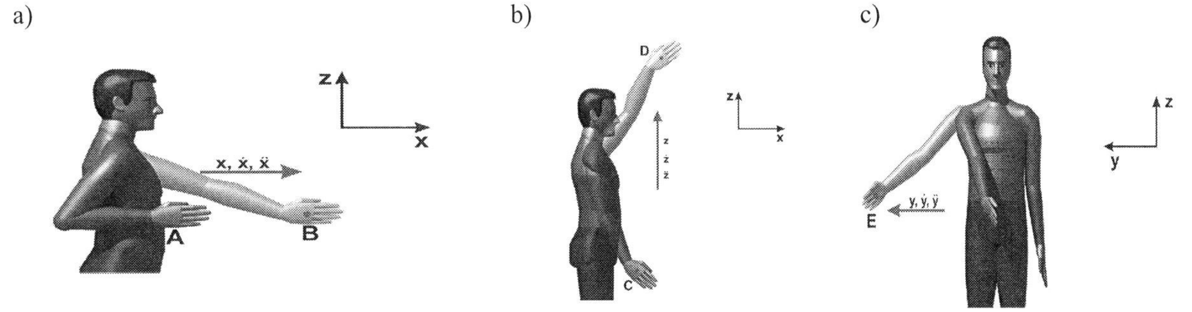

Fig. 3. Characteristic movements of human hand: a) horizontal movement; b) vertical movement; c) cross movement

Selected to execute characteristic hand's movements after connected allow to use the whole hand's work area. The anthropomorphic manipulator's model which copping hand's movement also use the whole work area. The research object was a potential operator of the anthropomorphic manipulator. During the study to the hand were glued tags/markers which change position was recorded. The research was carried out by using the measuring apparatus based on a system of digital image correlation. The devices which were used in research were stereovision cameras, special tripod and device for data acquisition (Fig. 4).

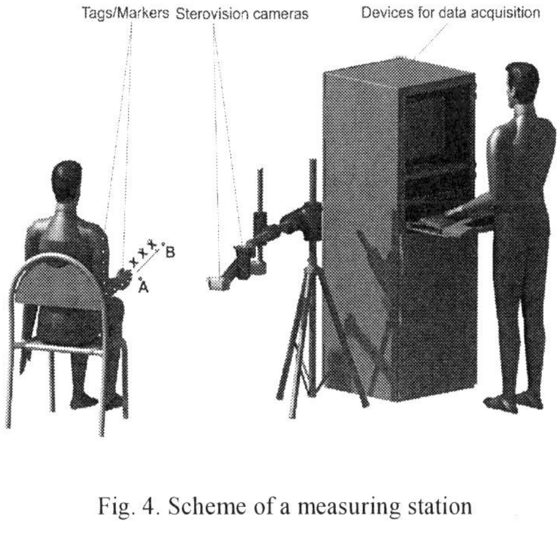

Fig. 4. Scheme of a measuring station

Used equipment has allowed the registration of changes of the markers position which were glued to hand with an accuracy of 0.1 mm. During the research the displacement, velocity and acceleration wrist's marker was recorded. This marker directly reflects the require manipulator effector's location.

2.2. Research results

The research of the operator hand's movement has helped to identify the real value of speed and acceleration during the execution the three characteristic motion. For the horizontal movement of the operator's hand (Fig. 3, *a*) course of changes the displacement, velocity and acceleration is given in Figure 5.

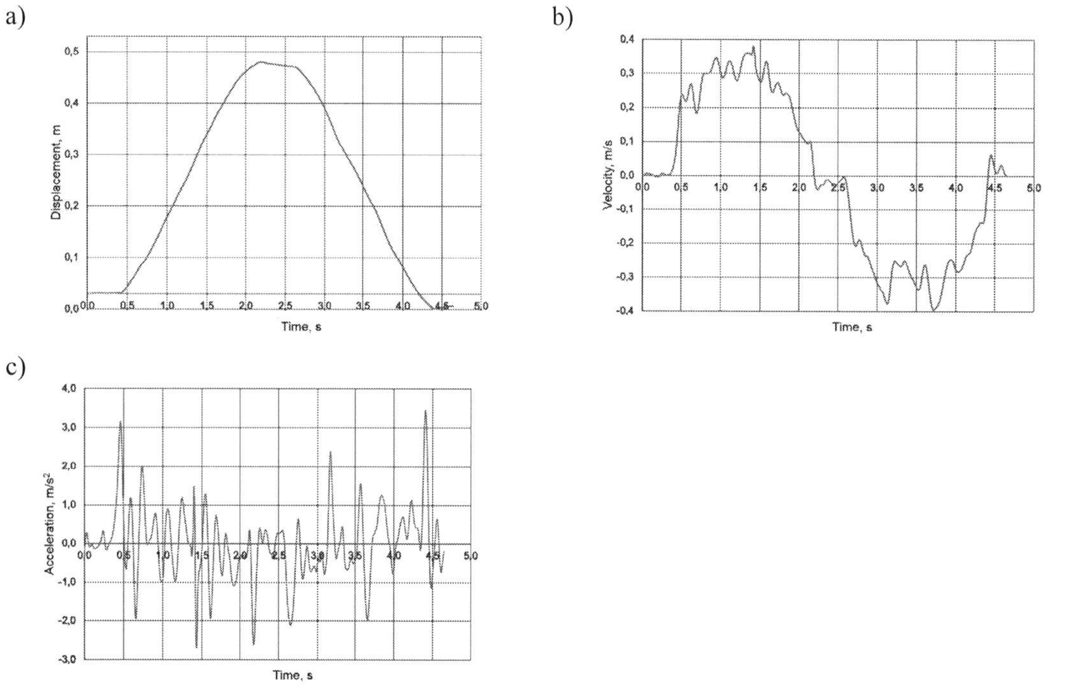

Fig. 5. Waveforms describing the motion wrist's marker for horizontal movement: a) displacement; b) velocity; c) acceleration

For the vertical movement of the operator's hand (Figure 3, *b*) course of changes the displacement, velocity and acceleration is given in Figure 6.

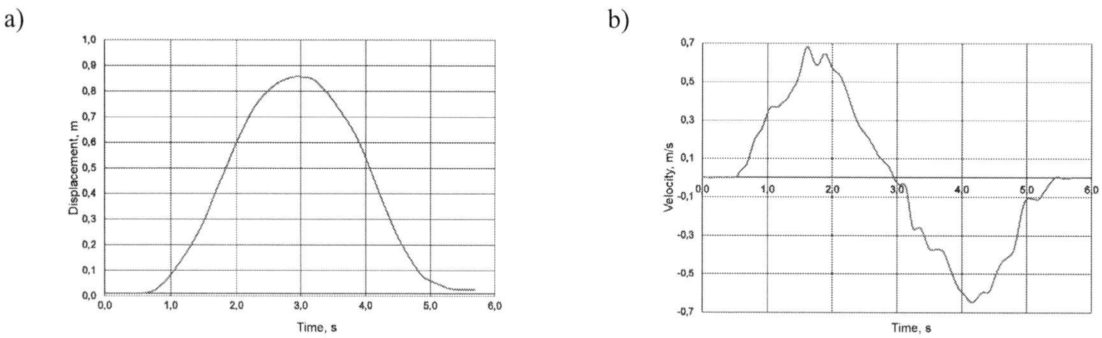

c)

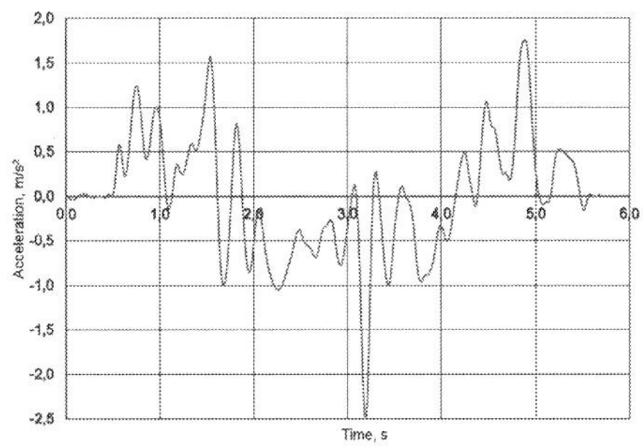

Fig. 6. Waveforms describing the motion wrist's marker for vertical movement: a) displacement, b) velocity, c) acceleration

For the cross movement of the operator's hand (Figure 3, *c*) course of changes the displacement, velocity and acceleration is given in Figure 7.

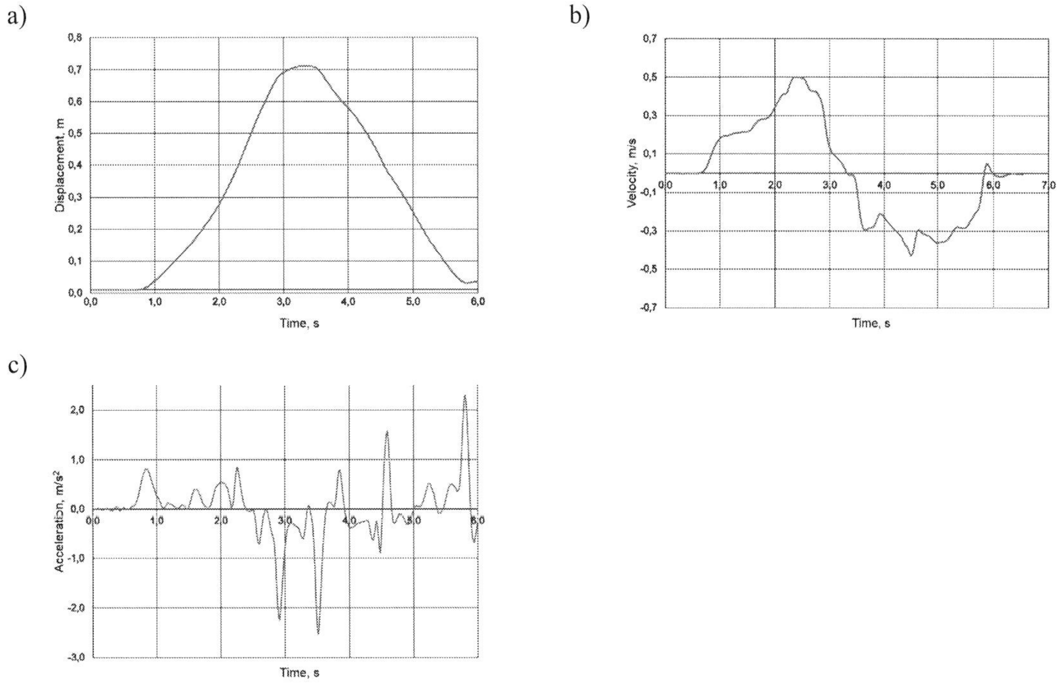

Fig. 7. Waveforms describing the motion wrist's marker for cross movement: a) displacement, b) velocity, c) acceleration

The maximum running time for each of the analysed cases does not exceed 5 s. The displacement of wrist's marker is from the 0.48 m for the horizontal movement to the 0.88 m for the vertical movement. The maximum value of the marker's velocity does not exceed 0.7 m/s and acceleration 3.5 m/s^2.

3. Simulation research

3.1. Anthropomorphic manipulator's model

The anthropomorphic manipulator's model used for the simulation research is functional and anatomical equivalent of the human hand. This construction has seven degrees of freedom (DOF). The manipulator's model degree of the magnification relative to the human arm is 4. The kinematic scheme of the manipulator's model given in Figure 8.

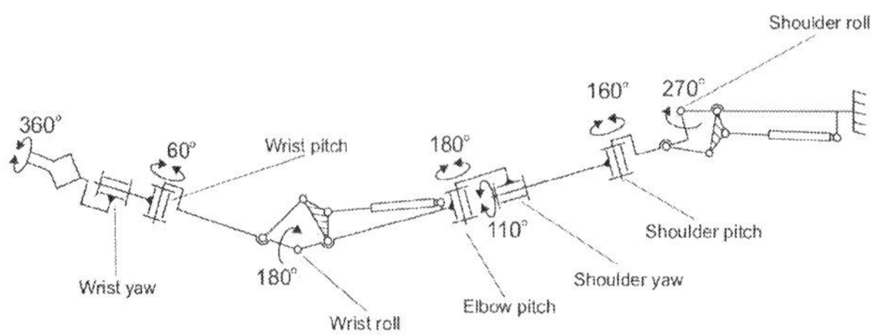

Fig. 8. Kinematic scheme of the anthropomorphic manipulator's model

The manipulator uses a hydraulic drive system. Model was made in ADAMS Software [14]. 3D model of the manipulator given in Figure 9.

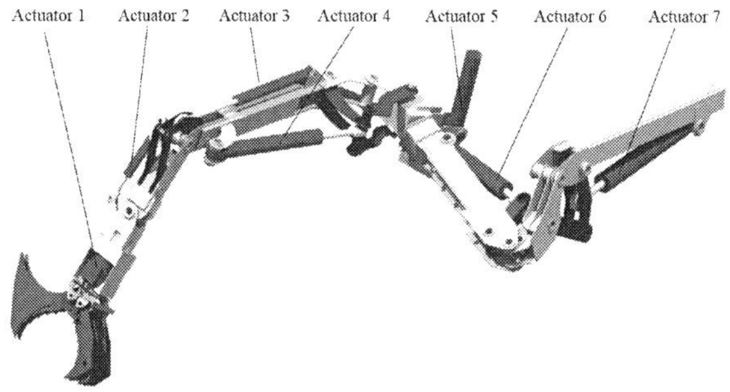

Fig. 9. Anthropomorphic manipulator's 3D model

3.2. The program and methodology of research

The aim of the simulation studies was to check the possibility transfer the operator hand's movement identified during the research by the anthropomorphic manipulator's model. The scheme of the simulation research is given in Figure 10.

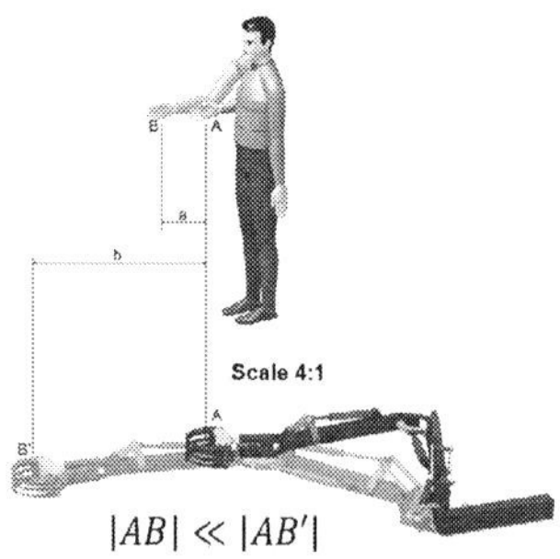

Fig. 10. Simulation research's scheme

During the study were identified the velocities and accelerations of the anthropomorphic manipulator's effector while copying the characteristic hand movements with the 4-times reinforcement. The reinforcement is equal to the manipulator's magnification in relation to the hand. The obtained values of velocity and acceleration referenced to the existing in working machines. The research was carried out for the two cases. First was assumed introduction to the model literature value of stiffness and damping for susceptible elements of hydrostatic drive system taking into account the working fluid in the actuator (Fig. 11).

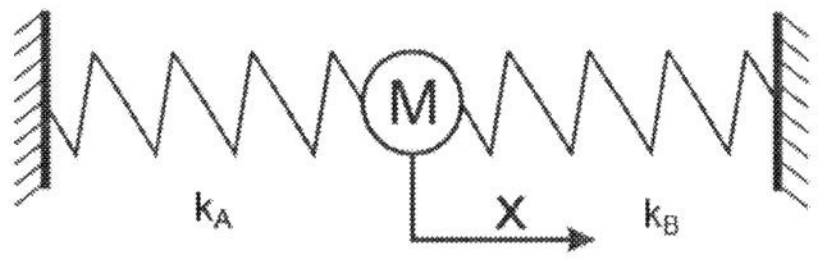

Fig. 11. Hydraulic actuator's model; kA – stiffness of working fluid on piston side, kB – stiffness of working fluid on rod side

The equals which used to determinate of the working fluid's stiffness in actuator is given by [15]:

$$k_A = A_K^2 \cdot \frac{E_{oil}}{A_K \cdot h_K}, \frac{N}{m}; \tag{1}$$

$$k_B = A_K^2 \cdot \frac{E_{oil}}{A_K \cdot (h_K - h)}, \frac{N}{m}; \tag{2}$$

where h – actual actuator stroke, m;

h_K – maximum actuator stroke, m;

A_K – piston area, m^2;

A_R – piston area on rod side, m^2;

E_{oil} – bulk module of the working fluid, Pa.

In the second case to the manipulator's model implemented the real values of stiffness and damping values of the hydrostatic drive system's elements [16]. A scheme of the anthropomorphic manipulator's movement is given in Figure 12.

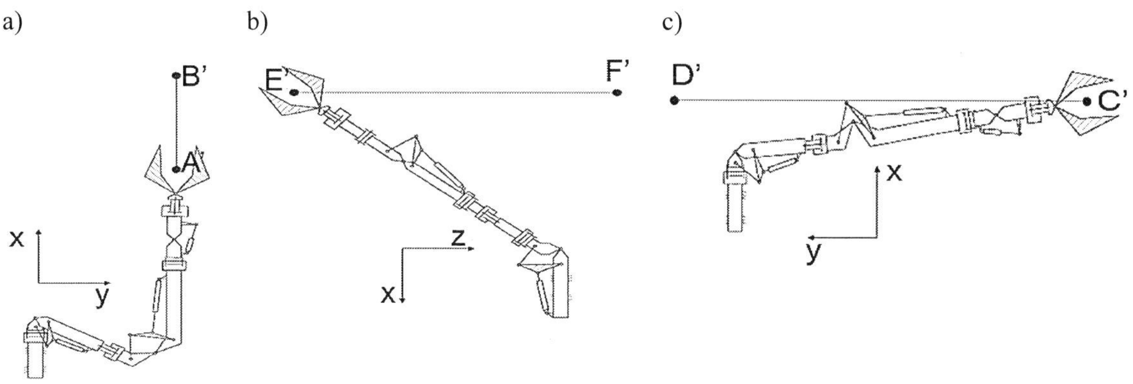

Fig. 12. Anthropomorphic manipulator's movements: a) horizontal movement; b) vertical movement; c) cross movement

The second part of manipulator's research was concern to possible to continue execution of the assumed movement given in Figure 12 by minimal actuator number.

3.3. *Results of the simulation research*

The examples of velocity and acceleration waveforms of the manipulator's effector for the case which copy horizontal movement are given in Figure 13. Waveform shows the velocity and acceleration changes of the manipulator's effector in three axes x, y, z for the first stage of research.

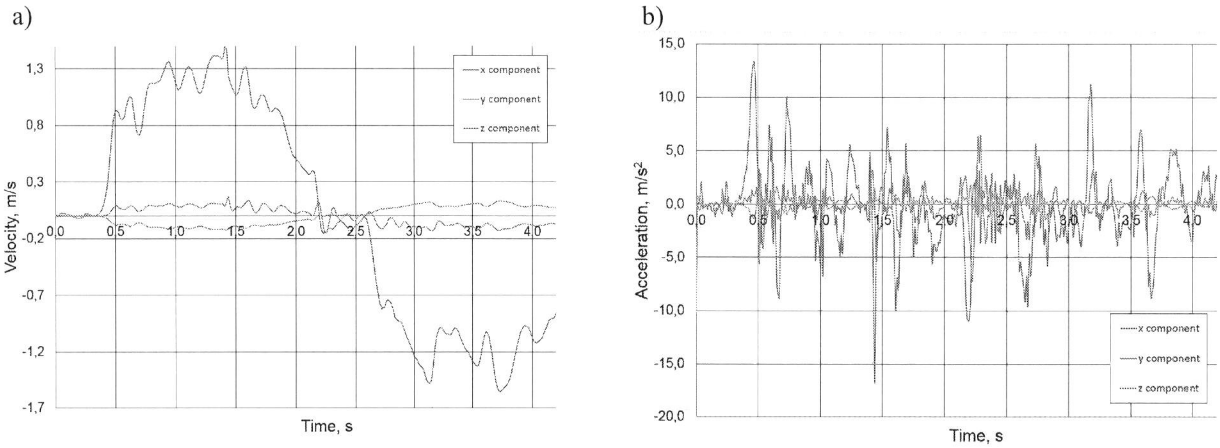

Fig. 13. Waveforms obtained while copying the horizontal movement of by the manipulator: a) effector's velocity; b) effector's acceleration

The maximum velocity value of the effector while copying the horizontal movement is 1.5 m/s and the maximum acceleration is 17 m/s². The value of root mean square acceleration is 3.8 m/s². In the second phase of the research considered the real values of the stiffness and damping parameters of the hydrostatic drive system's susceptible elements. The waveforms of the effector's acceleration for this case is given in Figure 14.

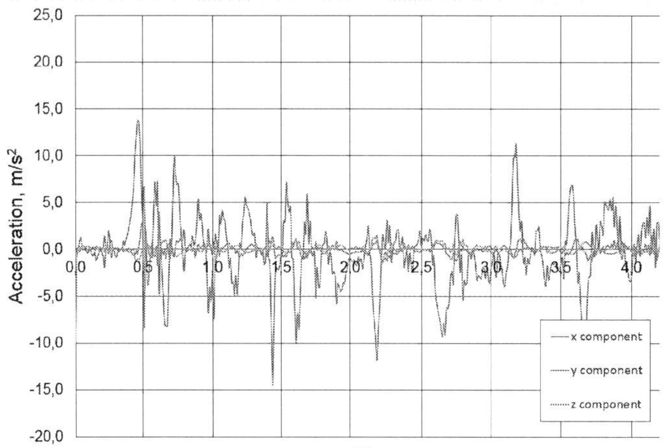

Fig. 14. Waveforms of the effector's acceleration for the second stage of research

The maximum effector's acceleration value for second steps of study is 14 m/s² and the root mean square acceleration's value is 3.6 m/s^2. On diagrams given in Figure 13 and 14 are a many of effector's oscillations. Their occurrence is related to the stiffness and damping parameters of the susceptible elements of hydrostatic drive system. Effector's oscillations in practice may difficult realization of the manipulation function by the operator. During the simulation research of the anthropomorphic manipulator were identified velocity and acceleration of hydraulic actuator need to realize the manipulator movements. The maximum velocity value achieved by the hydraulic actuators does not exceed 0.25 m/s and the acceleration value is less than 2.5 m/s^2. The obtained values of velocity and acceleration are similar to those in working tools of loaders, excavators and robots manipulators [17]. During the second part of research there were performed almost 100 simulation tests for every type of movement. The result of the research was presented in table 1 and they concern only the minimal number of hydraulic actuators which allowed to execution of movement. The actuator in a table was name as on Figure 9.

Table 1. results of the research of minimal number hydraulic actuators needed to execution of movement

O. N.	Name of the actuator	Movement along x axis	Movement along y axis	Movement along z axis
1	Actuator 1			
2	Actuator 2			
3	Actuator 3			
4	Actuator 4	✓	✓	✓
5	Actuator 5			
6	Actuator 6	✓	✓	✓
7	Actuator 7		✓	✓

According to research only tree hydraulic actuator of the manipulator is necessary to do every assumed movement. However, it concerns only quick localization the griper in the manipulator work area. Nevertheless, the precision manipulation of the gripper is possible when every hydraulic actuators of the manipulator are used.

Conclusions

The simulation studies were performed to concerning the possibility of direct copy the movement of the operator's hand by the anthropomorphic manipulator with hydrostatic drive system. To the manipulator's model was introduced literature and real stiffness and damping parameters of the susceptible elements of hydraulic drive system. Considered the values of stiffness and damping parameters significantly changes the load the manipulator. During the study on the manipulator's construction influence the acceleration values more than 1g. This may be a serious problem for the structural strength of the construction. In the anthropomorphic manipulator's model, the effector's acceleration values were over 5 times higher than those occurring in the operator's hand.

Achieved during research velocity and acceleration value are obtainable by the hydraulic actuators and are not a problem for hydrostatic drive system. Simulation research did not include the impact of control element - hydraulic valve.

Acknowledgements

The article is based on the works also carried out and financed within the project PBS 936 on: Development of design and manufacturing technology and effectiveness of teleoperation systems of Unmanned Ground Vehicles.

References

[1] Morecki A., Knapczyk J., Kędzior K. Teoria mechanizmów i manipulatorów, WNT, Warszawa 2002.
[2] Morecki A., Knapczyk J. Podstawy robotyki teoria i elementy manipulatorów i robotów, WNT, Warszawa 1999.
[3] Friedman D. C. W., Dosher J., Kowalewski T., Hannaford R. J. Automated toot handling for the trauma pod surgical robot, EEE International Conference on Robotics and Automation, Apr 10–14, 2007, Rome, Italy, pp. 1936–1941.
[4] University of Southern California. Computational Learning and Motor ControlLab. Access via internet: http://www-clmc.usc.edu/Research/ExperimentalEquipment [Accessed: 2017-11-08].
[5] La Tour Medical Group. Image gallery of robot da Vinci Xi. Access via internet: http://www.la-tour.ch/en/doctors-db-en/medical-specialties/robotic-surgery/image-gallery-of-robot-da-vinci-xi/ [Accessed: 2017-11-08].

[6] Engadged. Access via internet: http://japanese.engadget.com/gallery/3-t-53/#!slide=2588078 [Accessed: 2017-11-08].

[7] Sarcos. Access via internet: http://www.sarcos.com/?page_id=1409 [Accessed: 2017-11-08].

[8] Konopka S., Lopaka M. J., Krogul P. Simulation Research of Hydrostatic Power System Control of Engineer Robot Manipulator, Proceedings of 8th International Conference on Intelligent Technologies in Logistics and Mechatronics Systems ITELMS 2013, Kaunas University Technology Press, Kaunas 2013, pp. 126–135.

[9] Tomczyk J. Modelowanie dynamiczne elementów i ukladów napędów hydrostatycznych. WNT, Warsaw, Poland, 1999.

[10] Bartnicki A., Łopatka M. J. and Muszyński T. Stiffness evaluation of Fire Rescue Robot suspension with hydropneumatic components, Mechatronic systems, mechanics and materials II. Solid State Phenomena, Vol. 210, Trans Tech Publications Ltd., pp. 301–308, Zurich 2014.

[11] Konopka S., Łopatka M. J. and P. Krogul. Simulation Research of Hydrostatic Power System Control of Engineer Robot Manipulator, Proceedings of 8th International Conference on Intelligent Technologies in Logistics and Mechatronics Systems ITELMS 2013, Kaunas University Technology Press, pp. 126–135, Panevezys 2013.

[12] Przybysz M., Rubiec A. Simulation research of kinematic discrepancy in multiaxis hydrostatic drive system, 11th International Conference On Intelligent Technologies In Logistics And Mechatronics Systems (ITELMS'2016), pp. 131–139, Panevezys, Lithuania, Apr 28–29, 2016.

[13] Krogul P., Typiak R. Issues with controling hydrostaticly driven manipulator using an intuitive human - machine interface, Proceedings of 8th International Conference on Intelligent Technologies in Logistics and Mechatronics Systems ITELMS 2016, Kaunas University Technology Press, pp. 101–109, Panevezys 2016.

[14] Cieślik K., Konopka S. Concept and preliminary research of anthropomorphic manipulator with hydrostatic drive system for mobile robot, 21st International Conference on Methods and Models in Automation and Robotics, pp. 1154–1159, Miedzyzdroje, Poland 2016.

[15] Ewald E., Hutter J., Kretz D., Liedhegener F., Schenkel W., Schmitt A., Reik M. Proportional and Servo Valve technology, The hydraulic Trainer, Volume 2', Bosch Rexroth AG, 2003.

[16] Cieślik K., Łopatka M. J., Muszyński T. Stiffness identification of hydraulic hoses, 11th Intelligent Technologies in Logistics and Mechatronics Systems, pp. 53–59, Panevezys, Lithuania 2016.

[17] Bartnicki A., Krogul P., Spadło K. Influence of an EOD Engineer Robot manipulator structure on the effector's accuracy using intuitive control system, 21st International Conference on Methods and Models in Automation and Robotics, pp. 1148–1153, Miedzyzdroje, Poland 2016.

The 12th International Scientific Conference Intelligent Technologies in Logistics and Mechatronics Systems (ITELMS'2018), 26–27 April 2018, Panevėžys, Lithuania

Preliminary Analysis of Influence of Ground-Support System on Specified Operational Parameters of Teleoperated Ground Vehicles

Marcin Dejewski[a], Dariusz Kalinko[a], Józef Wrona[a]

[a]*Military University of Technology, Urbanowicza 2, Warsaw 00-908, Poland*

Abstract

There are presented the results of selected works related to influence of ground-support systems on visualization of working area and on load of executive mechanisms of chosen platforms. This paper considers importance of research question whether there is an influence of the ground-support system on the structure and characteristics of ground vehicles suspension systems travelling cross-country.

© 2018 M. Dejewski, D. Kalinko, J. Wrona
Peer-review under responsibility of the Kaunas University of Technology, Panevėžys Faculty of Technologies and Business

Keywords: soil models, ground-support/suspensionsystem,vision system, executive mechanisms

1. Introduction

Mobile platforms that are used by humans execute their functionalities in on-road or off-road conditions. When we consider on-road mobility there are some works that were done [1, 5, 9]. More complex issue is off-road, cross-country mobility which is characteristic for military vehicles and civilian ground platforms like construction, heavy duty machinery. It is exactly the natural environment of such a vehicles that execute their tasks in specific environmental conditions that ground plays an important role. The ground-support systems are crucial if we consider mobility of such a platforms. There are some models that are used for example to predict mobility of platforms in cross-country conditions. In the military domain there is the NRMM tool [10] which is an engineering model that accommodate the engineering characterization of the vehicle (power train, running gear, suspension, mass and inertial properties, and geometry) with the engineering characterization of the terrain (slopes, soil strength, obstacle geometry, surface roughness, and vegetation density). It is an analytical simulation model which predicts the vehicle's maximum attainable speed cross-country and on-road for a specified geographic region and environmental condition (wet season, dry season, snow, etc.), and defines within the geographic region areas of immobilization and

© 2018 M. Dejewski, D. Kalinko, J. Wrona **B427H0010** . Ed*Learning*.it
Peer-review under responsibility of the Kaunas University of Technology, Panevėžys Faculty of Technologies and Business

the reasons for such (Fig. 1). It makes possible to predict the maximum automotive performance of a vehicle travelling cross-country and on-road. There are works to find new such a model to take on board new technologies that allow to figure out new physics based rather than empirical platforms ground-support systems models and also their environmental components models. It is possible to solve this problem in the future when we find the soil model being part of the ground-support model to be used even in the designing process of such a platforms and then in their mobility algorithm also in case of unmanned ground vehicles. Such a works are conducted within NATO activity on: Next-Generation NATO Reference Mobility Model (NG NRMM) Development [4].

This topic is important not only on the Earth. According to NASA and other space exploration roadmaps, it is planned to send a manned lunar exploration mission and build infrastructures in the Moon. This requires multiple landings of rockets causing fine dust on the Moon which has harmful effect on human and infrastructures [6]. It is also a reason why it is important to find the methodology and tools to respond to scientific question whether there is an influence of ground-support system on specified operational parameters of teleoperated ground vehicles and to ask a research question whether there is influence of the ground-support system on the structure and characteristics of ground vehicles suspension systems travelling cross-country and being used for space explorations. It concerns not only manned but after all unmanned platforms.

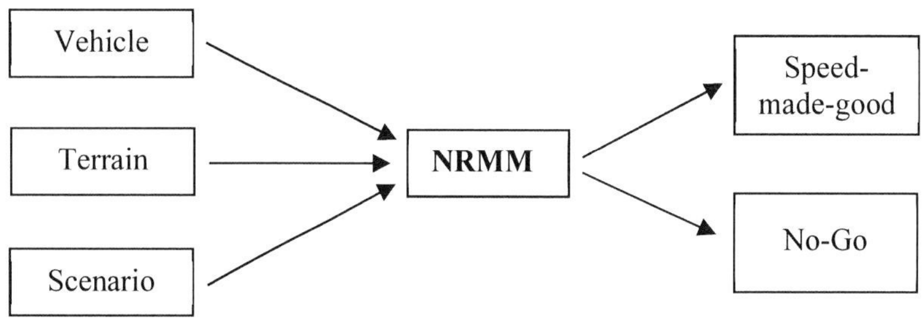

Fig. 1. NRMM Main Module Schematic

This problem is important for the ground vehicles industry to fulfill all end users' requirements both in civilian and military applications. In case of teleoperated ground vehicles the accuracy of these systems is the issue that producers should be focus on. It is important in designing process to know the impact of ground-support systems on the load of executive mechanisms. It is also crucial for military applications to improve travelling of military ground manned and unmanned platforms in off-road conditions.

2. Preliminary literature analysis of influence of ground-support systems on chosen specified operational parameters of teleoperated ground vehicles

There are many works focused on the prediction of mobility (for example [3], [11], [7]), but the preliminary analysis that are presented in this paper were focused on finding some efforts to describe influence of ground-support systems on accuracy of visual systems and load of executive mechanisms, which are typical problem when we talk about teleoperated ground vehicles. The ground-support systems are considered as a ground-wheel in case of accuracy of visual system and as a ground-stiffness leg (Fig. 4) in case of load of executive mechanisms. There will be used qualitative methods of analysis to respond to research question whether there is an influence of ground-support systems on chosen specified operational parameters of teleoperated ground vehicles and to find importance to ask the research question whether there is an influence of the ground-support systems on the structure and characteristics of suspension systems of the mobile ground vehicles travelling cross-country and on-road.

2.1. Influence of ground-support systems on accuracy of visual systems

Teleoperated ground vehicles work is based on transmission of the view from platform to operator's remote control unit and giving the feedback signal from control unit to vehicle. Regarding to unmanned vehicles moving in off-road conditions, quality of signal transmitted to operator has the main influence on their effectiveness. There is work describing effectiveness of vision system of unmanned teleoperated ground vehicles [2]. The results of the work show, that effectiveness of such a systems primarily depends on vibrations affecting on cameras. One of the main factors that generate vibrations is suspension system and cooperation between wheel and ground. Analyzes of vision systems effectiveness were based on analyzing relationship between effectiveness of detection and identification of markers (Fig. 3) by platform vision system of the operator, and vibrations affecting on cameras. The test was carried out by passing a special test vehicle on a prepared track at different speed (Fig. 2). During the tests, there were recorded vibrations affecting on the cameras and effectiveness of detection and identification of markers. The example results are shown in Figure 3.

In the development of these studies included in [8], a model of influence of suspension structure on the cameras vibrations was created. This model allows to create of suspension structure that minimizes the negative impact of vibration on the effectiveness of vision system of teleoperated ground vehicles. However, there is no model that allows taking into account the influence of soil deformation on the effectiveness of the vision system. Therefore creation of a tool allowing for a model representation of the ground-support system (ground-platform track system) would allow for further development and more efficient use of teleoperated land platforms.

Fig. 2. Test track view with located markers [8]

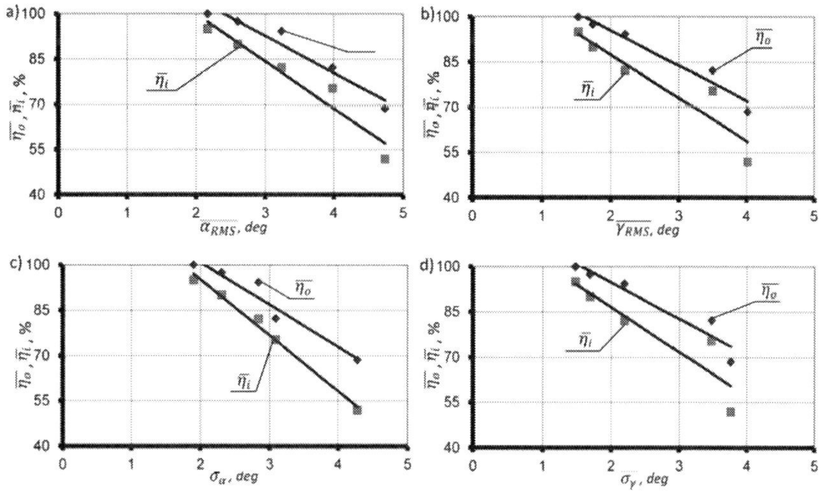

Fig. 3. Markers finding and identification efficiency ratio in function of: a) RMS values of the cameras angular displacement along carrier longitudinal axis; b) RMS values of the cameras angular displacement along carrier transverse axis; c) standard deviation values of the cameras angular displacement along carrier longitudinal axis; d) standard deviation values of the cameras angular displacement along carrier transverse axis [8]

2.2. Influence of ground-support systems on load of executive mechanisms

There are presented some exemplary research results analysis collected during the tests of single bucket excavator being equipped with experimental support system [12].

The aim of this research was to make a qualitative assessment of the influence of support on the displacement and load of working tool of selected hydraulic single-bucket excavator. During the tests, the character and size of the displacement of the suspension (displacement test method shown in Figure 4), and load of the tool's hydraulic cylinders for different support system were researched.

Fig. 4. View of the sensor for measuring the vertical displacement of the excavator's platform [12]

a)

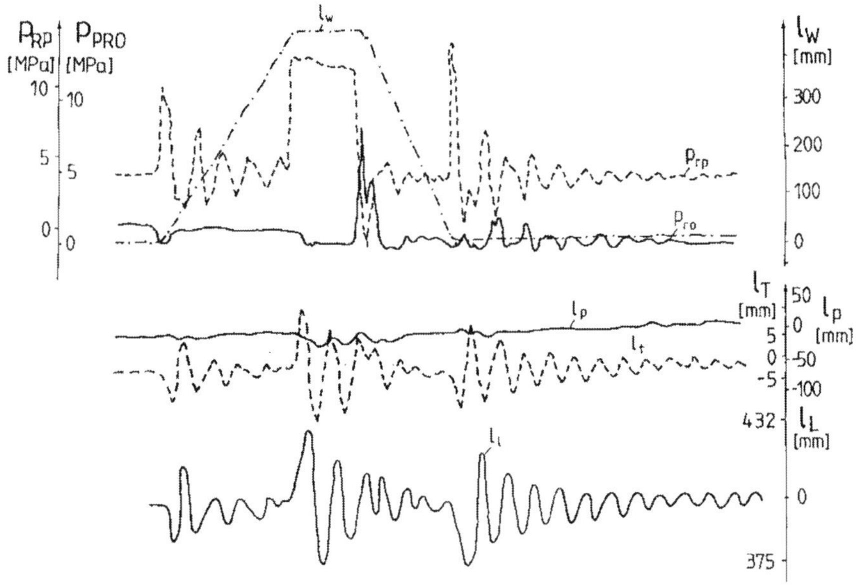

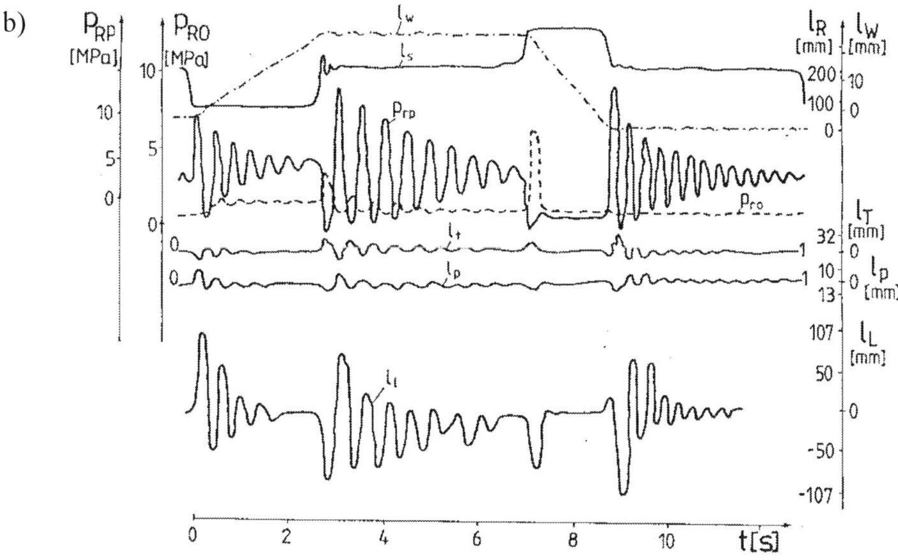

Fig. 5. Sample charts of registered pressures and displacements during forced movement of work tools by excavator arm's hydraulic cylinder [12]: a) wheeled single bucket excavator K–607; b) tracked single bucket excavator K–408; l_S – control signal; l_W – displacement of arm's piston rod; p_{RP}, p_{RO} – pressure above and below the piston of the arm's cylinder; l_P – vertical displacement of the front of platform; l_T – vertical displacement of the rear of platform; l_L – vertical displacement of the cutting edge of bucket

Excavator tests have been divided into two stages:

- measurement of excavator parameters during the movement of work tools with a bucket loaded with 750 kg;
- measurement of excavator parameters during excavation.

The exemplary research results are shown in Figure 5.

Analysis of load and displacement of working system and displacement of excavator allowed to draw the following conclusions:

- the end point of acceleration and deceleration phase is the beginning of the occurrence of large loads of arm's cylinder (p_{RP}, p_{RO}) and large displacements of unsupported part of platform, (l_T). There is clear analogy between the character (amplitude, frequency) of the arm's hydraulic cylinder and displacement of supports at the points of contact with a ground (support);
- load of arm's hydraulic cylinder can be reduced by developing a proper control signal or by changing the support properties, which should be determined by further simulation and experimental tests; this requires the creation of a simulation support model and an excavator simulation model with support model as an integral part;
- due to the fact, that a research did not concern the same excavator, differently supported, determined on the basis of their regularity results are only hypotheses, not strict regularities.

Conclusions

These works that are presented in the article allowed to determine directions of the follow-up research to be focused on elaboration of soil model that will consider both the type of soil and its moisture being a main determinant of the strength of the soil. That is why the models that have been in place until now should be considered and a new one should be developed. It is necessary to determine the model of interaction between such a new soil model and elaborated model of suspension system. In case of Lunar exploration, soil moisture is not an issue.

The original character of such an approach will be expressed by specific methodology of tests of soil and then ground-suspension systems. There will be verified existing tools and used a new one to measure strength of the soil and its moisture to find influence of the moisture on the strength of the soil. The elaborated ground-suspension system model considering different types of soils, different moisture coefficients and different types of suspension systems will be used as a component of mobile platform model. It will be possible to use it for different applications: military and civilian, for manned and unmanned ground vehicles.

The results of analysis show the importance to respond to scientific question whether there is an influence of the ground-support (wheels, tracks) systems on the structure and characteristics of suspension systems of the mobile ground vehicles travelling cross-country and on-road.

The article is based on the works also carried out and financed within the project PBS 936 on: Development of design and manufacturing technology and effectiveness of teleoperation systems of Unmanned Ground Vehicles.

References

[1] Bernard J., Clover C. L. Tire Modeling forLow-Speed and High-Speed Calculations, Society of Automotive Engineers, Paper 950311, 1995.
[2] Dąbrowska A., Jaskółowski M. B., Rubiec A. Cameras vibrations influence on efficiency of teleoperated Unmanned Ground Vehicle.
[3] Dudziński P., Stefanow D. New test device for determination of dynamical shearing resistance of soils, 2017.
[4] Final ReportET-148 Next-GenerationNATO Reference Mobility Model (NRMM), 2015.
[5] Garcia-Pozuelo D., Diaz V., Boada M. J. L. New tyre-road contact model for applications at low speed, International Journal of Automotive Technology, June 2014.
[6] Lee J., Chang B. C., Sangah L., Lee T. S. Automation System For Lunar Landing Pad, International Symposium on Automation and Robotics in Construction, 2011.
[7] Pytka J. Teoretyczno-doświadczalne studium układu element jezdny – podłoże odkształcalne, Politechnika Lubelska, Lublin 2011.
[8] Rubiec A. Forming properties of suspension systems of teleoperated engineering missions support robots, Doctoral Thesis, Warsaw, Poland 2018.
[9] Schramm D., Hiller M., Bardini R. Modelling of the Road-Tire-Contact, Vehicle Dynamics, 2012.
[10] Vong T., Haas G. A., Caledonia L. H. NATO Reference Mobility Model (NRMM) Modeling of, the DEMO III Experimental Unmanned, Ground Vehicle (XUV), Army Research Laboratory, 1999.
[11] Wong J. Y. Theory of Ground Vehicles, Fourth Edition, 2008.
[12] Wrona J. Influence of support system on load of working equipment and undercarriage movements of single bucket excavator, Doctoral Thesis, Warsaw, Poland, 1999.

ITMS'2018

panevėžys faculty
of technologies
and business

The 12th International Scientific Conference Intelligent Technologies in Logistics and Mechatronics Systems (ITELMS'2018), 26–27 April 2018, Panevėžys, Lithuania

Investigation of the Circle Fractal Structure Interaction with Gigahertz Frequency Electromagnetic Waves

Dainius Jasaitis[a], Vaida Vasiliauskienė[a], Paulius Miškinis[a], Jovita Damauskaitė[a], Artūras Jukna[a*], Aleksandr Kopyltsov[b**], Genady Lukyanov[c], Konstantin Korshunov[d], Igor Serov[d]

[a]Vilnius Gediminas Technical University, Sauletekio av. 11, Vilnius 10223, Lithuania
[b]Saint-Petersburg Electrotechnical University, Professor Popov str. 5, Staint-Petersburg 197376, Russia
[c]ITMO University, Kronverksky av. 49, Saint-Petersburg 197198, Russia
[d]AIRES Human Genome Research Foundation, 61, Vyborgskaya nab., Saint-Petersburg 194044, Russia

---

Abstract

We present results of investigations of Si crystal (100) with a circularly periodic structure formed on its surface (resonator-converter) interaction with low power radiation which can be attributed to electromagnetic pollution (EP). Due EP interaction with circularly periodic (diffraction grating) structure of the front/rear antennas and periodic structure of Si crystal our device in the regime of optical reflection can efficiently damp EP if incident power is ≥ 2 W. The efficiency of EP damping by means of our device is noticeably higher than that one of measured for a regular conducting plate of identical dimensions and orientation in the space. The results of our experimental measurements and model of power attenuation by the resonator-converter is presented and discussed.

© 2018 D. Jasaitis, V. Vasiliauskienė, P. Miškinis, J. Damauskaitė, A. Jukna, A. Kopyltsov, G. Lukyanov, K. Korshunov, I. Serov
Peer-review under responsibility of the Kaunas University of Technology, Panevėžys Faculty of Technologies and Business

Keywords: fractal antenna, electromagnetic polution, optical reflection, optical transmission, diffraction grating

---

* Corresponding author. Tel.: +370-5-274-4833; fax: +370-5-274-4844
E-mail address: arturas.jukna@vgtu.lt
** Corresponding author. Tel.: +7-921-401-9427; fax: +7-812-233-7720
E-mail address: kopyl2001@mail.ru

© 2018 D. Jasaitis, V. Vasiliauskienė, P. Miškinis, J. Damauskaitė, A. Jukna, A. Kopyltsov, G. Lukyanov, K. Korshunov, I. Serov B427H0011
Peer-review under responsibility of the Kaunas University of Technology, Panevėžys Faculty of Technologies and Business

1. Introduction

All organic and inorganic objects in nature are constantly being exposed to a natural electromagnetic radiation (EMR) produced by the cosmic radiation [1] (the biggest part of it is absorbed by the Earth atmosphere), emitted by the Earth's crust (due to nuclear reactions of radioactive elements in it), and coming from artificial sources in the highly urbanized and/or domestic industry environment. In fact, humankind, animals and plants use EMR for a variety of their living activities, like those of telecommunication, control and regulation of their various behavioral and physiological functions. However, though essential for living objects, exposure to excess low frequency (equal or higher than 50 Hz) [2, 3] and/or high frequency (below 2.5 GHz) EMR beyond the naturally evolved tolerance limits can cause various human illnesses [2] and/or harmful biological effects in humans' bodies [4, 5].

The artificial sources of ambient non-ionizing electromagnetic pollution (EP) are power supply lines (the higher the voltage, the higher radiation power), microwaves and radars, telecommunication equipment/devices and electrical appliances, radio and TV signals transmitters, even computers and their monitors, electric clocks, heated waterbeds, blankets *etc*. All together these sources create surrounding us EP with frequency of EMR ranging from 0 Hz (static electromagnetic field) to 300 GHz (microwaves and millimeter waves). The effects of extremely low-frequency EMR are dependent on dose and duration of exposure and are cumulative. Therefore, todays humankind start asking questions how to assess the exposures to the human body from EP and to protect them from excess of EMR which can lead to cells and/or neurons damage or even damage to chromosomes altering the structure of our DNR [6].

Protecting themselves from EP one can simply limit time period of direct contact with electronic appliances, but technically it is difficult to accomplish in nowadays living/working environment, be aware of high power radiation sources what is almost impossible in highly urbanized/industrial areas, or use electromagnetic shields which potentially can increase immunity of the shielded device and decreases the power of undesirable radiation from it [7]. Moreover, the electronic appliances emit power from a number of primary and secondary sources and their emissions become an uncorrelated broadband EP which can be totally suppressed only by means of the signals phasor addition for all contributions resulting the radiation field.

Our work aims on experimental testing of a silicon crystalline (100) resonator-converter interaction with low power 0.9–2.5 GHz frequency continuous wave signals which frequency can be attributed to frequency range of ambient non-ionizing EP. The resonator-converter (device) is shaped in a form of circular diffraction grating with a variable period starting with 0.001 mm wide "slits" at the device's centre and ending with 0.1 mm wide "slits" at the Si crystal edges and consists of front and rear antennas which are also circularly periodic diffraction gratings spaced by a 1-mm-thick insulator (ε = 3.5@10 GHz – 4@2.5 GHz) [8] located in between them. Our results of experimental investigation show that the resonator-converter suppresses power of incident EMR and, due to incident wave interference with a wave reflected from device's surface, it retransforms the incident EMR in terms of frequency/wavelength and phase. Our experimental results of incident power suppression when our device is operating at room temperature in the regime of optical reflection and in the regime of optical transmission are presented and discussed.

2. Samples and measurement setup

Our tested resonator-converter (device) contains a 500-µm-thick round shaped Si (100) crystal plate glued onto surface of a front antenna (Fig 1 *on the left*) which physical shape defined as a fourth iteration circle fractal. The surface of Si crystal plate modified by means of plasma-chemical etching of 600-nm-deep and 600-nm-wide groves which produce a net of intercrossing rings (Fig. 1 *on the right*). Since etched groves on the silicon surface affects a periodic change in crystal's thickness, the crystal behaves like a diffraction grating with a variable period produced by intercrossing rings of the fourth iteration circle fractal with a diameter of D_1 = 7.5 mm. It means that for every further step of iteration the diameter of rings decreases proportionally to ratio $D_2 = D_1/2$, $D_3 = D_2/2$, and $D_4 = D_3/2$ in such a way producing a diffraction grating of a variable period (Fig. 1 *on the right*).

The circles of groves on the surface of Si plate form a circularly periodic topology of the Si resonator-converter, which is fixed on a top of front antenna (Fig. 1. *on the left*). Two identical antennas, the front and rear, in our device are used for receiving ultra-wide band frequency continuous-wave EMR with the frequency band ranging between 0.9 and 2.5 GHz frequencies, which can be attributed to frequency range of ambient non-ionizing EP. The front and rear conducting (metallic) antennas are also manufactured as diffraction gratings produced by 0.1-mm-thick and 0.1-mm-wide intercrossing rings of metal fourth iteration fractal of the circle with a diameter of D_1 = 12.5 mm with a diameter of rings of every further step of iteration following ratio $D_2 = D_1/2$, $D_3 = D_2/2$, and $D_4 = D_3/2$ in such a way producing a diffraction grating (Fig. 1 *on the right*) of a variable period ranging from several millimeters down to tens of micrometers.

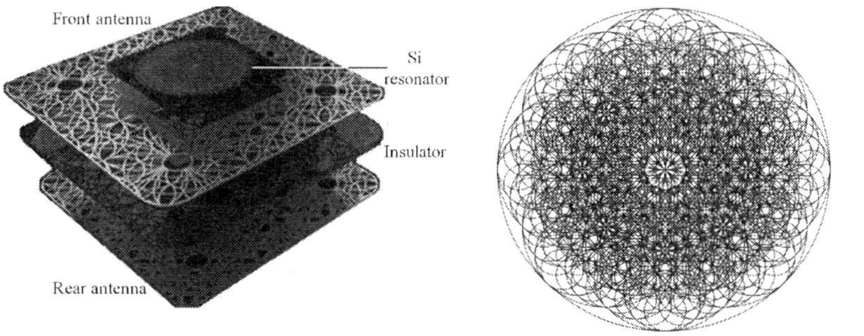

Fig. 1. Schematics of the resonator-converter consisting of a silicon resonator fixed on the top of the front antenna (*on the left*) and a schematic sketch of plasma-chemically etched groves on its surface (*on the right*). Similar design of the fourth iteration circle fractal (on the right) has been applied for the front and rear antennas of this device

The equivalent circuit of the resonator-converter is a capacitor comprising two conductive electrodes separated by an insulator. The interaction of the capacitor with incident EM wave results in charging up of the capacitor C and the device's antenna gain depends of antenna impedance given by $(L_p/C)^{1/2}$ where L_p stands for the parasitic inductance of both metallic antennas and of the Si resonator.

Our measurement setup consists of continuous-wave signals generators (for this purpose we have been using different radiation sources of household appliances), emitting wide band signals in frequency band ranging between 0.9 GHz and 2.5 GHz and of resonator-converter. For signals detection we used the Signal Hound Spectrum analyzer receiving signals in the 0–8 GHz frequency band. The signals were detected in the near-field (the resonator-converter located close to radiation source) and in the far-field (the resonator-converter located at distance $L > 10\lambda$) zones. Here λ is the wavelength of the central frequency of the signal received by the signals detector.

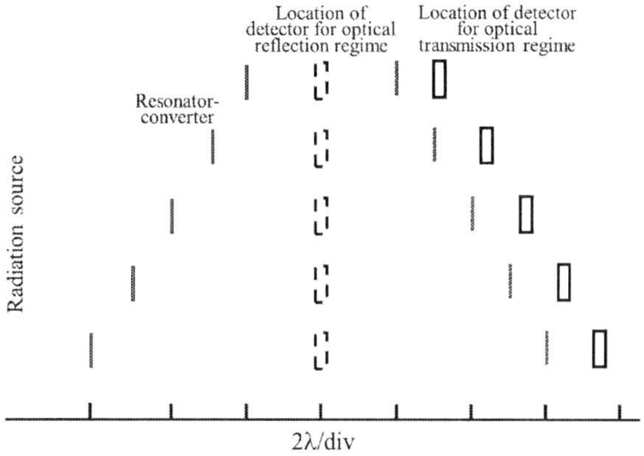

Fig. 2. A scheme explaining experimental setup for detection of signals when the resonator-converter is used in a regime of optical reflection (dashed lines) and in the regime of optical transmission (solid lines) and distances between radiation source, detector, and the resonator-converter are given in λ-as. Here λ is the wavelength of the central frequency of the source radiated wave

We measured power of 0.9–2.5 GHz electromagnetic waves (in dBm) transmitted through the resonator-converter (the regime of optical transmission (Fig. 2)) and that of reflected waves from our device (the regime of optical reflection) in the near-field and far-field zones of the resonator-converter and then the power values were converted into electric field amplitude, presuming that radiation sources are ideal electric dipoles with antenna gain of 2.15 dBi. For the optical reflection measurements, our detector of radiation we located in between radiation source and resonator-converter in the near-field and far-field zones. The power of detected signal was analyzed by means of the FFT method sampling the detected signal over a period of time and measuring amplitudes of its frequency components.

To calculate experimental mean values all measurements either under laboratory or field conditions were repeated for several times at fixed distances between the resonator-converter and the detector given in λ-as. Four different radiation sources transmitting 0.5 W @ 0.9 GHz (No. 1), 2 W @ 0.9 GHz (No. 2), 400 W @ 2.5 GHz (No. 3), and 800 W @ 2.5 GHz (No. 4) we used for our measurements. However, to minimize errors of our experimental measurements, in current report we will mainly focus on experimental results using the most powerful radiation source No. 4.

3. Measurement results and discussion

3.1. Experimental results

EMR interaction with the resonator-converter has been studied by means of detection of residual signal power versus distance between the detector and a radiation source in the case when the resonator-converter operates in the regime of optical transmission. Figure 3 represents difference in electric field amplitudes E_0–E_1 versus distance of the detector from mentioned above four types of radiation sources. Here E_0 stands for electric field amplitude of radiation source emitted signal and E_1 is the residual electric field amplitude transmitted through the resonator-converter which was located in the near-field zone of radiation source.

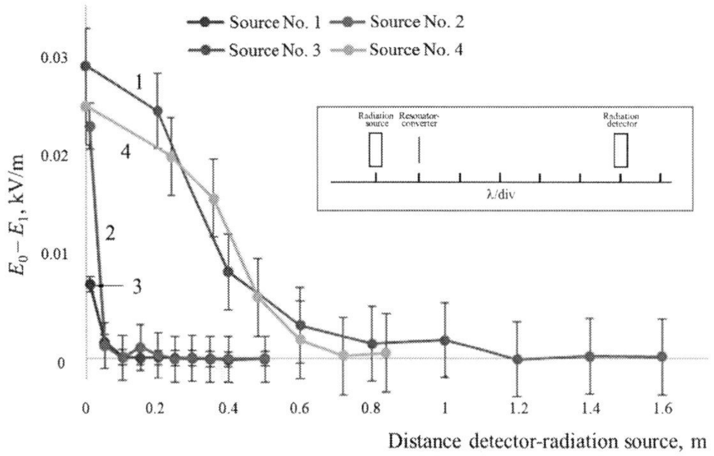

Fig. 3. A plot of difference of electric field amplitudes (E_0–E_1) recorded by the signals detector versus distance detector-radiation source. Here E_0 is electric field amplitude of the source emitted signal and E_1 is electric field amplitude of the signal transmitted through the resonator-converter which was fixed in the near-field zone of the radiation source. Numbers and colors of curves in the graph represent results of experimental measurements using four different sources of EMR. Inset represents a schematic of our measurement setup

The electric field amplitude of the transmitted through the resonator-converter gigahertz frequency EMR decreased by ~27 % (i.e. electric field amplitude damping ratio -2.07 dB) level on average in whole range of our tested distances from the radiation source No. 4 (Fig. 3, curve 1 (grey)). Here and in all other experimental cases the resonator-converter is attached directly to the frame of a radiation source, *i.e.* located in its near-field zone.

Increasing distance from the radiation source, the parameter E_0–E_1 gradually decreased and finally vanished at distance 1.2 m (curve 1), 0.85 m (curve 4) and at distances below 0.2 m (curves 2 and 3).

All curves in Figure 3 show strongly nonlinear behavior with increasing detector's distance from the radiation source and this feature let us to predict that the lever of power damping by the resonator-converter should depend on power of incident EMR. Since our tested radiation, sources are not zero-dimensional and in all cases amplitude and frequency range of radiation is a result of superposition of radiation of several primary and secondary sources, the electric field amplitude of 800 W radiation source does not decrease with increasing distance squared. However, the variation of the parameter E_0–E_1 show that the resonator-converter turns-on at high power and turns-off when the incident power appears to be below some threshold value P_{min}. As it follows from our measurement results the minimum power required for turning-on the resonator-converter is of order of P_{min} ~ 2 W (curve 2 (red)). Our preliminary results also show that P_{min} is a function of frequency/wavelength of an incident electromagnetic wave and P_{min} decreasing with increasing central frequency of an incident EMR.

The electric field amplitude of 800 W power radiation source versus distance to the signals detector is shown in Fig 4 *on the left* (here the resonator-converter used in the regime of optical transmission) and in Figure 4 *on the right* (the regime of optical reflection). In both experimental cases, the distance between the resonator-converter and signals detector was kept constant while the detector moves away from the radiation source.

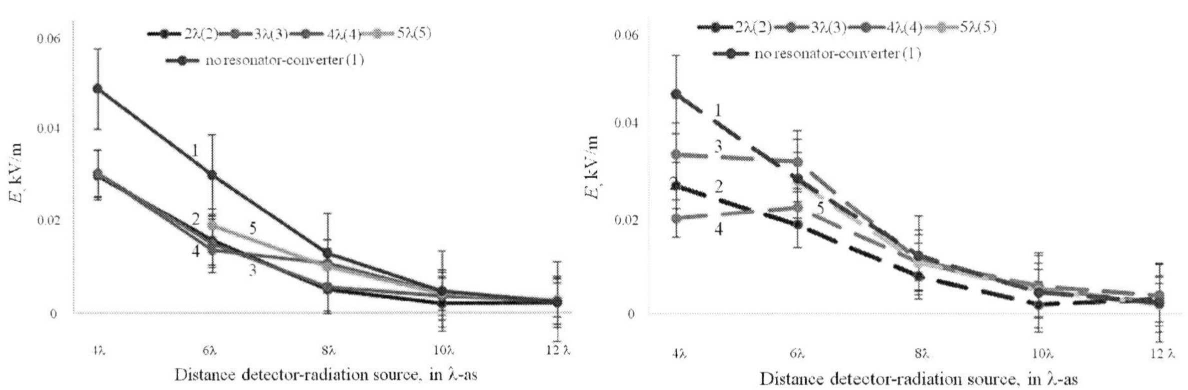

Fig. 4. The electric field strength detected at various distances (in λ-as) from an 800 W radiation source. Here figure (on *the left*) represents the case when our tested resonator-converter used in the regime of optical transmission and figure (*on the right*) when the resonator-converter used in the regime of optical reflection. In both experimental cases the distance between the resonator-converter and detector is fixed at 2λ (curve No. 2), 3λ(3), 4λ(4), and 5λ(5) (see figure legends) while the detector moves away from the radiation source. Here λ = 12 cm is the wavelength of the central frequency of the source radiated signals

The electric field amplitude of the transmitted through the resonator-converter EMR decreases considerably (Fig. 4 *on the left*), however the parameter E_0–E_1 (it was calculated by subtracting curves Nos. 2-5 from curve No. 1 in figure) does not depend on the resonator-converter distance from the detector. Slightly weaker damping of the electric field amplitude, but very similar behavior of the parameter E_0–E_1 with increasing distance from the detector demonstrates the conductive (metallic) plate when we substituted it for the resonator converter. Having same dimensions as the resonator-converter the metallic plate in our measurement setup exhibited effect of electromagnetic screening of EMR electric field component demonstrating 7–8 percent lower level of EMR damping if compare it with that one of the resonator-converter.

When the resonator-converter operated in the regime of optical reflection (Fig. 4 *on the right*), the maximal damping of electric field amplitude we observed for the distances between the resonator-converter and the detector lower or equal to 2λ or when the resonator-converter attached to the frame of the detector. These results let us to conclude that most efficient damping of the EP could be expected when the resonator-converter operates in regime of the optical reflection and when incident EMR power is greater or equal to P_{min}.

3.2. Results of numerical simulation

Our results of numerical simulation of the resonator-converter interaction with an incident monochromatic wave of 550 THz frequency showed that 5.75 cm in diameter and 1 mm thick Si crystal modified surface by means of plasma-chemically etched 0.5-µm-wide and 0.5-µm-deep grooves undergoes non-uniform distribution of free carriers. An incident EMR charges up Si crystal producing stronger electric field (*i.e.* inducing larger concentration of free carriers) between the bottom of grooves located on a top of the silicon crystal and crystal's bottom surfaces. We simulated both the redistribution free-carrier concentration along the Si crystal surface and a strength of the electric field in the near-field zone of this crystal. Our results showed that high concentration of free-carriers collected in the grooves of the Si crystal can initiate a spontaneous electric discharge causing the electric field strength redistribution and flow of electric current along the resonator-converter surface [9]. It means that for the most efficient damping of EP of specific frequency range the doping of Si crystal (*i.e.* its electric conductivity) of the resonator-converter has to be optimized. It is also demonstrated that efficiency of the EMR damping by the resonator-converter is a function of Si crystal and metallic antennas topology as well as characteristic frequency/wavelength of an incident EMR.

Due to EMR interaction with both antennas and signals interaction with diffraction grating built on the surface of Si crystal, the signal is back scattered in a form of coherent radiation with the maximal amplitude of the electric field located in the centre of the Si crystal (Fig. 5).

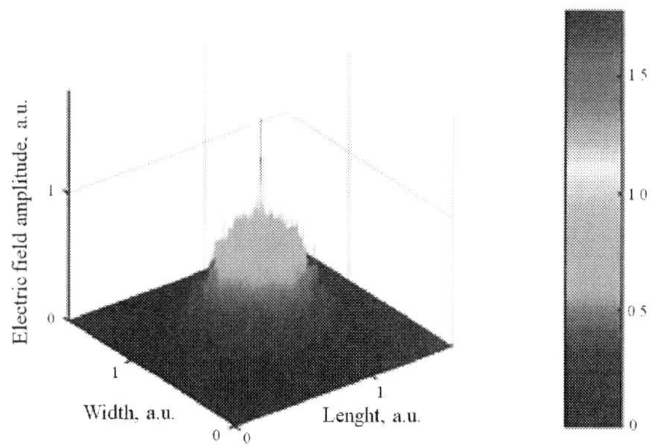

Fig. 5. The three-dimensional (3D) view of the electric field strength distribution in near-field zone of the 1 mm thick, round (5.75 cm in diameter) shaped Si crystal with plasma-chemically etched 0.5-µm-wide and 0.5-µm-deep grooves of rectangular cross-section on its surface

Increasing power of an incident EMR, a noticeable field amplitude appears also at the crystal edges in such a way producing a bell-like shape of field distribution in the near-field zone of the resonator-converter. The backscattered EMR interfere with incident EMR and the electric field distribution around the resonator-converter surface turns into an intricate three-dimensional (3D) structure [9]. A backscattered signal affects redistribution of electric field amplitude recorded by the signals detected, exhibiting sharp non-linear growth of electric field amplitude starting with the device's antenna edges and spreading towards its center as well as towards the centre of the Si crystal. Our results of numeric simulation show that due to backscattered wave interference with incident EMR, the frequency of resulting wave could increase for several times and for our tested geometry of Si crystal it increases by 3.2 times [9] in respect to 550 THz frequency of monochromatic EMR from the radiation source.

Conclusions

The resonator-converter partly suppresses power of incident on it electromagnetic radiation (EMR). The maximal efficiency of signals suppression can be achieved when the resonator-converter is attached directly on a frame or located in a near-field zone of the radiation source, since in the near-field zone is easier to reach and exceed the threshold power P_{min} necessary for the device's onset.

The minimal power needed for the resonator-converter's excitation is $P_{min} \geq 2$ W measured for the 2.5 GHz frequency radiation source. Our preliminary results also suggest that P_{min} is a function of antenna and Si crystal surface topology, doping of the Si crystal, and frequency/wavelength of an incident EMR.

For the case of damping of electromagnetic pollution (EP) (i.e. far-field zone of a radiation source), the maximal damping of EMR power we achieved when the resonator-converter is directly attached to the frame of the signals detector or located at distance shorter or equal to 2λ away from it. Therefore, as a potential protector against EP, the resonator-converter should be used in the near-field zone of signals detector (*i.e.* human body, shielding electronic device *etc.*) and/or radiation source (*i.e.* source of potential EP).

Acknowledgements

Authors A. J and D. J. acknowledges company "Aireslita" UAB and head of the company Mr. Darius Višinskas for the Aires Defenders provided for our experimental measurements.

References

[1] Nguyen VD., Bouisset P., Kerlau G., Parmentier N., Akatov YA., Archangelsky VV., Smirenniy LN., Siegrist M. A new experimental approach in real time determination of the total quality factor in the stratosphere. *Rad. Prot. Dos.* 1993, 48(1): 41–46.

[2] Johansen Ch. Electromagnetic fields and health effects – epidemiologic studies of cancer, diseases of the central nervous system and arrhythmia-related heart disease. *Scand J Work Environ Health* 2004, 30 Suppl 1: 1–80.

[3] Hardell L., Sage C. Biological effects from electromagnetic field exposure and public exposure standards. *Biomedicine & Pharmacotherapy* 2008, 62: 104–109.

[4] Terzia M., Ozberka B., Denizb OG., Kaplanb K. The role of electromagnetic fields in neurological disorders. *J. Chem. Neuroanatomy.* 2016, 75: 77–84.

[5] Repacholi MH., Basten A., Gebski V., Noonan D., Finnie J., Harris AW. Lymphomas in E mu-Pim1 transgenic mice exposed to pulsed 900 MHz electromagnetic fields. *Radiation Research.*1997, 147(5): 631–640.

[6] Phillips JL, Singh NP, Lai H. Electromagnetic fields and DNA damage. *Pathophysiology* 2009, 16(2-3): 79–88.

[7] Marvin AC., Dawson JF., Ward S., Dawson L., Clegg J., Weissenfeld A. A proposed new definition and measurement of the shielding effect of equipment enclosures. *IEEE Trans. Electromag. Compat.* 2004, 46(3): 459–468.

[8] Land SO., Tereshchenko O., Ramdani M., Leferink F., Perdriau R. Proceedings of the International Symposium on Electromagnetic Compatibility (EMC'14/Tokyo), 2014: 777–780.

[9] Kopyltsov AV., Korshunov KA., Lukyanov GN., Serov IN. Distributed Calculations of the Interaction of Electromagnetic Radiation with a Structured Surface//Regional Informatics and Information Security. Collected Papers. Saint Petersburg Society of Computer Science, Computer Hardware, and Communication and Control Systems 2016, 2: 383–387.

ITMS'2018

The 12th International Scientific Conference Intelligent Technologies in Logistics and Mechatronics Systems (ITELMS'2018), 26–27 April 2018, Panevėžys, Lithuania

Efficiency of the Wall-Window Linear Thermal Bridge in the Net-Zero Energy Building

Jovita Kaupienė[a,b*], Donatas Aviža[a,b], Rita Baltušnikienė[a,b], Zita Kasperiūnienė[b]

[a]*Kaunas University of Technology, Nemuno g. 33, LT–37164 Panevėžys, Lithuania*
[b]*Panevėžys university of applied sciences, Laisvės a. 23, Panevėžys*

Abstract

In the buildings of the energy efficiency class A++ (net-zero energy); thermal loss occurs through partitions and linear thermal bridges. A wall-window connection node is highly relevant in the buildings of this type. The size of the linear thermal bridge directly depends upon the position of the window in the wall. Therefore, the article co-authors carried out the applied research of the window installation in the wall of 1–2 flats in a residential building of the A++ energy efficiency class. The study was based on the principles of energy and economic efficiency. There were eight-window installation in the wall positions analysed in the research. Having done the empirical study, it was found out that the most efficient windows of the A++ class buildings are those, which are installed at the middle of the outer thermal insulation layer (T3 position). In this area, the wall-window linear thermal bridge is the smallest and it is equal to 0.0232 W/(mK). The annual cost for one window connection is 0.87 €. The highest thermal losses occur through the T8 junction and might reach up to 81.06 kWh during the heating season because of one incorrectly installed window. The annual cost of such a solution might increase to 5.7 times; therefore, windows for the class A++ buildings must be installed in the thermal insulation layer.

© 2018 D. Aviža, R. Baltušnikienė, J. Kaupienė, Z. Kasperiūnienė
Peer-review under responsibility of the Kaunas University of Technology, Panevėžys Faculty of Technologies and Business

Keywords: linear thermal bridge, A++ energy efficiency class, thermal losses, net-zero energy building

* Corresponding author. Tel.: +370-652-14307
E-mail address: jovita.kaupiene@ktu.lt

© 2018 D. Aviža, R. Baltušnikienė, J. Kaupienė, Z. Kasperiūnienė B427H0012 . EdLearning.it
Peer-review under responsibility of the Kaunas University of Technology, Panevėžys Faculty of Technologies and Business

1. Introduction

The objective of the work is to carry out the study of the efficiency of the wall-window linear thermal bridge of the A++ class building.

The tasks of the work:

1. Perform the scientific and technical literature analysis of the requirements for linear thermal bridges and for the A++ energy efficiency class building;
2. Create a theoretical model of a wall-window junction;
3. Create the methodology of calculations;
4. Simulate the model and compare the outcomes.

Methodology of the research: scientific and technical literature analysis, empirical study, model simulation, comparison of the calculation outcomes obtained.

From December 31, 2020, according to the requirements of the European Directive 2010/31/EU, the Member States of the EU must ensure that all newly constructed structures shall be net-zero energy buildings [2]. There is a three year period provided to achieve this purpose.

Residential one family net-zero energy houses are the buildings of high energy efficiency with the consumption of a zero or very low energy: most of the energy consumed consists of renewable energy sources, including the energy generated locally or nearby produced energy from renewable sources. In Lithuania such a category would belong to the "A++" class buildings (Fig. 1).

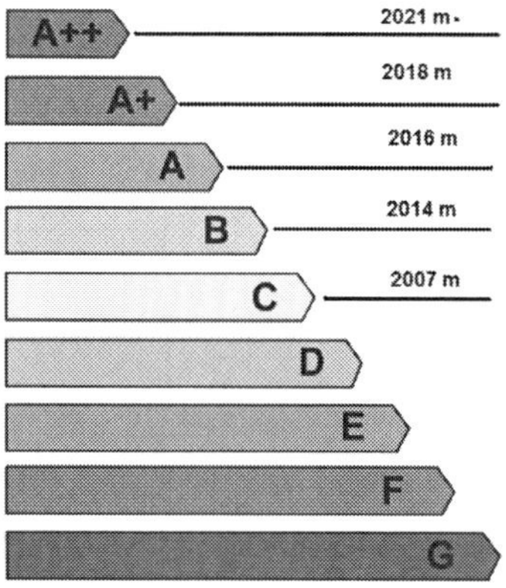

Fig. 1. Classification of the energy efficiency of buildings into classes: class A++ is the highest class, indicating that the building uses net-zero energy, class G indicates not an efficient building energetically [1]

According to the Technical Regulation of Construction STR 2.01.09:2012 "Energy efficiency of buildings; Certification of energy efficiency" [9], the thermal transfer coefficient values for linear thermal bridges of the A, A+, A++ energy efficiency class buildings (their parts) must be based on the calculations in accordance with the LST EN ISO 10211:2008 requirements: "Thermal bridges for building constructions. Thermal flows and surface temperatures. Detailed calculations" [4]. Thermal losses of a typical house are demonstrated in Figure 2.

Windows are the most heat conductible part in building partitions. Therefore, when designing buildings, the properties, size and also orientation of the windows in relation to the countries of the world ought to be evaluated. Heat transfer coefficient, the value of U (W/m^{2}K), is used to define the level of the thermal flow that is conducted through windows and isn't related to solar energy. U values have been established according to the European standards and reflect the properties of the whole window, including the window frame and a small dividing frame. The proper installation of windows has a highly significant effect on energy efficiency. Modern building and thermal insulating materials make it possible to significantly reduce the heat escape from buildings [5].

Freeze bridges might form due to the properties of the thermal conductivity of different materials. In the mounting seam, because of the packings of the materials of other thermal properties, of which the thermal conductivity is higher than the surrounding materials, there is a great probability of freeze bridges being formed.

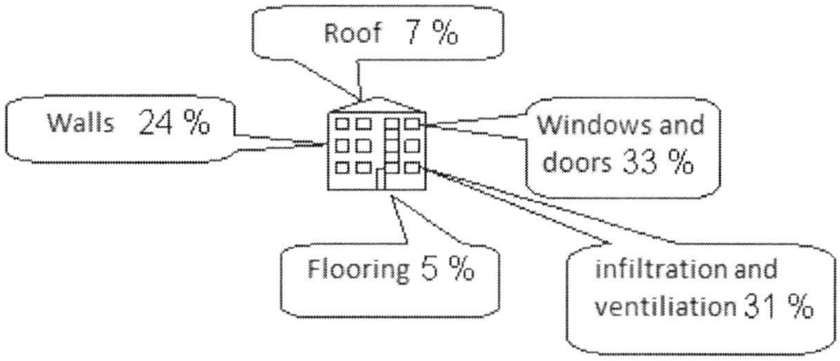

Fig. 2. Heat losses of the building [12]

Such packings make an isothermal displacement and a significant temperature change, therefore, this leads to the formation of a dew point and also condensate, what might be the cause of erosion of the assembly unit [8].

The linear thermal bridge is the areas of increased heat flows in partition structures. They have a significant effect on thermal losses and on the structure surface temperatures. The heat (cold) bridge, if improperly designed, will not only be conductive to heat, but also due to low temperatures will provide conditions for condensation to moisture and grow mould. The areas of heat bridges is often the testing result of the tightness of the building [11]. Improperly designed wall-window interconnection units increase the influence of thermal bridges and worsen the testing results of the tightness of the building.

The window withstands a lot of effects, it functions in different environments and conditions simultaneously. Therefore the location of the window installation is extremely important and is also an integral part of the thermal parameters of the window [3].

To design a building of A++ energy efficiency class it is necessary to calculate linear thermal bridges in the house. On average, there are eight areas in the building where losses are possible through linear thermal bridges (Table 1). In more complex objects their number might even be higher.

Table 1. Standard values for linear thermal bridges (STR 2.05.01:2013 [10])

Linear thermal bridges	The values of heat transfer coefficients for linear thermal bridges $\Psi_{(A++)}$ (W/(m*K))
Between the foundation and exterior walls of the building	0
Around the window openings in the walls	0,05
Around the openings of the entrance door in the walls	0,05
Between the walls and roof of the building	0
In outer and inner corners of the facade	0
In the balcony floor intersections with exterior walls	0,01
Between the ceilings that border with the outside and the walls	0
Between gable windows, bay windows and the perimeter of other transparent openings of partitions	0,05

The linear thermal bridges of the window-wall intersection and the area of the most efficient window installation will be analysed in this research below.

2. Model of the research

The A++ energy efficiency class residential building was selected while calculating the efficiency of the wall-window linear thermal bridge of the building. The research model consists of a two-component (external wall and window) junction with eight window installation positions.

The detail of the wall and the layers are shown in Table 2. The wall is the masonry of two layers, which is insulated with foam polsterene EPS70 boards. The heat transfer coefficient of the wall is U=0,1 W/(m2·K).

The window selected for the A++ class building with the following parameters: heat transfer coefficient– U_w=0,7 W/(m^2·K); window frame – PVC; width 85 mm; three pane package. Dimensions of the window being tested 1.5x1.5 m.

Table 2. The wall detail (ST 2124555837.01:2013 [7]) with the names of layers and thicknesses

Type of walls	No.	Name of wall layers	Thickness in mm
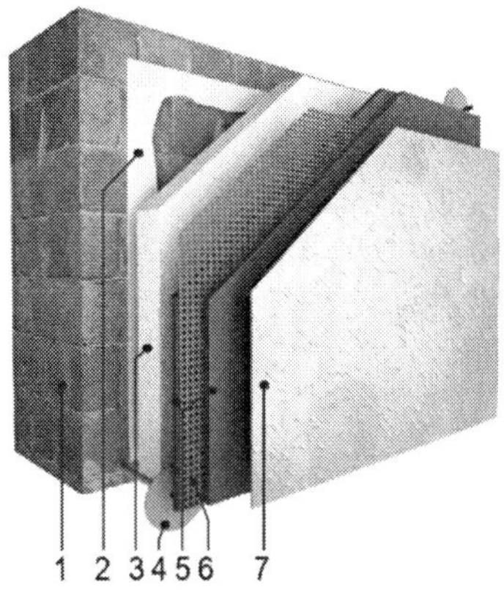	*	Lime and sand plaster	15
	1	Cellular bricks (500 kg/m^3)	250
	2	Adhesive layer	5
	3	**Polystyrène foam boards (EPS70)**	**340**
	4	Plastic studs	-
	5	Armored plaster	
	6	Armature mesh	5
	7	Facing plaster	

To determine the most efficient location of the window installation in the wall, the positions of eight windows have been selected for the research study (Fig. 3).

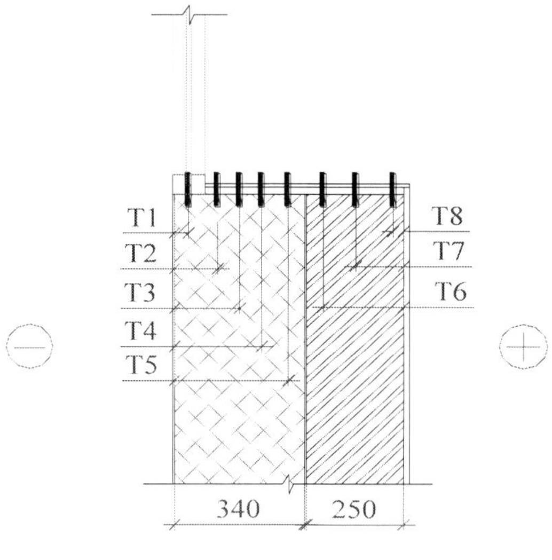

Fig. 3. Determination of the location scheme and testing positions of the wall-window junction of the calculated linear thermal bridges: T1 – at the edge of the external thermal insulation; T2 – 1/3 of thermal insulation from the outer edge; T3 – 1/2 of thermal insulation; T4 – 1/3 of thermal insulation from the inner edge; T5 – at the inner edge of thermal insulation; T6 – at the edge of the outer masonry; T7 – at the masonry wall; T8 – at the inner edge of the masonry

The coefficient of heat transfer of the linear thermal bridge formed at the junction of the wall and window shows how much the total heat transfer through the window and the wall differs from the heat transfer through the window and the wall (Fig. 4).

Generally, the coefficient of the total heat transfer is 7–9 times higher than the heat transfer coefficient of the wall. The difference of the heat transfer coefficient between the window and the wall is even higher [3].

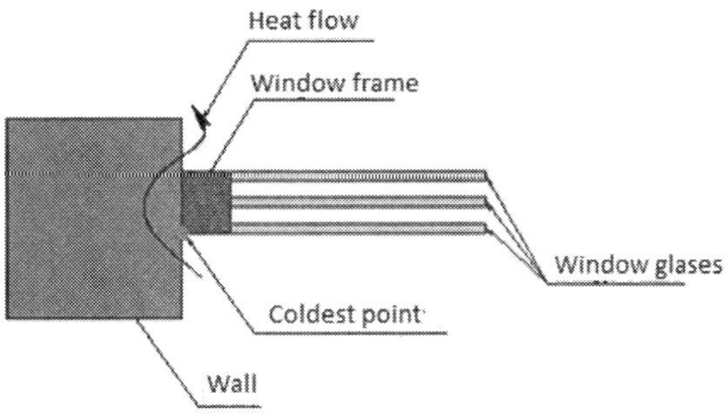

Fig. 4. The scheme for the formation of linear thermal bridges between the walls and the window

Therefore, to reduce the impact of linear thermal bridges, it is important to analyse where the installation of the window is most efficient.

2. Methodology of calculations

Basing upon the requirements of standard LST EN ISO 10211:2008 "Thermal bridges of building structures. Heat flows and surface temperatures. Detailed calculations" [4], the heat transfer coefficient of the linear thermal bridge was calculated according to the formula (1):

$$\psi = L_{2D} - U_1 \cdot l_1 - U_2 \cdot l_2; \tag{1}$$

where Ψ – the heat transfer coefficient of the wall-window linear thermal bridge W/(m·K);
L_{2D} – the specific heat loss of the linear thermal bridge having been indicated by calculating the two-dimensional temperature field for a component that separates two environments (inside and outside);
U_1 – the heat transfer coefficient of the wall has been calculated in one-dimensional temperature field, which separates two environments;
L_1 – the length of the two-dimensional geometric module for which the heat transfer coefficient value has been calculated U_1 (Fig. 6));
U_2 – the heat transfer coefficient of the window has been calculated in one-dimensional temperature field, which separates two environments;
L_2 – the length of two-dimensional geometric module for which the value of the coefficient of heat transfer U_2 has been calculated.

The specific heat loss of the linear thermal bridge L_{2D} is calculated according to the formula (2):

$$L_{2D} = \frac{\Phi_{lj}}{(\theta_i - \theta_e)}; \tag{2}$$

where Φlj – heat flow of the linear wall-window thermal bridge (calculated using (THERM 7.4 or Htflux program), W/m;
θ_i – outside temperature, K;
θ_e – inside temperature, K.

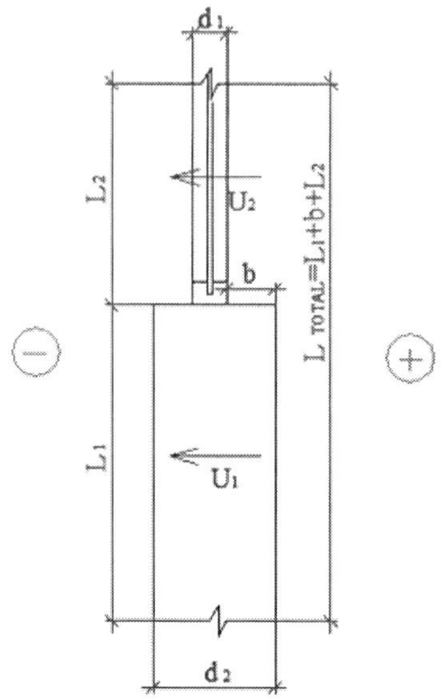

Fig. 5. The calculation scheme of the window and the wall junction of the linear thermal bridge: L1 – not less than 3d2 and not less than 1 m; L2 – not less than 3d1 and not less than 1 m [4]

Heat losses (Q) and annual costs (I) due to the impact of the window-wall linear thermal bridge are calculated according to the formulas: (3) and (4).

$$Q = \psi \cdot l \cdot (\theta_i - \theta_e) \cdot 24 \cdot t; \tag{3}$$

$$I = Q \cdot E. \tag{4}$$

The formulas (3) and (4) are presented in Table 3.

Table 3. Technical - economic indicators of the selected model

Name of the indicator and units of measurement	Numeric value
Perimeter of the window l, in metres	6
Outdoor temperature in winter θ_e, in degrees	0,6
Indoor temperature in winter θ_i, in degrees	20
Heating season t, in days	220
Heating time per day, in hours	24
Price of thermal energy, E (CŠT), €/kWh	0,061

The same model (wall-window unit) has been selected for further calculations only the window placement positions have been changed. Therefore, it might be assumed that the linear thermal bridge is the same throughout the window perimeter.

3. Research study outcomes

According to the model composed, simulative calculations of linear thermal bridges have been performed using Htflux software. The outcomes are presented in Figures 6 and 7.

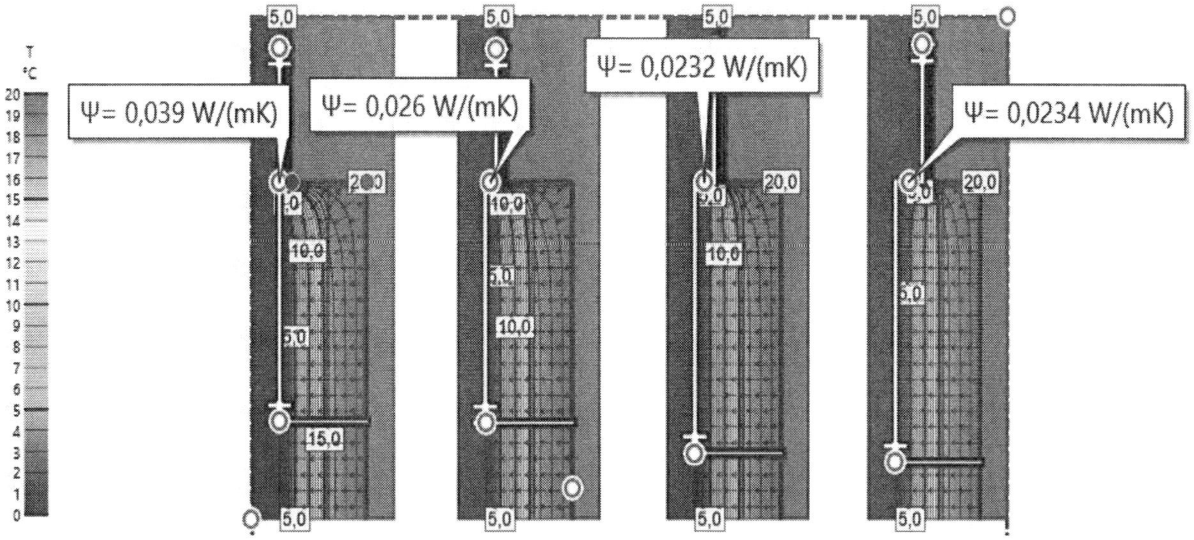

Fig. 6. Window-wall linear thermal bridges T1–T2–T3–T4

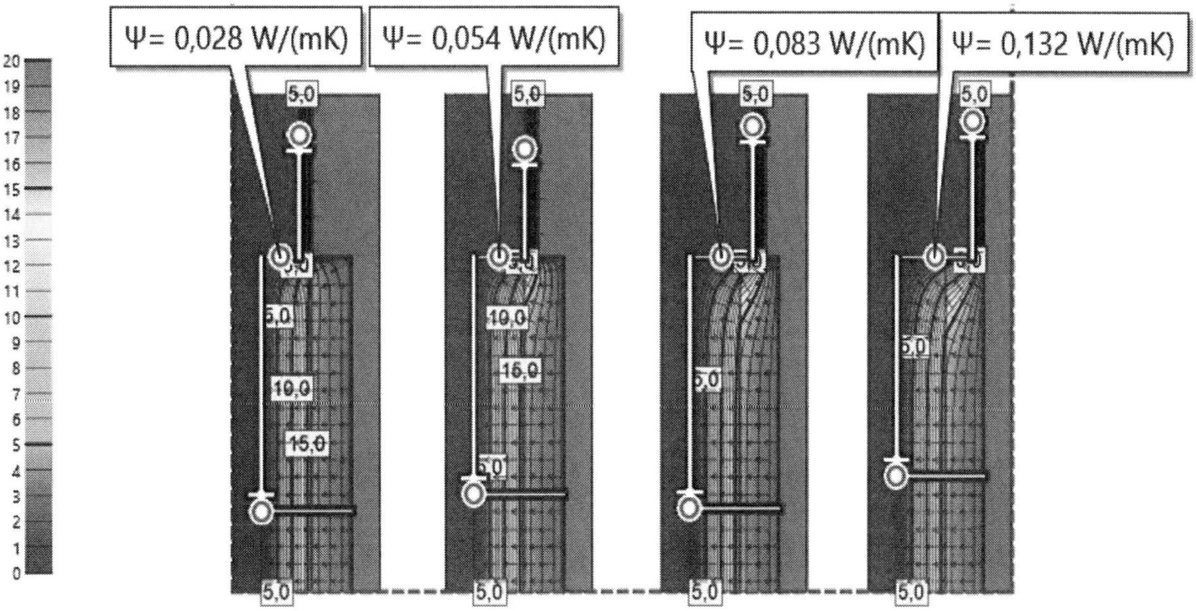

Fig. 7. Window-wall linear thermal bridges T5–T6–T7–T8

According to the outcomes of the linear thermal bridges, the annual thermal losses due to the linear bridge have been calculated in one test of PVC window, the dimensions of which are 1.5m x 1.5m (Table 4).

Table 4. Heat losses during the heating season

Bridge number	Linear bridge ψ, W/(mK)	Thermal losses, kWh/1 vnt.
T1	0,0390	23,97
T2	0,0260	15,98
T3	0,0232	14,26
T4	0,0234	14,38
T5	0,0282	17,33
T6	0,0538	33,07
T7	0,0829	50,95
T8	0,1319	81,06

Having performed the model simulation, it was determined that the most efficient wall-window junction is at the middle of the thermal insulation layer, which corresponds to T3 position. The worst option is the T8 position where the window is installed at the edge of the masonry.

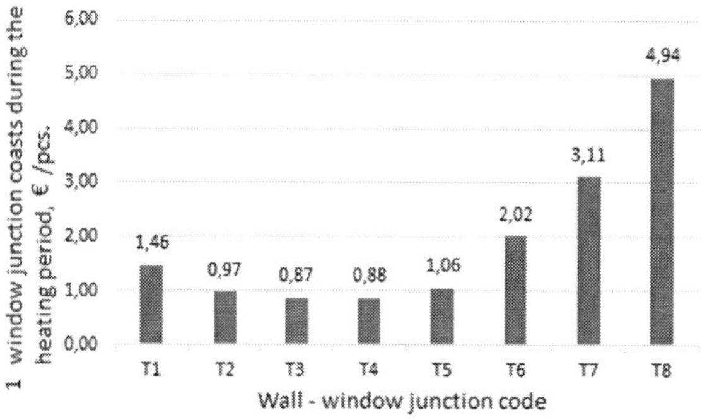

Fig. 8. Comparison of the costs during the heating season

Having performed the economic evaluation (Fig. 8), it is estimated that the lowest costs due to linear thermal bridges are then, when the window is designed at T3 position – 0.87 €. Having designed the window installation at T8 position – annual costs might increase to 5.7 times, i.e. up to 4.94 €.

Conclusions

The most efficient installation of the window of A++ class buildings is in the middle of the outer thermal insulation layer (centre). In this case (T3 position) the window-wall linear thermal bridge will be the lowest and equal to 0.0232 W/(mK). The annual costs – 0.87 €. The highest thermal losses at T8 position may increase up to 81.06 kWh during the heating season because of one improperly installed window, and annual costs might even increase to 5.7 times.

References

[1] Aviža D., Turskis Z., Kaklauskas A. A Multiple criteria decision support system for analyzing the correlation between the thickness of a thermo-insulation layer and its payback period of the external wall. *Journal of Civil Engineering and Management*, 2015, 20(6): 827–835.

[2] Deng S., Wang. R. Z., Dai I. Y. How to evaluate performance of net zero energy building – A literature research, *Energy*, 2014, 71: 1–16.

[3] Langų montavimo apšiltinimo sluoksnyje sistema: paprasčiau ir kokybiškiau. Access via internet: www.asa.lt [Accessed: 2016-10-11].

[4] LST EN ISO 10211:2008 Statinių konstrukcijų šiluminiai tilteliai. Šilumos srautai ir paviršiaus temperatūros. Detalieji skaičiavimai. Vilnius, 2008.

[5] Paroc: Atitvaros: šiluma, sandarumas ir drėgmė. Access via internet: http://www.paroc.lt/verta-zinoti/energinis-efektyvumas/pastatu-projektavimas/atitvaros [Accessed: 2016-10-17].
[6] Praktiški patarimai renkantis langus: TIKS LANGAI. Access via internet: http://www.apstatyba.lt/praktiski-patarimai-renkantis-langus-tiks-langai-1225 [Accessed: 2016-10-17].
[7] ST 2124555837.01:2013 Atitvarų šiltinimas polistireniniu putplasčiu. Vilnius, 2013.
[8] ST 2491109.01:20015 Langų, durų ir jų konstrukcijų montavimas. Vilnius, 2015.
[9] STR 2.01.09:2012 Pastatų energinis naudingumas. Energinio naudingumo sertifikavimas. Vilnius, 2012.
[10] STR 2.05.01:2013 Pastatų energinio naudingumo projektavimas. Vilnius, 2013.
[11] Šalčio (šilumos) tiltelių projektavimas. Access via internet: http://energinissertifikavimas.lt/silumostilteliu-projektavimas-55 [Accessed: 2016-10-08].
[12] Šilumos nuostoliai. Access via internet: http://www.ena.lt/pat_sil_nuost.htm [Accessed: 2016-10-21].

ITMS'2018

panevėžys faculty of technologies and business
1922

The 12th International Scientific Conference Intelligent Technologies in Logistics and Mechatronics Systems (ITELMS'2018), 26–27 April 2018, Panevėžys, Lithuania

Renovation of Buildings in Lithuania: Energy Savings and Environmental Impact

Jovita Kaupienė[a,b*], Andrius Stasiškis[a], Donatas Aviža[a,b], Elvyra Zacharovienė[a]

[a]*Kaunas University of Technology, Nemuno g. 33, LT–37164 Panevėžys, Lithuania*
[b]*Panevėžys university of applied sciences, Laisvės a. 23, Panevėžys*

Abstract

A lot of heat energy is consumed for the heating of old houses, because of poor building maintenance and material insulating properties. Therefore, the renewal of such houses becomes especially important and actual due to the deterioration of building energetic properties and the increase in energy price. Residential buildings energy consumption is about 40% of the total European Union final energy balance and the same amount of greenhouse gas (GHG) emissions. This situation creates a significant potential for the implementation of cost-effective energy saving measures in the building sector. To emphasize and raise demands to EU members, the European Commission has legislated to promote energy saving measures. One of these laws is the Energy Performance of Building Directive, which set requirements for Nearly Zero Energy Buildings for all newly built buildings from 2020 and for all public buildings from 2018. In order to implement this directive, energy saving is important not only in the new buildings, but also in renovation of existing ones.

© 2018 J. Kaupienė, A. Stasiškis, D. Aviža, E. Zacharovienė
Peer-review under responsibility of the Kaunas University of Technology, Panevėžys Faculty of Technologies and Business

Keywords: construction, renovation, impact, environment, energy, savings, residential buildings

1. Introduction

Construction is an important fragment of the state's national economy. Each country cares about it, regardless of its prosperity level, geopolitical situation, natural resources and other factors. Today about 1.64 million residents

* Corresponding author. Tel.: +370-652-14307
E-mail address: jovita.kaupiene@ktu.lt

© 2018 J. Kaupienė, A. Stasiškis, D. Aviža, E. Zacharovienė
Peer-review under responsibility of the Kaunas University of Technology, Panevėžys Faculty of Technologies and Business

(approx. 547 thousand apartments) live in multi-apartment buildings. 38435 residential blocks of flats (i. e. three or more apartments) with a total living area of 47894459 m^2 or almost 48 km^2 are registered in the Register of Legal Entities of Lithuania republic.

The distribution of materials used for the building walls (except auxiliary farm) registered in the Real Estate Registry is as follows: the majority buildings are from bricks and blocks – 47.9%, 36.3% buildings from logs, 2.9% from reinforced concrete panels, from monolithic concrete – 0.7%, from other materials – 12.2%. According to statistics data, 64% of all residential houses in Lithuania are built up to 1975. Even 75% of all residential buildings are multi-apartment buildings, i.e. three and more apartments.

National and even international efforts are required to mitigate climate change. The largest potential for reducing greenhouse gases exhaust has been found in the construction sector, where residential buildings use about 27% of energy. The Energy Efficiency Directive and the Energy Performance of Buildings Directive (EPBD) are the main laws in the European Union aimed at improving the energy efficiency of new and existing buildings. They are implemented in national policies and programs, the main objective of which is to reduce energy consumption. Improving energy efficiency can also affect the quality of the internal environment and the health and well-being of the population.

Approximately half of the Lithuania population (the clear majority in cities – about 80%) lives in multi-apartment buildings. A lot of these houses are depreciated; their condition is constantly deteriorating due to insufficient supervision and too little investments to renew them. The energy consumption does not meet current standards and exceeds them by 1.5–2 times. Apartment house renewal (modernization) program is the main measure to implement one of the Lithuanian housing strategy goals, i.e. to ensure the efficient use, maintenance and renewal of existing residential buildings, and the rational use of energy resources. Modernization is an essential component of building exploitation and maintenance system, which enables to restore the physical and moral value of the building, improve energy efficiency; thermal comfort and reduce negative impact on the environment. However, the need and scope of buildings modernization are currently assessed solely based on economic and physical depreciation criteria. Such an approach is insufficient to maintain the long-term value of a building and to ensure that the buildings meet one of the essential requirements – energy saving and heat preservation [12]. One of the main goals of energy efficiency improvement is to increase thermal comfort. The insufficient level of thermal comfort often is the main reason to make decision to renew the building. Building modernization allows restoring the normal temperature in the building also reducing the moisture level.

2. Literature review

There is always dilemma: what is more profitable to renovate a house or demolish it and build a new one. Some researchers' studies show that, in the economic, environmental and social aspects, it is more beneficial to renovate a house [3, 4, 8]. However, modernization of a building is not always a better solution. The construction of a new building in comparison with the renovation has much more potential to reduce energy consumption. Renewal of the building is limited due to the building location, used materials and already made engineering decisions. Although the new building has many benefits, it is very important to consider two non-economic factors:

- the environmental impact of renovation and construction of a new building;
- the social-economic impact.

Firstly, the renovation will always cause less damage to the environment than the demolition of the old building and the building new one. The reason is that the creation of new materials leads to additional CO_2 emissions during their production. The demolition of the building and the disposal of the remaining waste issues additional CO_2 emissions. Also, the waste is hard to decompose and therefore occupies large volumes of landfill.

The thermal renovation of the building is considered to restore building's value. If depreciated value of the building is fully or partially recovered during the renewal, the building's value could become higher than the initial building's value. However, both the reconstruction and the thermal renovation costs must be effective and well-founded [10].

The renovation of the house provides not only direct benefits to the building residents, but also indirect, which affects all the residents of the region. Therefore, in assessing the benefits of renovation, it is necessary to evaluate not only the micro-level benefits, but also the macro [7] split macro-financial benefits into three parts:

- environmental benefits (less contamination, limited conservation of natural resources);
- social benefits (growing social justice, improving human health);
- economic benefits (increased employment, wealth creation, etc.).

All these factors are closely related, complementing each other, and thus affecting the usefulness of the renovation. It is important to emphasize that renovation should be considered taking into account these factors. Each of these factors should be analyzed separately.

Renovation at country level is an ambitious goal without any doubts. From the perspective of the country and business, the importance of renovation can be based on the country's energy consumption balance, re-distribution of revenues, direct fuel saving effect and local economy. This process should also stimulate creation of new work places; increase the number of taxpayers etc. The benefits of renovation were explored by several researchers [1, 5, 9, 12], they analyzed the technical and economic aspects of this process.

The environmental benefits of renovation are analyzed by [6, 11]. The main conclusion is that after the implementation of the renovation projects, pollution and fossil fuel demand will be reduced.

3. Methodology

The empirical research structure is following: firstly, the recent changes of Lithuanian population quantity and structure will be analyzed, and the economic dependency ratio will be calculated. Secondly, the changes of Lithuanian population will be statistically predicted and the expected economic dependency ratio in 2040 will be calculated. Finally, the impact of Lithuanian population changes on the State's Social Insurance Fund in 25 years perspective will be estimated. The linear, exponential and logarithmic regression models will be developed for the prediction of population changes. The correlation coefficient, determination coefficient, mean squared error (MSE), mean absolute deviation (MAD), and mean absolute percentage error (MAPE) will be used to assess the validity of developed models. The data sources for the statistical analysis are the World Bank, OECD, Lithuania Statistics and Lithuanian Social Insurance Fund databases.

4. Energy consumption in multi-apartment buildings

The designed cost of heat energy, expressed in kilowatt hours per square meter of living space in non-renovated houses is 160–180 kWh/m^2 per year, and for the newer houses built after 1993 – 80–90 kWh/m^2 per year. Thermal energy consumption in multi-apartment buildings built up to 1993 according to technical standards – about 5,000 GWh per year.

The multi-apartment house sector has the largest energy saving potential in the country – about 3 TWh per year. Thus, modernization of multi-apartment buildings will allow Lithuania to implement energy saving obligations, reduce heating costs, achieve better quality of life, use of rational resources and reduce budget expenditures for heating costs compensation.

Today An "House Renewal/Modernization" program is being implemented. It aims to renew or modernize 70% (about 24,000) residential buildings (constructed up to 1993) till 2020; to reduce the thermal energy consumption of modernized apartment buildings by 30 per cent in comparison with the thermal energy costs before the renovation; reduce up to 400 thousand tons of carbon dioxide emissions per year [2]. In the program, calculations are made for the total heat consumption for heating, per standard year and one square meter of useful area. Houses with 1500 m^2 and more useful floor area consuming least quantity of heat per year, i. e. 147 kWh/m^2, after implementation of the modernization, it is planned to use 74 kWh/m^2. Relatively more energy is consumed in houses with a useful area from 501 to 1500 m^2, i. e. 198 kWh/m^2, after modernization energy consumption should be reduced to 95 kWh/m^2. Relatively the biggest part of energy is consumed in homes with a useful floor area of less than 500 m^2, i. e. 243 kWh/m^2, and after modernization 113 kWh/m^2. The difference in heat consumption between the smallest (buildings with a useful floor area up to 500 m^2) and biggest buildings (buildings with a useful floor area more than 1500 m^2) is 96 kWh/m^2 per year, after the modernization, the difference should be 39 kWh/m^2. Small houses, especially up to 500 m^2 of useful floor area, are subjected to higher heat losses, as relatively there are more external walls (Fig. 1–2).

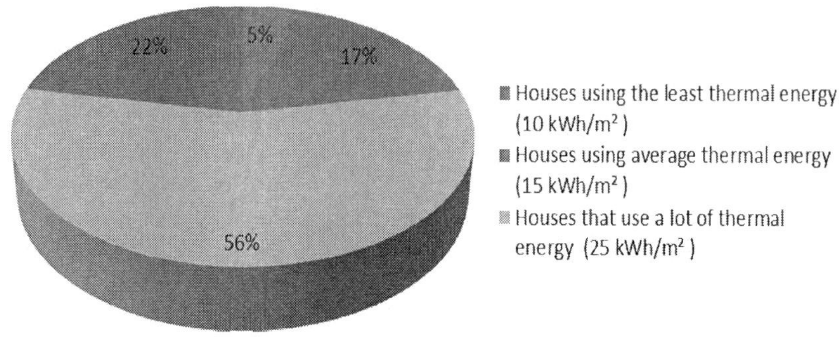

Fig. 1. Multi-apartment building distribution according to energy consumption categories (60 m^2)

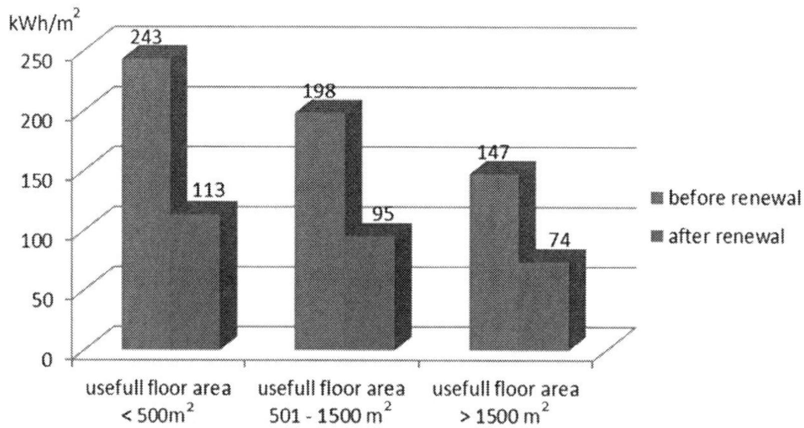

Fig. 2. Energy consumption in buildings with different useful floor area

The year of building construction does not have a significant impact on energy consumption in an analogous group of buildings, because the coefficients of thermal conductivity of external walls are similar. The technical condition of the house (timely repair of partitions, building engineering systems and qualified exploitation) has a greater influence on energy consumption. Thus, energy consumption depends directly on the quality of building maintenance.

Energy savings throughout the building life cycle depends on a large extent of the decisions made in design or project preparation phase. It is valid for both a new building construction and renovation of an existing one. The main strategic directions for saving energy in the buildings:

- reduce energy needs;
- use of renewable energy resources;
- rational use of energy.

The loss of the heat is very clearly visible in thermovision measurements of panels, i.e. the investigation of seams or surfaces by thermovision.

The thermographic photo (Fig. 3) clearly shows the places where most of the heat is lost.

Heat losses can be reduced and energy efficiency increased through using modern solutions of heat insulation:

- increase energy efficiency of buildings using heat insulation solutions for isolation of external partitions. Current solutions for thermal insulation are used both for the construction of new ones and for renovation (upgrading) of previously built buildings;
- reduce heat losses in thermal pipes, i. e. use of durable and efficient heat insulating materials to repair existing heating and hot water supply pipelines in buildings.

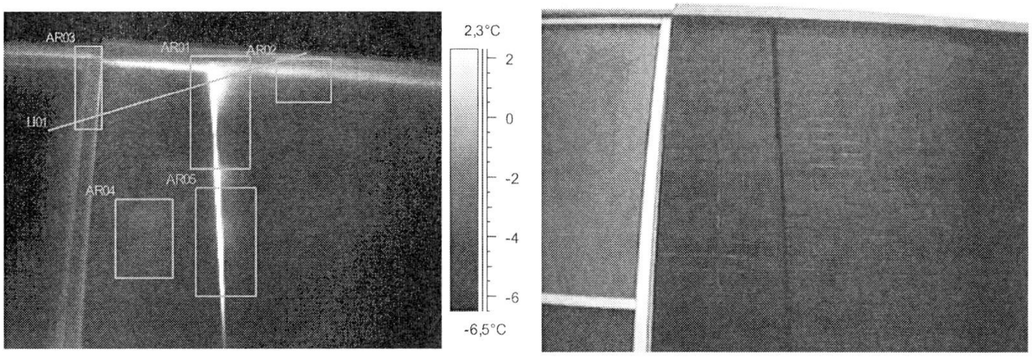

Fig. 3. Thermographic photo of building wall

It is often assumed that a new building consumes less energy than a renovated one, and has a lower negative impact on the environment during lifetime. Such effect is shown in Figure 4. There are two important effects: new buildings use less energy after an intersection point, which can be shifted to the right by improving the energy efficiency by renovation. The figure also shows that if the intersection point is behind the emission reduction target, the reduction of CO_2 throughout the life cycle is negligible and renovation is the best option.

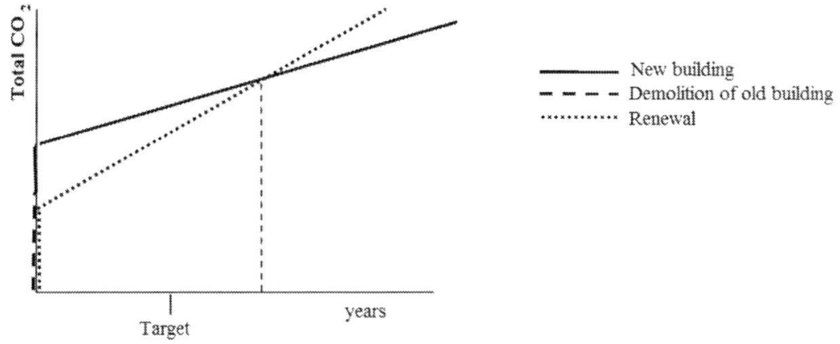

Fig. 4. CO2 emission during the construction of new building and renewing old one [1]

Building gets lower costs for heating, better indoor climate, lower CO_2 emissions, better acoustic comfort, higher fire safety and also higher market value by investing into improving the energy efficiency.

5. Measures to increase the energy efficiency of multi-apartment houses

The renovation / modernization plan for multi-apartment buildings describes how each of the measures should be implemented. It covers: the insulation of walls and roof, the replacement of unmodified windows and exterior doors, the insulation of pedestal and floor above cellar, the modernization of electric and heating systems, the renovation of the water supply and sewage engineering system, the renewal of rain collector and drainage from the roof and installation of fire and security alarms.

The complex renewal of the multi-apartment building (insulating walls, roof, replacing windows, glazing the balconies, renovating the heating system) can save 50% and more energy.

During the modernization of the building, the construction elements and engineering systems are renewed, what restores the designed building properties and reduces the degree of depreciation. It should be noted that modernization of building structures and engineering systems also restores the thermal insulation properties to projected level. According to the calculations of the National Audit Office's auditors which are based on the statistical data of the Lithuanian heat suppliers' association, after restoring the thermal insulating properties of buildings to the designed level and proper maintenance, approximately 20% of heat energy could be saved.

Conclusions

Renovation (modernization) of multi apartment buildings is one of the most important field of housing policy, which is currently being given special attention. One of the main strategic goals of European Union countries (including Lithuania) in housing policy is reducing energy consumption in buildings. To achieve this goal, various programs are being implemented in Lithuania to increase energy efficiency in public buildings and multi-apartment buildings.

Building renewal allows to reduce heating costs, bring in new quality of life and comfort, raise the value of the property. The greatest economic benefits (heat savings) are achieved through the implementation of complex renovation of buildings.

Favorable financing models are developed for the renovation, but this process is still very slow. Residents of multi-apartment buildings do not receive enough information on the economic benefits of modernizing their buildings. Also, information about the benefits of people who have already participated in the program is not shared. Thus, this lack of information does not encourage the residents to participate in the program for renewing (modernizing) multi-apartment buildings.

References

[1] Baker N. V. The handbook of sustainable refurbishment: non-domestic buildings. Access via internet: http://books.google.com/books?id=FTQoAZighDwC&lpg=PP1&dq=The%20handbook%20of%20sustainable%20refurbishment%3A%20non-domestic%20buildings.&pg=PP1#v=onepage&q&f=false [Accessed: 2018-01-10].

[2] Biekša D., Jaraminienė E., Martinaitis V. Daugiabučių namų renovacijos vertinimas atsižvelgiant į trejopą naudą. Aplinkos apsaugos inžinerija, Vilniaus Gedimino technikos universitetas, pp. 98–104. Access via internet: http://www.mla.vgtu.lt/index.php/mla/article/viewFile/mla.2011.093/pdf_1 [Accessed: 2017-12-18].

[3] Dong B., Kennedy Ch., Pressnail K. Comparing life cycle implication of building retrofit and replacement options. Canadian Journal of Civil Engineering December 32(6). Access via EBSCO Publishing [Accessed: 2017-12-14].

[4] Erlandsson M., Levin P. Environmental assessment of rebuilding and possible performance improvements effect on a national scale, Building and Environment November 40(11). Access via EBSCO Publishing [Accessed: 2017-12-14].

[5] González A. B. R., Díaz J. J. V., Caamaño, A. J., Wilby, M. R. Towards a universal energy efficiency index for buildings, Energy and Buildings 43(4). Access via EBSCO Publishing [Accessed: 2017-11-25].

[6] Janukonis A. Lithuanian heat sector: today based on imported fossil fuel, tomorrow – on local biofuel and wastes. World energy congress: Montreal 2010: Issue 4.1. Access via internet: http://www.worldenergy.org/documents/congresspapers/239.pdf [Accessed: 2017-12-14].

[7] Mickaitytė A., Zavadskas E. K., Kaklauskas A., Tupėnaitė L. The Concept Model Of Sustainable Buildings Refurbishment. International Journal Of Strategic Property Management March 12(1). Access via EBSCO Publishing [Accessed: 2017-12-19].

[8] Power A. Does demolition or refurbishment of old and inefficient homes to help to increase our environmental, social and economic, viability? Energy Policy. Access via internet: http://www.bis.gov.uk/assets/bispartners/foresight/docs/energy/energy%20final/power%20paper-section%205.pdf [Accessed: 2017-12-19].

[9] Rogoža A., Šiupšinskas G., Martinaitis V. Daugiabučių namų modernizavimo programos monitoringas. Vilnius: Housing and Urban Development Agency.

[10] Serbent, V. Daugiabučių pastatų energinio efektyvumo didinimo programos. Access via internet: http://www.am.lt/VI/files/0.764302001362137879.pdf [Accessed: 2017-12-19].

[11] Užšilaitytė L., Martinaitis V. Search for optimal solution of public building renovation in terms of life cycle. Journal of environmental engineering and landscape management. Access via EBSCO Publishing [Accessed: 2017-12-19]..

[12] Zavadskas E. K., Raslanas S., Kaklauskas A. The selection of effective retrofit scenarios for panel houses in urban neighbourhoods based on expected energy savings and increase in market value: the Vilnius case, Energy and Buildings 40(4). Access via EBSCO Publishing [Accessed: 2017-12-19].

ITMS'2018

ktu 1922 panevėžys faculty of technologies and business

The 12th International Scientific Conference Intelligent Technologies in Logistics and Mechatronics Systems (ITELMS'2018), 26–27 April 2018, Panevėžys, Lithuania

Mathematical Simulation of Spatial Oscillations of the "UNDERFRAME-TRACK" System Interaction

Irina Klimenko[a], Laura Černiauskaitė[b], Larisa Neduzha[a], Oleksandr Ochkasov[a*]

[a] Dnipropetrovsk National University of Railway Transport named after Academician V. Lazaryan, 2 Lazarian St., 49010 Dnipro, Ukraine
[b] Vilnius College of Technologies and Design, Petras Vileisis Railway Transport Faculty, K. Kalinausko str. 7, Vilnius, Lithuania

Abstract

The article presents mathematical simulation of spatial oscillations of the "underframe-track" system interaction. It describes a design scheme of the freight-passenger electric locomotive, taking into account details of construction and the mathematical model of its spatial oscillations. The connections imposed on the system, the forces arising in them, as well as the mutual displacements of all the bodies in the system are considered. Differential equations of system oscillations are composed on the basis of Lagrange's second order equations taking into account details of construction of the electric locomotive. The paper presents some results of calculations for the cases of an electric locomotive movement on the straight sections of the track, curves of small and mean radius.

© 2018 I. Klimenko, L. Černiauskaitė, L. Neduzha, O. Ochkasov
Peer-review under responsibility of the Kaunas University of Technology, Panevėžys Faculty of Technologies and Business

Keywords: "underframe-track" system, rolling stock, interaction, locomotive, mathematical model, design scheme, dynamic coefficients, stresses

The activation of international carriage by rail in the direction of international transport corridors, the running of the international container train "Viking", prospects for organizing passenger transportations between Lithuania and Ukraine – all this requires an appropriate scientific support.

One of the crucial tasks, in this case, is the task of increasing the speed of rolling stock. The solution of this problem requires carrying out both theoretical research and practical tests. When designing the rolling stock and

* Corresponding author. Tel.: +380677452071
E-mail address: abochkasov@gmail.com

© 2018 I. Klimenko, L. Černiauskaitė, L. Neduzha, O. Ochkasov
B427H0014 . EdLearning.it
Peer-review under responsibility of the Kaunas University of Technology, Panevėžys Faculty of Technologies and Business

improving the design of the track, significant attention is paid to the interaction of the elements of the "underframe-track" system.

The railway track and the rolling stock is a single system in which they interact, they are in close connection and interdependence. Currently, the locomotive, as one of the main representatives of the rolling stock, is the most reliable and economically advantageous vehicle. While increasing the mass of freight trains and axle loads, in addition to strength and durability, it has also to meet the following conditions [1–4]:

- design of the running gear – to ensure the safety of traffic on railways at the governed speed in straight and curved sections of the track;
- to have a sparing effect on the track – the load of wheel set should not exceed the established norms, depending on the type of the permanent way;
- when driving at maximum permissible speeds, it must ensure the safe operation of the railway;
- to have sufficient stability to prevent accidents;
- to have a quiet running for comfortable transportation of passengers and qualitative cargo safety condition.

Therefore, improvement, modernization and creation of new types of traction rolling stock is one of the important tasks involving efforts of a large team of scientists, designers, researchers. This requires a powerful experimental base, which is not always possible and it is associated with significant expenses, an increase in the cost of the vehicle.

In connection with this, mathematical simulation can help greatly [1, 2, 5–7]. It should be noted that the locomotive, as an object of simulation, is a complex system, and the development of its mathematical model involves the use of a generalized model, taking into account details of its construction at the design stage, which greatly facilitates the solution of the defined objectivities.

Main part. Let us consider the mathematical simulation of the "underframe-track" system on the example of DS3 electric locomotive. A number of scientific and production organizations together with Simens, including Dnipropetrovsk National University of Railway Transport named after Academician V. Lazaryan (DNURT), resolved the task of creating a four-axle freight-passenger electric locomotive DS3 of alternating current with asynchronous traction motors and traction transmission, class II [1, 4, 10, 11].

For the effective achievement of this objective, it is necessary to solve a set of tasks, among which there is the development of an experimental-theoretical method for predicting the dynamic qualities of an electric locomotive, which includes: the choice of the constructive scheme of the bogie and the concept definition of its rational scheme; a proposal of an algorithm for the mathematical model of spatial oscillations in the electric locomotive; the development of the computer program using a PC for the rational parameters of its running gears, research of the "underframe-track" interaction.

Before full-scale tests operation, it is expedient to simulate those processes that occur in separate nodes of the created device and the whole structure in general. This allows giving recommendations upon definition of rational parameters of the system that meets the normative documents at the design stage [8, 9].

For carrying out of theoretical researches the mathematical model of spatial oscillations of an electric locomotive was developed, based on the accepted design scheme (Fig. 1) [1].

The model takes into account the details of construction in the locomotive: suspension of the body – swing, eight-point with parallel located hydraulic snubbers, operating in a vertical and horizontal transverse directions; transmission of traction efforts – through inclined rod; axle-box suspension springs, end springs, elastic rods; suspension of TEM – support-and-frame with elastic coupling [1, 4].

Each of the two axle bolster-less bogies is connected in the longitudinal direction with the body of one inclined rod, working on tension and compression. The connectivity node of the inclined rod with the body has elastic rubber elements, and inclined rod is ball-jointed with bogies. In the body-bogie support system, intermediate beams (two for each bogie) are used, the inner ends of which are ball-jointed with the bogie frame, and loads from the body mass are attached to the outer ends through bolster hangers. In the intermediate section, the beam bears on spring kits that transmit the vertical load to the bogie frame. In the longitudinal direction, the intermediate beams are connected to the bogie frame by longitudinal bars with uniform elements (reactive rod).

The application of intermediate beams makes it possible to eliminate a unit that is not very reliable in operation, in which the body suspension springs are in direct contact with the bolster hangers, as well as to increase the static deflection of the suspension. Wheel set axle-boxes are connected with the bogie frame with the help of axlebox link bar, and in the vertical direction – by means of spring coils and hydraulic shock absorbers. The transverse flexible connection between the body and bogies is carried out with the help of a swing suspension, and shock absorbers using inclined hydraulic dampers.

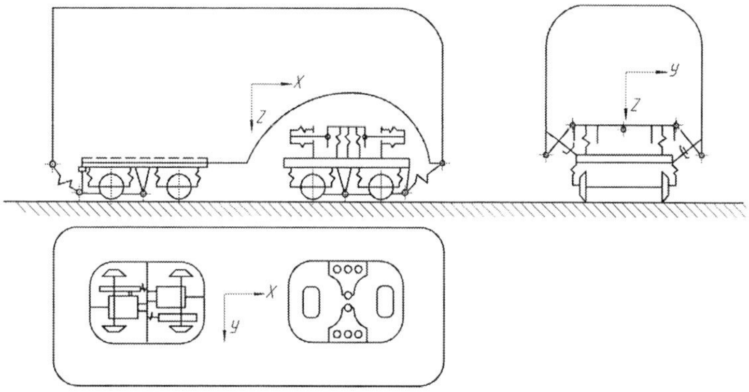

Fig. 1. Design scheme

The springs, receiving the vertical load, are located separately between the two intermediate beams and the bogie frame. The body through the bolster hangers is suspended to intermediate beams, which can be connected to the bogie frame in the horizontal longitudinal and transverse directions rigidly or with the help of hinged-rod connections with flexible-dissipative elements of one-sided action, working only on compression. The advantage of the scheme is that elements are separated in which the restoring forces arise in the vertical planes (springs) and horizontal (bolster hangers) planes [1, 4].

To determine the number of generalized coordinates describing the spatial oscillations of the system, we denoted by x, y, z – displacements along the corresponding axes (recoiling, swaying, bouncing), by θ, φ, ψ – rotation angle relative to the corresponding axes (side roll, pitching, wabbling). Positive linear displacements are along the corresponding axes, positive angular displacements – counterclockwise, viewing from the positive direction of the corresponding axis. Further on, without an index we will mark the displacements of the body, the index i ($i = 1.2...$ – bogie number) – bogie frames, index I_{ik} – intermediate beams, index i_m – ($m = 1.2...$ – number of wheelset in the bogie) – wheelset, index d_{im} – traction motors, index r_{im} – gearbox of traction motors, index p_{imk} – reduced masses ($k = 1$ – left in the course of movement, $k = 2$ – right side of an electric locomotive). In total, the system under consideration consists of fifteen solids (body, two bogie frames, four intermediate beams, four gearboxes of traction motors, four wheel set) and track mass reduced to the wheel-rail contact points that move in two directions. The total number of displacements $15 \cdot 6 + 18 \cdot 2 = 106$ (their symbols are given in Table 1).

Table 1. Symbols of system bodies displacements

Systems bodies	Displacements					
	linear along the axes			angular relative to the axes		
	x	y	z	x	y	z
Body	x	y	z	θ	φ	ψ
Bogies frames	x_i	y_i	z_i	θ_i	φ_i	ψ_i
Intermediate beams	$x_{I_{ik}}$	$y_{I_{ik}}$	$z_{I_{ik}}$	$\theta_{I_{ik}}$	$\varphi_{I_{ik}}$	$\psi_{I_{ik}}$
Wheelsets	x_{im}	y_{im}	z_{im}	θ_{im}	φ_{im}	ψ_{im}
Traction motors	$x_{d_{im}}$	$y_{d_{im}}$	$z_{d_{im}}$	$\theta_{d_{im}}$	$\varphi_{d_{im}}$	$\psi_{d_{im}}$
TEM gearboxes	$x_{r_{im}}$	$y_{r_{im}}$	$z_{r_{im}}$	$\theta_{r_{im}}$	$\varphi_{r_{im}}$	$\psi_{r_{im}}$
Rails in contact points		$y_{p_{im}}$	$z_{p_{im}}$			

Additional symbols are introduced: displacements and forces arising at each point of the swing suspension – $K_{i_{mk}}$. In flexible elements between the intermediate beams and the bogies frames – $I_{i_{mk}}$. Between the bogies frames and the axle-boxes – $B_{i_{mk}}$. Between the bogies frames and TEM gearboxes – r_{i_m}. In inclined rod – R_i, in flexible couplings – C_{i_m}. Displacements and forces in the corresponding directions are additionally supplemented by the indices x, y, z or ψ (at angular displacements), in the flexible elements – by the index "0", in dissipative – "d" [1].

The body and intermediate beams in the vertical direction are connected rigidly:

$$z_{I_{ik}} + (-1)^k (b_1 - b_{C1})\theta_{1ik} = z + (-1)^k (b_{10} - h_{10} tg\delta_0)\theta; \ \phi_{1ik} = \phi; \tag{1}$$

and in the horizontal plane the restoring moments between these bodies arise in connection with the change in the position of the center of the body mass (more details of the obtained expressions are described below).

Links between the intermediate beams and the bogie frame with vertical and horizontal transverse displacements are as follows:

$$z_{I_{ik}} - (-1)^k b_{C1}\theta_{I_{ik}} = z_i, \ y_{I_{ik}} = y_i, \ x_{I_{ik}} = x_i; \ \varphi_{I_{ik}} = \varphi_i. \tag{2}$$

Between the bogies frames and the axle-boxes, links allow the bodies to make mutual displacements in all directions.

On mutual displacements of the TEM gearboxes and wheel sets of the bogies links in all directions are imposed, except the gearboxes pitching, that is:

$$x_{r_{im}} = x_{im}, \ y_{r_{im}} = y_{im} + (-1)^m (l_C - a_{Dk})\psi_{ik}, \ z_{r_{im}} = z_{im} - (-1)^m l_C \varphi_{r_{im}};$$

$$\theta_{r_{im}} = \theta_{i_m}, \ \psi_{r_{im}} = \psi_{i_m}. \tag{3}$$

It is considered that the wheels move without slipping (they are taken into account in pseudo-slip theory):

$$\varphi_{im} = \frac{x_{im}}{r}; \tag{4}$$

and without lift-off from the rails:

$$z_{Pim_k} = z_{im} + (-1)^k b_2\theta_{im} + \Delta r_{im_k} - \eta_{im_k}. \tag{5}$$

Thus, 56 equations of constraints have been introduced, in total there are $106 - 56 = 50$ degrees of freedom.

The generalized coordinates (without masses of rails in the horizontal plane) are written as follows:

- body displacement: $q_1 = z, q_2 = \mathrm{j}, \ q_3 = \mathrm{q}, \ q_4 = y, \ q_5 = \mathrm{y}$;
- bogies frames displacements: $q_n = z_i, \ (n = 6;\ 7), q_n = \mathrm{j}_i \,(n = 8;\ 9)$,
- $q_n = \mathrm{q}_i, \ (n = 10,\ 11), q_n = y_i, \ (n = 12;\ 13), q_n = \mathrm{y}_i \,(n = 14;\ 15)$;
- wheelset displacements:

$$q_n = z_{im},\ (n=16,...,19), q_n = \text{q}_{im}(n=20,...,23),$$
$$q_n = y_{im},\ (n=24,...,27),\ q_n = \text{y}_{im}(n=28,...,31); \tag{6}$$

- bodies recoiling $q_n = x_i,\ (n=32,\ 33), q_n = x_{im}(n=34,...,37), q_{38} = x$;
- TEM pitching $q_n = \text{j}_{r_{im}}(n=39,\dots,42)$.

In the model mutual displacements of the system bodies were considered and the generalized forces, acting in the system were determined; a number of significant factors were considered and taken into account [1].

To compile the differential equations of oscillations [1, 5, 7], the kinetic energy of the system was determined as the sum of the kinetic energies of all structures – the body, bogies frames (inclusive of the mass of the traction motor and the part of the gearbox mass and the corresponding moments of inertia), intermediate beams, wheelset (it is taking into consideration that part of the gearbox mass is attached to the wheelset), reduced masses of the rails at the contact points:

$$T = T_0 + \sum_{i=1}^{2} T_{\text{T}_i} + \sum_{i=1}^{2}\sum_{k=1}^{2} T_{\text{п}_{ik}} + \sum_{i=1}^{2}\sum_{m=1}^{2} T_{\text{к}_{im}} + \sum_{i=1}^{2}\sum_{m=1}^{2} T_{\text{p}_{im}} \sum_{i=1}^{2}\sum_{m=1}^{2}\sum_{k=1}^{2} T_{\text{p}i_{mk}}\ . \tag{7}$$

After substitution the sum of the kinetic energies of all structures, the general expression for the kinetic energy (7) has the form:

$$\begin{aligned} T = \frac{1}{2}\Big[& a_z\dot{z}^2 + a_\varphi\dot{\varphi}^2 + a_\theta\dot{\theta}^2 + a_y\dot{y}^2 + a_\psi\dot{\psi}^2 + a_x\dot{x}^2 + \\ & + \sum_{i=1}^{2}\left(a_{x_i}\dot{x}_i^2 + a_{z_i}\dot{z}_i^2 + a_{\varphi_i}\dot{\varphi}_i^2 + a_{\theta_i}\dot{\theta}_i^2 + a_{y_i}\dot{y}_i^2 + a_{\psi_i}\dot{\psi}_i^2\right) + \\ & + \sum_{i=1}^{2}\left(a_{x_i}\dot{x}_i^2 + a_{z_i}\dot{z}_i^2 + a_{\varphi_i}\dot{\varphi}_i^2 + a_{\theta_i}\dot{\theta}_i^2 + a_{y_i}\dot{y}_i^2 + a_{\psi_i}\dot{\psi}_i^2\right) + \\ & + \sum_{i=1}^{2}\sum_{m=1}^{2}\left(a_{z_{im}}\dot{z}_{im}^2 + a_{\theta_{im}}\dot{\theta}_{im}^2 + a_{y_{im}}\dot{y}_{im}^2 + a_{\psi_{im}}\dot{\psi}_{im}^2 + a_{x_{im}}\dot{x}_{im}^2 + a_{\varphi r_{im}}\dot{\varphi}_{r_{im}}^2\right) + \\ & \sum_{i=1}^{2}\sum_{m=1}^{2}\left(a_{y_{im}\psi_{im}}\dot{y}_{im}\dot{\psi}_{im} + a_{z_{im}}\varphi_{r_{im}}\dot{z}_{im}\dot{\varphi}_{r_{im}}\right); \end{aligned} \tag{8}$$

where a_z, …, $a_{z_{im}}\varphi_{d_{im}}$ – inertial coefficients.

The resulting equitation for the kinetic energy of the system (8) will substituted in the Lagrange second order equations [1, 2, 6, 7]

$$\frac{d}{dt}\left(\frac{\partial T}{\partial q_n}\right) = Q_n,\quad (n=1,...,42). \tag{9}$$

As a result, there is the following system of differential equations [1, 6]

$$
\begin{aligned}
& a_n \ddot{q}_n = Q_n && (n = 1,...,15;\ 20,...,23;\ 32,...,38), \\
& a_{z_{im}} \ddot{q}_n + a_{z_{im}\varphi_{r_{im}}} \ddot{q}_{n+23} = Q_n && (n = 16,...,19), \\
& a_{y_{im}} \ddot{q}_n + a_{y_{im}\psi_{r_{im}}} \ddot{q}_{n+4} = Q_n && (n = 24,...,27), \\
& a_{\psi_{im}} \ddot{q}_n + a_{y_{im}\psi_{r_{im}}} \ddot{q}_{n-4} = Q_n && (n = 28,...,31), \\
& a_{\varphi r_{im}} \ddot{q}_n + a_{z_{im}\varphi_{r_{im}}} \ddot{q}_{n-23} = Q_n && (n = 39,...,42).
\end{aligned}
\tag{10}
$$

Results of the research. The solution of the obtained system of differential equations (10) was carried out by numerical integration using a PC. Calculations are made for the cases of the electric locomotive movement on the straight sections of the track, curves of the mean radius R = 600 m and a small radius R = 300 m at a cant of the outer rail of 0.1 m and 0.15 m respectively. When moving along straight sections of the track, the speeds range is from 60 up to 200 km/hour. When moving in curves, the speed range was chosen taking into account the unballanced acceleration constraint that is $[a_{un}] = 0.7\ \text{m/s}^2$. Practically for mean radius curves, the speed range was 60...100 km/h, and for small radius curves – 60...80 km/h.

In the course of calculations forces, accelerations, displacements of the elements in the "electric locomotive-track" system, and stresses in the elements of the track were determined (since the formation of defects in rails of tracks is a serious practice problem due to the danger of creating emergency situations). Accordingly to the obtained values of these sizes, the corresponding coefficients of vertical and horizontal dynamics were determined too.

Since the calculations were carried out in parallel with the design of the electric locomotive, a number of parameters were changed and refined. Calculations are performed for a number of options of the initial parameters values. In total six basic options were considered, for which the dynamic indicators and indicators of traffic stability were defined in parallel.

The main attention was paid to the normalized quantities, for which we give the notation and the largest permissible value [1, 8, 9]:

1. Coefficient of vertical dynamics in the axle-box stage of spring suspension – $[K_{VA}] = 0.4 / 0.35^*$.
2. Coefficient of vertical dynamics in the body stage of the spring suspension – $[K_{VB}] = 0.25 / 0.2^*$.
3. Coefficient of horizontal dynamics in the axle-box stage of spring suspension, determined accordingly to the value of the frame force, it is also a criterion of stability against shear of assembled rails and sleepers – $[K_{dh}^{A}] = 0.4$.
4. Coefficient of horizontal dynamics by forces in the body stage of spring suspension – $[K_{dh}^{c}] = 0.2$.
5. Vertical acceleration of the driver's seat in portions $g - [\ddot{z}_{km}] = 0.4$.
6. Horizontal transverse acceleration of the driver's seat in portions $g - [\ddot{y}_{km}] = 0.35$.
7. Stresses in the edges of the rail base – $[\sigma_e] = 240$ MPa.
8. Stress in the rail head – $[\sigma_h] = 260$ MPa.
9. Calculated stress in wooden sleepers for crushing action under the rail chair – $[\sigma_n] = 2.2$ MPa.
10. Calculated stress in the ballast under the slipper – $[\sigma_s] = 0.5$ MPa.

For an example, coefficient of vertical dynamics curves K_{VB} in the body stage of spring suspension is presented (Figures 2–4).

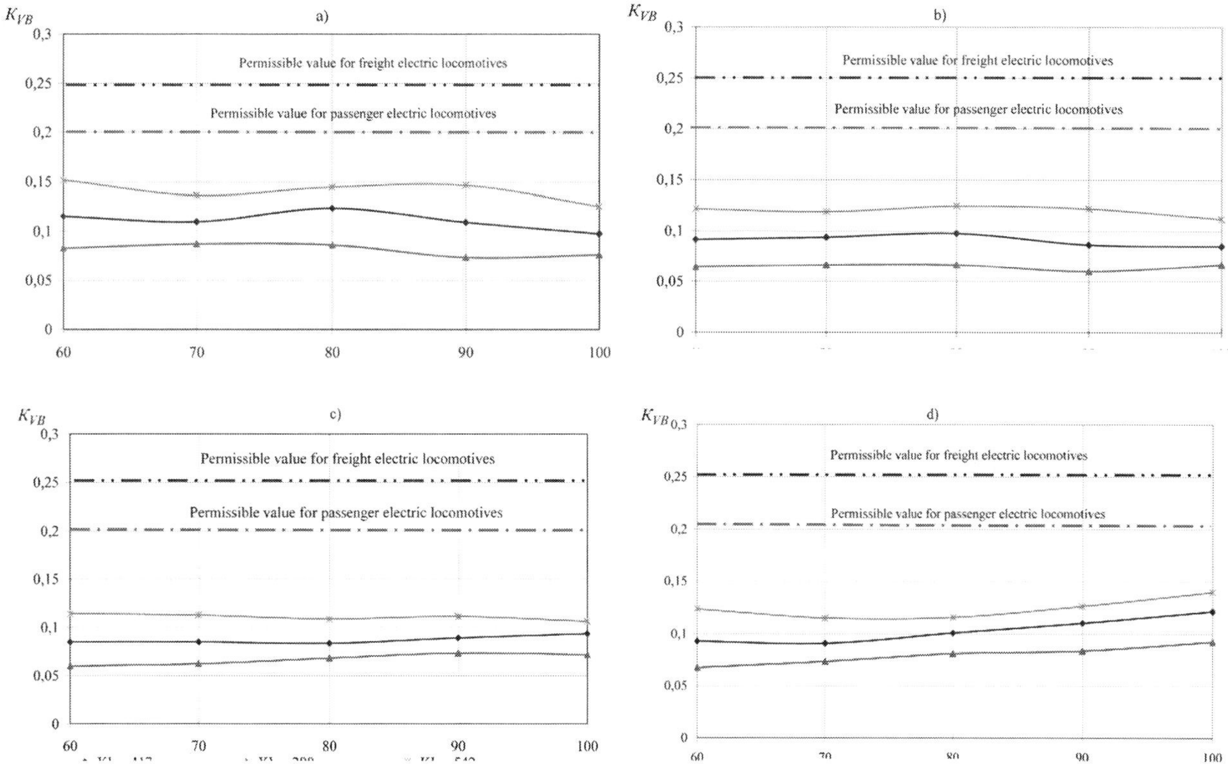

Fig. 2. Change in the coefficient of vertical dynamics in the body stage of spring suspension when moving in a curve R = 600 m from the movement speed: a) maximum values K_{VB}, random track irregularities; b) maximum probable values K_{VB}, random track irregularities; c) maximum values K_{VB}, deterministic track irregularities; d) maximum probable values K_{VB}, deterministic track irregularities

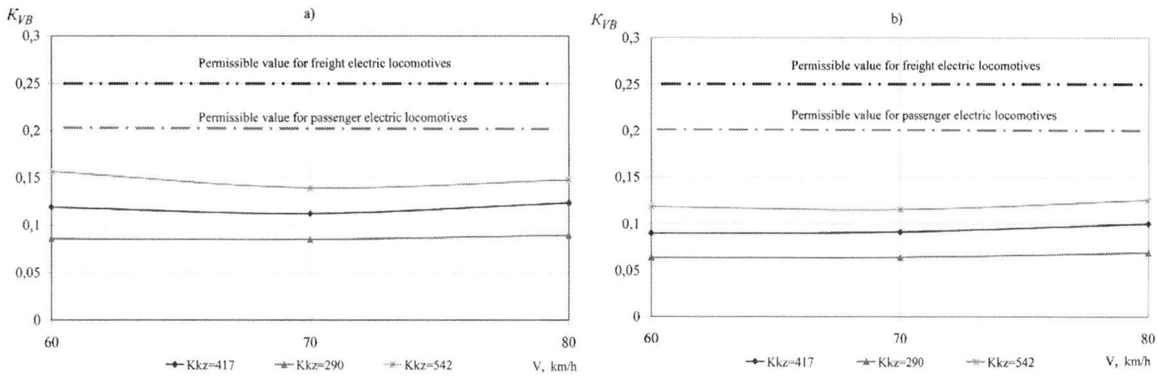

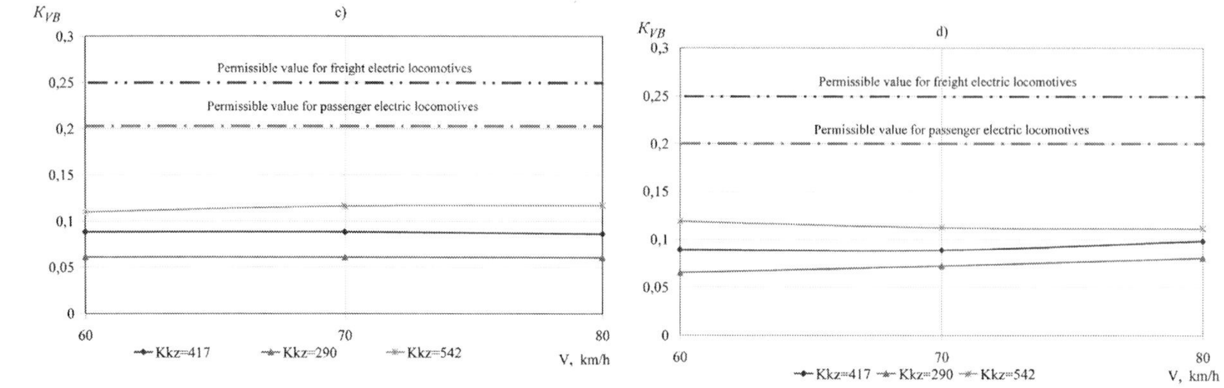

Fig. 3. Change in the coefficient of vertical dynamics in the body stage of spring suspension when moving in a curve R = 300 m from the movement speed: a) maximum values K_{VB}, random track irregularities; b) maximum probable values K_{VB}, random track irregularities; c) maximum values K_{VB}, deterministic track irregularities; d) maximum probable values K_{VB}, deterministic track irregularities

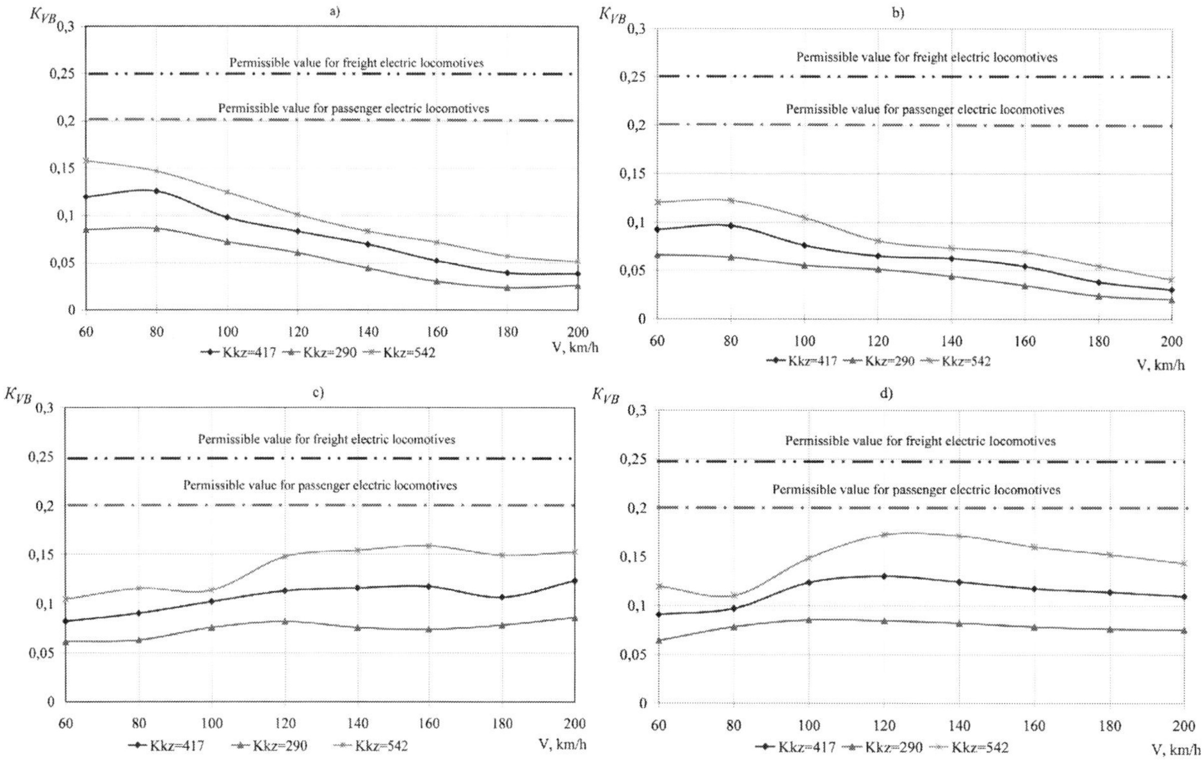

Fig. 4. Change in the coefficient of vertical dynamics in the body stage of spring suspension when driving in the straight section from the movement speed: a) maximum values K_{VB}, random track irregularities; b) maximum probable values K_{VB}, random track irregularity; c) maximum values K_{VB}, deterministic track irregularities; d) maximum probable values K_{VB}, deterministic track irregularities

Stresses in the rail head σ_h (Fig. 5) at the variation of the springs stiffness values of the central stage k_{kz} of spring suspension depending on the electric locomotive speed.

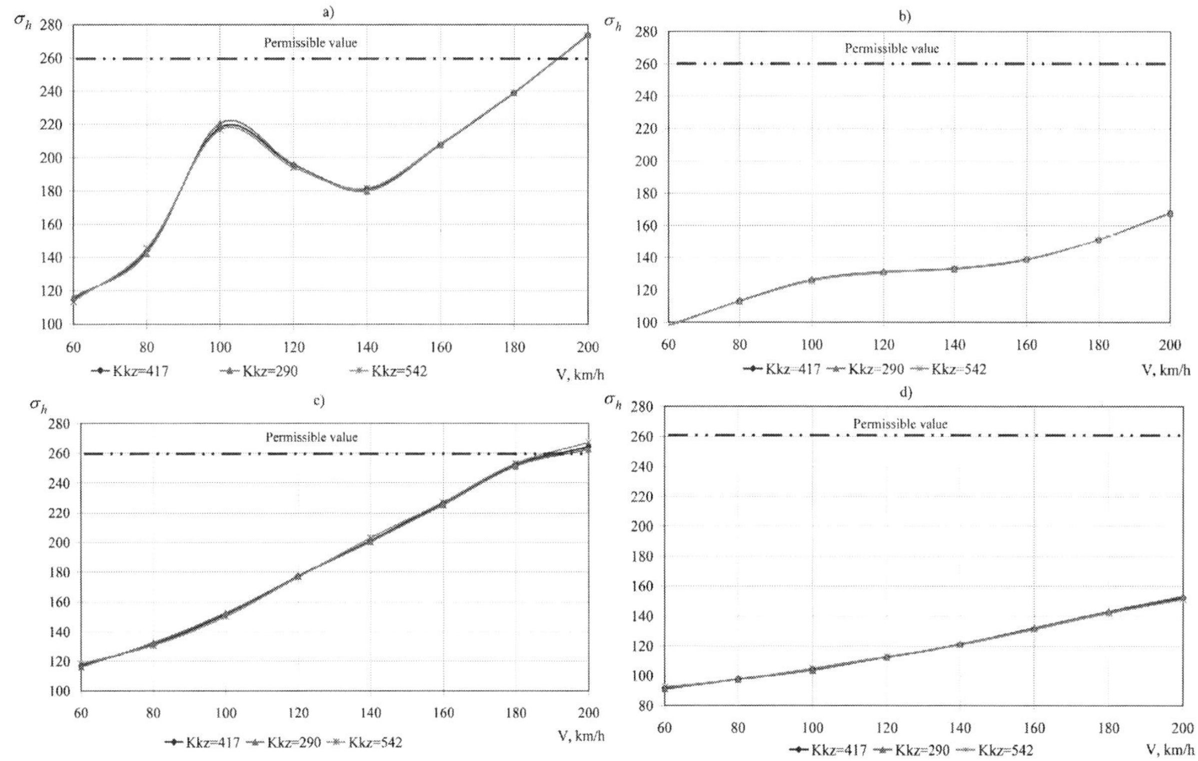

Fig. 5. Change of the stress in the rail head when driving in the straight section of the track from the movement speed: a) maximum values K_{VB}, random track irregularities; b) maximum probable values K_{VB}, random track irregularities; c) maximum values K_{VB}, deterministic track irregularities; d) maximum probable values K_{VB}, deterministic track irregularities

As resultant, the largest values from obtained ones for homogeneous elements of the electric locomotives or tracks were accepted. The comparison was carried out (on all graphs, Figures *a* and *b*) for the largest and largest probable values (in all the graphs, Figures c and d).

Minimum and maximum values are defined for each of the dynamic indicators. These values correspond to the abscissas of the track at which these values have been reached. The normal random values distribution law was adopted. The average value at the section, dispersion, the standard deviation was determined. The maximum possible values were determined from the condition of value deviation of the random quantity from an average one of not more than 2.5 r.m.s. deviations.

On the graphs the value $k_{kz} = 417$ corresponds to the variant with the nominal value of the variable parameter; reduced by 30 % to parameter value $k_{kz} = 290$; increased by 30% – $k_{kz} = 542$. The comparison was carried out for the maximum (max) and maximum probable (max prob) values of the coefficient of vertical dynamics in the body stage of spring suspension at the variation of the springs stiffness values of the of the central stage k_{kz} of spring suspension.

As can be seen from the graphs given above, in all cases the calculated figures of the regulatory values are less than the permissible ones for the movement along the straight lines up to speeds of 160 km/h and when moving along curves of mean and small radius with the above values of the highest unballanced acceleration.

Conclusions

The paper described the design scheme of a freight-passenger electric locomotive, taking into account the details of its construction and a mathematical model of spatial oscillations. The connections imposed on the system, the forces arising in them, as well as the mutual displacements of all the bodies in the system are considered.

Differential equations of system oscillations are composed on the basis of Lagrange's second order equations taking into account the details of construction in the electric locomotive. Analysis of calculations and subsequent good agreement of the theoretical and experimental studies results confirmed the correctness of the choice of the design scheme, parameters of the running gears of the locomotive. It has a regulatory-permissible effect on the railway track.

References

[1] Клименко И. В. Нагруженность рам тележек перспективных электровозов. Дисс. на соиск. ... к.т.н. 1994. 163 с.

[2] Myamlin S., Dailydka S., Neduzha L. 2012. Mathematical Modeling of a Cargo Locomotive / Proceedings of 16th International Conference "Transport Means 2012". Kaunas, pp. 310–312.

[3] Myamlin S., Neduzha L., Urbutis Ž. 2015. Research of Innovations of Diesel Locomotives and Bogies / Proc. of 9th Intern. Scientific Conf. Transbaltica 2015. Vilnius, pp. 470–475.

[4] Myamlin S. V. Peculiarities of running gear construction of rolling stock // Science and Transport Progress. 2017. Vol. 3 (69), pp. 130–146.

[5] Lazaryan V. Dynamics of Carriages (Динамика вагонов). Транспорт, 1964, 256 p.

[6] Медель В. Б. Подвижной состав электрических железных дорог. Транспорт, 1965. 280 с.

[7] Lagrange J. L. Oeuvres. T. 1–14. Paris, pp. 1867–1892.

[8] Нормы для расчета и оценки прочности несущих элементов, динамических качеств и воздействия на путь экипажной части локомотивов железных дорог МПС РФ колеи 1520 мм. МПС РФ, ВНИИЖТ. Москва, 1998.Standards for design and construction of railroads with 1520 mm. track cars (non-self-propelled)]. Moscow, GosNIIV-VNIIZhT Publ., 1996. 352 p.

[9] Типовая методика испытаний подвижного состава по воздействию на путь после изготовления или перед вводом в эксплуатацию. ИЦЖТ ВНИИЖТ, МПС. Москва, 1990.

[10] Bodnar B. Improving Operation and Maintenance of Locomotives of Ukrainian Railways, Technologijos ir Menas = Technology and Art. 2016. № 7, pp. 109–114.

[11] Очкасов А. Б. Усовершенствования бортовой системы диагностирования электровоза: авт. дис... к. т. н.: 05. 22. 07 / Днепропетровский нац. ун-т железнодорожного транспорта им. академ. В. Лазаряна. 2005.

ITMS'2018

panevėžys faculty of technologies and business

The 12th International Scientific Conference Intelligent Technologies in Logistics and Mechatronics Systems (ITELMS'2018), 26–27 April 2018, Panevėžys, Lithuania

The Influence of Layer Re-Melting on Tensile and Fatigue Strength of Selective Laser Melted 316L Steel

Janusz Kluczyński[a*], Lucjan Śnieżek[a], Krzysztof Grzelak[a], Janusz Torzewski[a]

[a]*Military University of Technology, Faculty of Mechanical Engineering, Institute of Machine Building, 2 Gen. Witolda Urbanowicza Street, 00-908 Warsaw 49, Poland*

Abstract

Additive manufacturing techniques such as Selective Laser Melting (SLM) allows for fabrication of complex components from different types of metal powders directly from CAD data. The process is based on selective melting metal powder layer by layer using ytterbium laser. It may be paused by operator or interrupted by machine software due to different reasons. Damaged metallic parts could be also regenerated by additive manufacturing missing geometry on original part. Continuing process or beginning regeneration of damaged part requires initial exposure which re-melts the top layer of the element. The paper contains results of microstructure, tensile strength and fatigue strength analysis made on specimens which were manufactured in specially prepared process. Building process has been programmed in such a way that it was interrupted in half job and continued after the machine has fully cooled down. The continuation of the process was preceded by re-melting the top layer of specimens. To check the influence of this operation as best as possible the specimens were manufactured vertically. With this research method the influence of exposure repeating on microstructure, tensile and fatigue, strength was taken into account. The results provides useful knowledge about fatigue strength of elements, which were manufactured in interrupted process. It is also a base for improving metallic parts regeneration by using powder based additive manufacturing systems.

© 2018 J. Kluczyński, L. Śnieżek, K. Grzelak, J. Torzewski
Peer-review under responsibility of the Kaunas University of Technology, Panevėžys Faculty of Technologies and Business

Keywords: additive manufacturing, Selective Laser Melting, stainless steel, mechanical properties, fatigue and tensile strength

* Corresponding author.
E-mail address: janusz.kluczynski@wat.edu.pl

© 2018 J. Kluczyński, L. Śnieżek, K. Grzelak, J. Torzewski
B427H0015 . Ed*Learning*.it
Peer-review under responsibility of the Kaunas University of Technology, Panevėžys Faculty of Technologies and Business

1. Introduction

Selective Laser Melting (SLM) is one of the most important 3D additive manufacturing (AM) technique because of its good alternatives for conventional manufacturing technologies. Selective laser melting is a powder – based process which allows obtaining three – dimensional parts for final usage, basing on a CAD model. During the process, each layer of metal powder is molten and consolidated with the base by the energy of a high intensity laser beam. The SLM technique allows to produce almost fully dense parts [1]. A process characterization is shown in Figure 1.

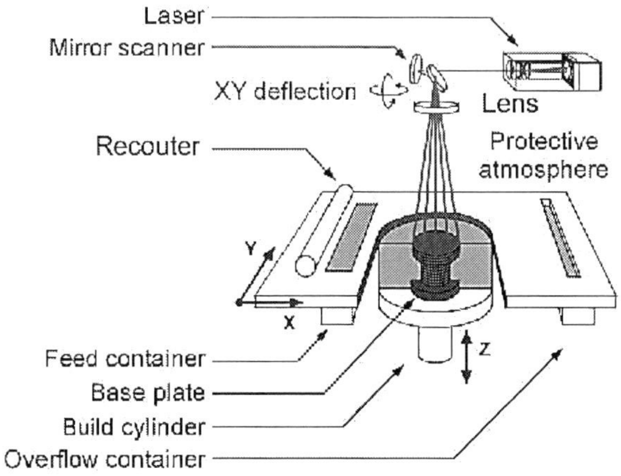

Fig. 1. SLM process characterization [1]

Exposing each layer in SLM process cause the shrinkage increasing, when 50 % powder density became almost 100 % solid [1]. It is caused by the short interaction of powder layer and heat source where there is high scanning speed of the high power laser beam. In conclusion, the heat transfer and melted metal flow have influence on the size and shape of the melt pool which is shown on Figure 2. The mechanical properties of the final product depends also from:

- the cooling rate;
- the transformation reactions in the melt pool;
- heat – affected zone;

there is also a dependency between the melt pool geometry and the grain growth in the part microstructure [2]. Growth of the SLM manufacturing systems is caused by real possibility of obtaining completely new geometries of parts which have significant application in many technical objects. Ability of reaching very common geometries could decrease product total weight and production cost. It can also raise machines efficiency and simplify production processes. As was mentioned – SLM technique is based on melting metal powder layers using the laser beam, so "*layer-by-layer*" process allows to continuing production after interruption caused by operator, machine software or lack of electric power supplies. This characteristic feature allows to combine materials during production of each part and also regenerate broken mechanical part by repair of damaged geometry on a serviceable basis.

Due to constant development of the SLM technique there are available many different materials:

- iron alloys (stainless steel, tool steel);
- titanium alloys;
- aluminumalloys;
- cobal-chrome alloys;
- nickel based alloys.

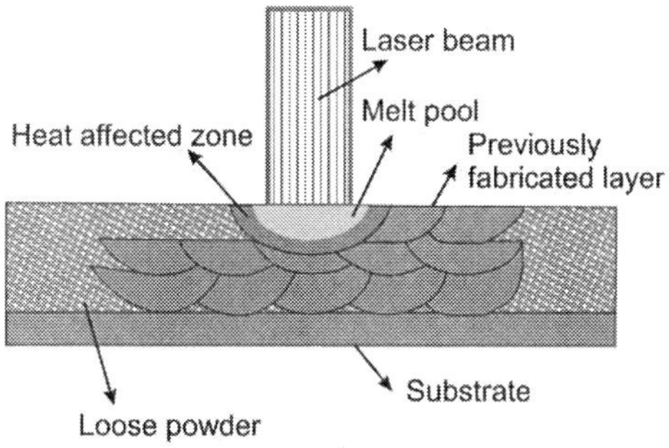

Fig. 2. Different zones during SLM process

One of the most common materials due to its usage and low shrinkage during additive manufacturing is 316L stainless steel. Biocompatibility of this material makes it one of the primary in medical applications. In addition, the corrosion resistance, small linear expansion coefficient, good resistance of oxidation [3] makes it a very good material for turbine machine vanes, chemical and marine applications.

Mechanical properties of SLM manufactured products are strongly dependent on the way they are arranged in the machine workspace [4]. Therefore there is significant anisotropy of manufactured parts which has to been taken into account during part designing and process preparation. Moreover, whenever the SLM process is interrupted after different issues it is important to expose the top layer once or twice before the process continuation. Every modification in SLM manufacturing process could cause final product properties change.

Presented above research results are connected with process interruption with repeated laser exposure and its influence on mechanical properties (microstructure, tensile strength, fatigue strength). Specimens manufacturing process was intentionally interrupted three times to find out how the interruption of the process and another exposure of last layer (before interruption) affects the mechanical properties.

2. Experimental materials and methods

2.1. Research description

Specimens which were used for research had been manufactured on the SLM 125HL machine. Research is based on the literature analysis of material microstructure, tensile and fatigue strength of own specimens analysis. Tensile and fatigue properties were determined with using Instron 8802 servohydraulic test system. Specimens were made of 316L austenitic stainless steel and they were manufactured vertically (Fig. 3). The results of fatigue tests conducted by a factor of asymmetry cycle of R=0.1 are shown in a Wöhler graph.

Fig. 3. ASTM E466 96 vertically orientated specimens

2.2. Processing parameters and material description

Metallic specimens were manufactured using SLM process on SLM 125HL. Main process parameters were similar to producers:

- laser beam power: 190 W;
- hatching distance: 0,12 mm;
- exposure speed: 900 mm/s;
- layer height: 0,03 mm.

Throughout the build process, the ambient temperature was maintained at 26 °C and oxygen level in machine chamber was minimized using an argon atmosphere to decrease the oxygen level to 0.2 %. All parameters was setup as machine producer recommends to check the typical influence of interrupting process on elements which were manufactured in the most unfavorable arrangement in machine workspace.

Grain size of the material used in research was from 15 to 60 μm. The 316L steel powder was manufactured via gas atomization. The composition of the material is shown in Table 1.

Table 1. Chemical composition of 316L steel

C	Mn	Si	P	S	N	Cr	Mo	Ni
% wt								
max. 0.03	max. 2.0	max. 0.75	max. 0.045	max. 0.03	max. 0.10	16.0–18.0	2.00–3.00	10.0–14.0

The manufacturing process was interrupted three times:

- during specimens measuring zone production;
- during specimens transition between the measuring zone and the handle zone production;
- during specimens handle zone production.

After each interruption there was about 3 hour break to let the material and machine "return to the starting point" including protective argon atmosphere reduction to normal conditions.

2.3. Microstructural characterization

The specimens were metallographically prepared using standard techniques dedicated for stainless steels. For etching was used 5% phosphoric acid solution. Images for each specimen were taken using Olympus LEXT confocal microscope.

2.4. Tensile and fatigue testing

In this study, all tensile and fatigue specimens were built according to programmed building process and some of the samples have been additionally machined. All specimen dimensions (rectangular bar) and test conditions were in accordance with ASTM E8 on an Instron Load Frame at room temperature. All tensile tests were conducted at a constant strain rate of 3 mm/min. During tensile test, the yield and ultimate strengths were recorded together with to the elongation to failure. Specimens were prepared via Additive Manufacturing (AM) directly to net dimensions as well as machined to net dimensions. Figure 4 shows fatigue specimens with marked re-melting areas for as-built and machined samples. The dimensions of the fatigue specimens were 105 mm in total length, the gage length was 25 mm in the center having a constant cross section with width of 9 mm and thickness of 3 mm.

Machining and polishing was done on all sides of the machined condition specimens to obtain a surface finish of 1.25 μm or better. Fatigue samples were tested at different constant maximum stress levels, from 141 to 235 MPa, to failure in order to generate an approximate S–N curve. Testing of the specimens was accomplished at a load ratio of R = 0.1, a frequency of 10 Hz, on an Instron load frame at room temperature.

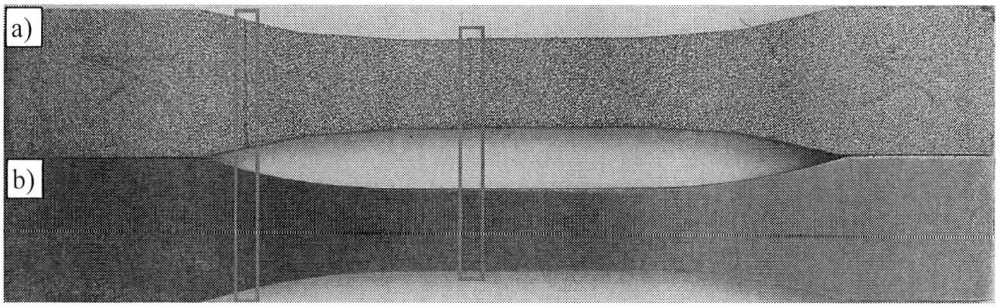

Fig. 4. Fatigue specimens with marked re-melting areas: a) as-built and b) machined

3. Results and discussion

3.1. Microstructure analysis

The total specimens thickness was 3mm so the microstructure is characterized by relatively high porosity in additive manufactured specimens (5.6 %). The porosity was determined using Gimp 2 software dedicated for image edition. Therefore, the porosity analysis basing on specimens surface measurements (number of pixels) – which is shown in Figure 5. Results of porosity measurements are shown in Table 2.

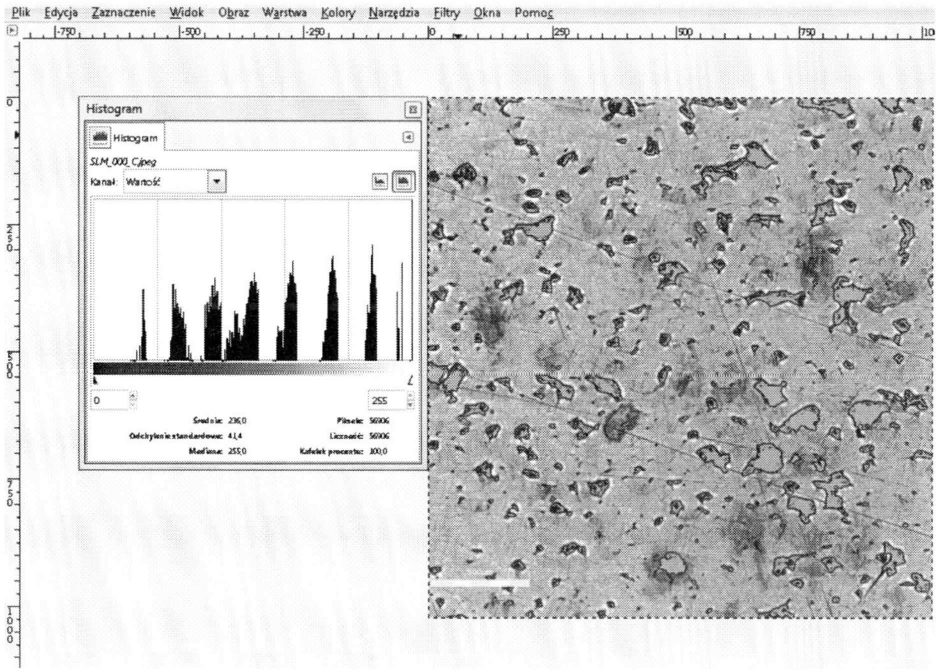

Fig. 5. Porosity analysis basing on histogram comparison in Gimp 2 software

Table 2. Porosity obtained from histogram analysis in Figure 5

	The number of pixels in pores	Total number of pixels	Porosity %
PV-ST1 (as-built)	56096	1048676	5,42

The microstructure (Fig. 6–7) has a typical, dual phase austenitic – ferritic structure constituted by molten pools, which depth-to-width ratio is 2.5–4.5. There is visible (area "A" in Fig. 6) zone of re-melting area after machine downtime, where depth-to-width ratio is 1.5–2 times bigger than other part of visible microstructure. There is also visible higher porosity (from 5.6 % to 7.3 %) on re – melting line which was marked by blue outline in the Figure 6. In the re-melted area porosity increase which is caused by top surface roughness change after another laser beam exposure. During repeated exposure of each layer its roughness is decreasing [5] so the amount of powder on the layer is smaller. The influence of this linear increase of porosity on mechanical properties is shown in the next subsection.

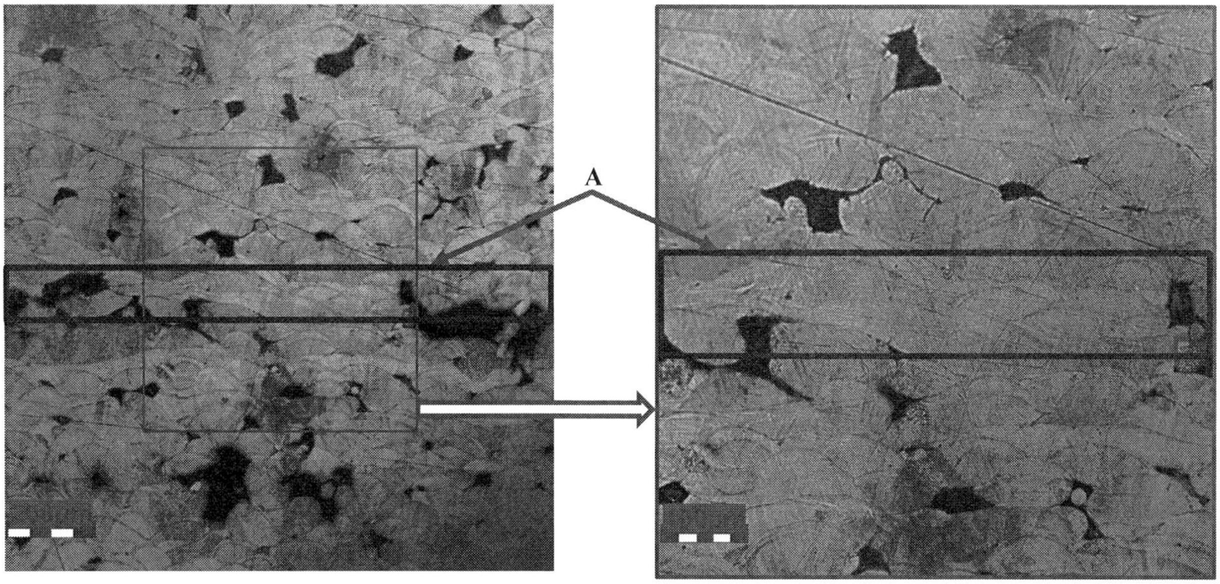

Fig. 6. Microstructure of the specimens with marked re-melted area (black outline)

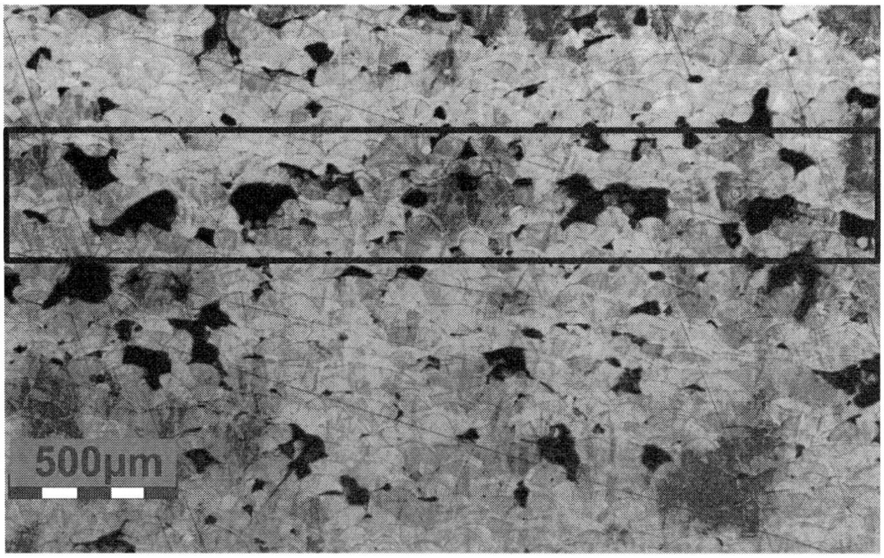

Fig. 7. Microstructure of the specimen with marked higher porosity in the re-melted area (black outline)

3.2. Tensile strength analysis

Stress-strain curves for the tensile specimens tested in this article are given in Figure 8. This diagram shows a comparison of three monotonic tensile tests, two of which were made according to programmed building process (layer re-melting) and last one was prepared using ordinary Additive Manufacturing process. Additionally, there is evaluation as-built sample with after machining ones on this graph.

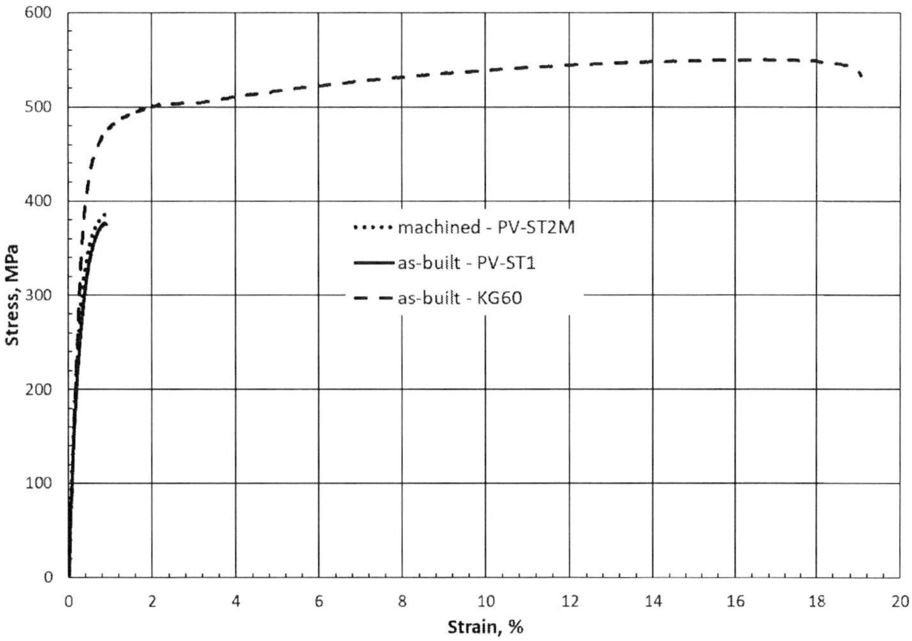

Fig. 8. Tensile stress-strain curves for Additve Manufactured 316L stainless steel

Table 3 lists their average monotonic properties including yield strength, ultimate tensile strength (UTS) and elongation.

Table 3. Tensile properties obtained from stress-strain curves in Figure 8

	Yield strength MPa	Ultimate tensile strength MPa	Elongation %
PV-ST1 (as-built)	320	375	1
PV-ST2M (machined)	340	385	1
KG60 (as-built)	440	550	19

It may be seen that the sample made without re-melting had a higher tensile properties than made according to programmed building process. The ultimate tensile strength (UTS) of the re-melted specimens is about 380 MPa and thereby about 30 % lower than the UTS of the specimens as SLM ordinary processed. Noticeable is also a decrease of Yield strength at approximately 25 % for specimens made of layer re-melting of AM 316L steel. Moreover, elongation plunge because of the unique thermal history inherent to manufacturing along Z-axis (which is the most unfavorable configuration in terms of the element tensile strength) and also re-melting SLM 316L steel specimens. This always causes the samples to crack at the place of resumption of the powder melting.

Additionally, the aspect considered in the article was the evaluation of the impact of a machining on the tensile properties of samples in re-melting. As can be seen, the yield strength and the ultimate tensile strength of the specimen with machined surface is slightly higher than that of the samples with the as-built surface.

As discussed in [6], machining removes the stress concentration effect of the rough surface and therefore a bit better tensile properties for the machined surface condition is not surprising.

3.3. Fatigue strength analysis

The fatigue properties were described on the basis of the carried out research for the as-built and machined specimens produced using SLM technique. The specimens were exposed to variable levels of maximum stress and stable cycle asymmetry ratio R = 0.1 (R = σ_{min} / σ_{max}). The load was sinusoidal with frequency f = 10 Hz.

The findings of the fatigue testing are shown in Figure 9 on a plot of maximum applied stress vs. number of cycles to failure.

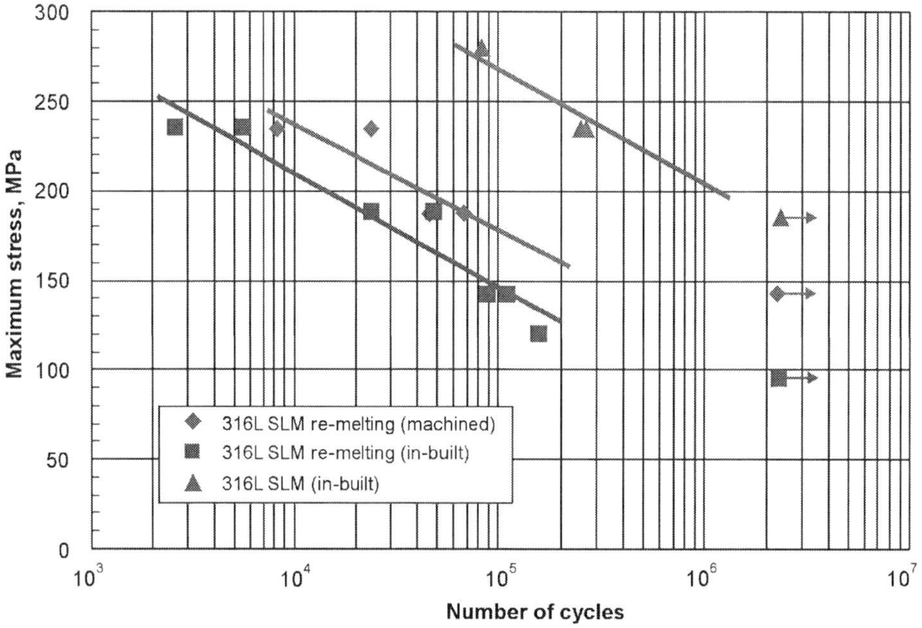

Fig. 9. S-N curves for Additve Manufactured 316L steel

The graph presents a comparison chart of fatigue life of samples made of 316L steel in three ways: according to the program in the initial state (square), according to the program after machining (diamond) and made without interrupting the manufacturing process in Y-X axis (triangle). The marks with arrows shows the level of maximum stress when specimens do not fail during fatigue tests for each type of specimens. Regression lines obtained by least square method was carried out by the experimental points. As can be seen, the fatigue life on the same level of maximum stress of the samples with machined surface is higher than that of the samples with the as built surface. As discussed in other work [7], machining removes the stress concentration effect of the rough surface, and therefore, the longer life surface life for the machined surface condition is not surprising. There are longer fatigue lives for tests on the specimens manufactured in Y-X axis, as to compared to the re-melting specimens, see figure 9 where there is limitation of samples number. A greater number of samples could lead to a higher generalization of our fatigue research.

Conclusions

SLM manufactured elements are characterized by high properties anisotropy connected with manufacturing direction. In this paper there was shown results of research which were specially prepared to include the worst case from the mechanical properties point of view. Manufactured along Z axis and re-melted specimens are characterized by:

- higher porosity – 5,6 %, when monolithic elements has porosity is lower than 1 %;
- increased porosity in re-melted zone – up to 7.3 %;
- 32 % lower ultimate tensile strength in comparison to specimens manufactured in Y-X axis;
- 95 % lower elongation in comparison to specimens manufactured in Y-X axis;
- 3 % higher ultimate tensile strength after machining;
- up to 10 times lower fatigue live for re-melting samples in comparison to specimens manufactured in Y-X axis.

Selective Laser Melting process makes material structure similar to structure after quenching, which decrease material elongation. Layer re-melting process intensifies "quenching effect" and also cause porosity increase which has been presented in this paper. To reduce the effect of elements unfavorable arrangement in machine workspace and also reduce the re-melting artefacts it is necessary to use proper heat treatment or hi-pressing process which can be useful to reduce the porosity and increase the material elongation. The effect of re-melting on fatigue performance was found to be significant. More testing is needed on re-melting specimens with different manufacturing orientation and post-process heat treatment.

References

[1] Kruth J.-P., Badrossamay M., Yasa E., Deckers J., Thijs L., Van Humbeeck J. *Part and material properties in selective laser melting of metals*, J. Mater. Process. Technol. 149: 616–622.

[2] Childs T. H. C., Hauser C., Badrossamay M. *Selective laser sintering (melting) of stainless and tool steel powders: experiments and modelling*, Proc. IMechE Part B: J. Mechanical Engineering Science, 219: 339–358.

[3] Cherry JA., Davies HM., Mehmood S., Lavery N. P., Brown, S. G. R. and J. Sienz. *Investigation into the effect of process parameters on microstructural and physical properties of stainless steel 316L parts by selective laser melting*, International Journal of Advanced Manufacturing Technology, Vol. 76 Nos 5/8, pp. 869–879.

[4] Ziółkowski G., Chlebus E., Szymczyk P., Kurzac J. Application of X-ray CT method for discontinuity and porosity detection in 316L stainless steel parts produced with SLM technology. *Archives of civil and mechanical engineering* 14 (2014), pp. 608–614.

[5] Di W., Liu Y., Yongqiang Y., Xiao D. *Theoretical and experimental study on surface roughness of 316L stainless steel metal parts obtained through selective laser melting*, Rapid Prototyping Journal, Vol. 22, Iss 4, pp. 706–716.

[6] Spierings A. B., Starr T. L., Wegener K. *Fatigue performance of additive manufactured metallic parts*, Rapid Prototyping Journal, Vol. 19, Iss 2, pp. 88–94.

[7] Frazier W. E. *Metal Additive Manufacturing: A Review*, Journal of Materials Engineering and Performance, Volume 23, Issue 6, pp. 1917–1928.

The 12th International Scientific Conference Intelligent Technologies in Logistics and Mechatronics Systems (ITELMS'2018), 26–27 April 2018, Panevėžys, Lithuania

Adaptation of Ukrainian Grain Elevators for Unit Train Loading

Dmytro Kozachenko[a], Volodymyr Bobrovsky[a], Oleksandr Ochkasov[a*], Anatolii Shepotenko[a], Yuliia Hermaniuk[a]

[a]*Dnipropetrovsk National University of Railway Transport named after Academician V. Lazaryan, 2 Lazarian St., 49010 Dnipro, Ukraine*

Abstract

The aim of the paper is to improve approaches to the design and operation of private sidings of grain elevators for loading of unit trains. The research is based on the methods of organizing of railway operation. One of the main methods of reducing costs in the logistics chain, which provides for the export of grain cargoes is the using of unit train technology for transportation. Introduction of this technology in Ukraine is limited by the low loading capacity of elevators. In the paper the basic schemes of track development of elevator private sidings and analyzes the use of different technical means for performing shunting operations are presented. The results of the performed research can be used for designing and reconstruction of the existing elevators and for building the new ones, as well as for estimating the costs associated with different logistic schemes for grain delivery to seaports. As result the work develops a methodology and obtains a technical-operational assessment of the technical equipment and the operation technology grain elevator private sidings for loading of unit trains.

© 2018 D. Kozachenko, V. Bobrovsky, O. Ochkasov, A. Shepotenko, Yu. Hermaniuk
Peer-review under responsibility of the Kaunas University of Technology, Panevėžys Faculty of Technologies and Business

Keywords: railway transport, grain elevator, grain transportation, unit train; private sidings

1. Introduction

Today, grain is one of the main export goods of the Ukrainian economy. In the 2016/2017 marketing year, Ukraine has exported 43.9 million tons of grain. This is the second result in the world after the United States (92 million tons.) and in monetary terms it is about 17 % of the country's total exports and. Lowering the logistics costs is one of the problems that needs to be solved in order to increase the competitiveness of Ukrainian grain in foreign

* Corresponding author. Tel.: +380677452071
E-mail address: abochkasov@gmail.com

© 2018 D. Kozachenko, V. Bobrovsky, O. Ochkasov, A. Shepotenko, Yu. Hermaniuk
Peer-review under responsibility of the Kaunas University of Technology, Panevėžys Faculty of Technologies and Business

B427H0016 . Ed*Learning*.it

markets. Thus, according to the World Bank assessment [1], the logistics costs for grain exports in Ukraine are 50...55 USD per ton, that is around 35 % of its final cost. At the same time in the USA these costs are at the level of 30 USD per ton, which is about 9 % of the grain cost [2].

Lowering the logistics costs associated with grain transportations can be achieved through the improvement of the technical means and technologies that provide them.

Analysis of the Ukrainian system of grain storage and transportation for export was made in the work [3]. The results presented in this paper show that one of the modern problems of logistics support of grain transportation is inadequacies in the interaction of grain elevators and railway transport. Grain loading in Ukraine is carried out by more than 600 stations, while for 67 % of stations the average daily volume of loading does not exceed 1 car. As result, grain elevators have low loading capacity, which in most cases can load 10...12 cars per day. This situation, together with the large dispersal of grain loading stations across the territory of the country, leads to the impossibility of grain cargoes shipment by unit trains. In sum, grain is the only type of bulk cargo, which is mainly transported by single car consignment, and, with an average level of good transportation by unit trains on the Ukrainian railway 40...45 %, this figure for grain is only 10 % [4, 5].

In this regard the paper is aimed to improve approaches to the design and operation of grain elevators private sidings for the possibility of loading direct unit trains.

2. Problem statement

It should be noted that significant part of elevators in Ukraine were constructed during the Soviet Union. Moreover, "The norms of process engineering of grain receiving stations and elevators" [6] remain unchanged from those times and new elevators are also built in accordance with outdated requirements. According to the Norms [6], typical designs of grain handling devices are developed based on the calculation of processing of one group of cars with capacity of 1000 tons, 1500 tons, 2000 tons, 3000 tons. For the enterprises with calculated daily loading volume of more than 1000 tons, the daily grain loading should be accepted not less than the carrying capacity of the unit train. The loading and unloading front of unit trains should be determined from the condition of loading or unloading the train no more than in two-three supplies. For certain items of construction, the number of supplies per day and their volume were set by the Ministry of Railways (MR, now Ukrzaliznytsia). The calculated volume of operations in accordance with the work [6] should be determined taking into account the irregularity coefficients of grain shipping using the formula

$$B_c = \frac{BK_m K_d}{330}; \tag{1}$$

where B – is the annual volume of grain loading, ton;

K_m, K_d – are the coefficients of monthly and daily transportation irregularities, respectively;

330 – is the estimated number of days during which the grain loading is carried out over the year.

According to the Norms [6], for the typical designs of elevators, the values of irregularity coefficients should be taken as equal to $K_m = 2{,}0$ and $K_d = 2{,}5$.

The Fig. 1 shows dependence of the correlation between the estimated and average daily volumes of elevator loading on the annual loading volumes.

In general, the Norms [6] concern only track infrastructure and grain handling devices of elevator. This is due to the fact that servicing of grain elevators in the USSR was carried out in most cases by the locomotives of railways, and therefore the requirements [6] are formulated so that the elevator infrastructure enables them to be used most effectively during shunting operations. At the same time, significant costs for creating and operating the loading capacity reserves of elevators are economically unjustified for their owners. Due to the fact that in Ukraine there are no requirements concerning correlation between the estimated loading capacity of elevators and the volume of transportation orders from them, then the requirements [6] regarding the correlation between the calculated and average loading capacity of elevators are actually not fulfilled.

It should be noted that the loading capacity of grain elevators is determined not only by the capacity of grain handling devices and track infrastructure, but also by the technology of performing shunting operations and the technical means that provide them. However, in the work [6], these issues are not considered.

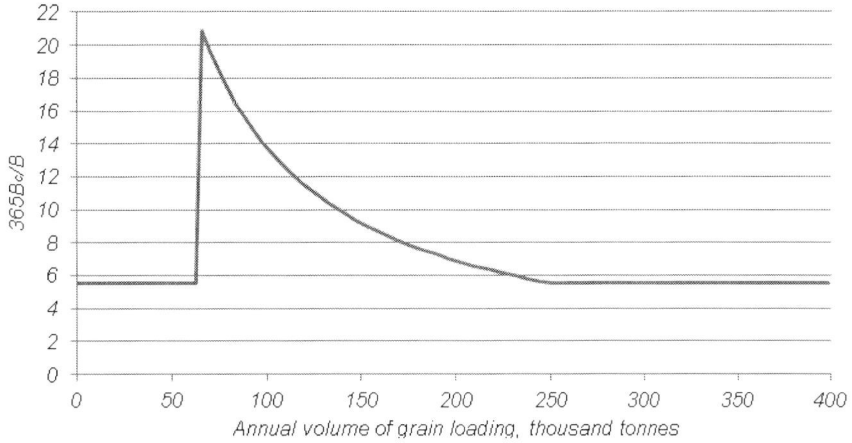

Fig. 1. Dependence of the correlation between the estimated and average daily loading volumes of grain elevator on the annual loading volumes

Propositions for improving infrastructure and technical supply of grain elevators are provided in the work [7]. This work provides performing shunting operations at elevator in the conditions of combination of both shunting locomotives of the railroad and shunting facilities of grain elevators. However, the problems of operation technology of grain elevator siding and its interaction with the mainline railway station are not developed in the work.

In order to reduce the mileage of vehicles transporting grain, the elevators are located in agricultural areas adjoining intermediate stations. At the same time, operation of the mainline railway transport of Ukraine is characterized by the drop in cargo transportation volumes by 60–70 % as compared to 1991 and considerable wear of the main technical means. In particular, wear of shunting locomotives has reached 99 %, and wear of grain carriers is 99.99 %. Combination of the above-mentioned factors leads to the fact that certain elevators cannot provide acceptable level of loading railway locomotives when loading direct unit trains. Therefore, their maintenance is carried out by locomotives of pickup trains and dispatching locomotives, together with other points of freight operation at the stations, which causes significant car demurrage. Formation of gradual routes [8], as well as technical routing, require coordination of loading at several stations and lead to increase in car demurrage as compared to formation of unit trains at one station.

Problems of interaction between grain elevators and railways are also typical for other countries of the world, first of all, for such large grain exporters as the USA and Canada.

In order to increase transportation efficiency of grain cargoes on the North American railways a shuttle train technology was introduced [9]. This technology involves the use of a special fare, which is by 46–52 % lower than that for the transportation of grain by collective dispatches. In this case, the shipper should be able to provide the loading of train with 75–120 cars for a limited time (about 15 hours [10]). Trains move between loading and unloading points on a rigid schedule in accordance with the contract for 6–9 months without re-forming and uncoupling train locomotives at the loading station [11]. Application of this technology requires increase in loading capacity of grain elevators, but its use makes it possible to reduce the costs for railway transportations. That is why it does not require involvement of shunting locomotives, reduces the circulation of cars and greatly reduces the need for use of technical facilities of railway stations.

It should be noted that domestic railways have a number of significant differences from the railways of North America. In particular, the share of electrified railways in the USA is about 0.6 %, while the share of electrified railways in Ukraine is 47 % of the operating length of the network. Moreover, taking into account the fact that the costs for energy resources at the electrified railways are 4 times lower than that of non-electrified ones, then the bulk of transportations in Ukraine is concentrated on electrified lines and the share of cargo turnover that is carried out with the use of electric traction is about 90 % of the total cargo turnover. In this regard, at the main directions the loading of grain unit trains with performing shunting operations at the elevators by train electric locomotives is technically impossible. In addition, the crisis in the economy of Ukraine, which took place after the Soviet Union collapse, had a minor influence on the volume of passenger traffic. In fact, current volumes of passenger transportations make up about 72 % of the passenger transportation volume in 1991. In that context, the requirements for railway infrastructure and organization of transportations even at low-powered lines are quite high, as they should ensure the safety of passenger train movement. Therefore, at the lines serviced by diesel traction organization of the grain unit train loading without its uncoupling requires significant capital investments in the development of railway infrastructure. The transportation technology of shuttle train in the USA is used mainly at the distances more than 1000 km. Grain transportations at the distance up to 500 km in a large number of cases is carried out by motor transport. Due to the fact that the USA area is 16.3 times bigger than Ukraine's area, the railway and motor transport may occupy different niches of the transportation market. At the same time, average distance of grain transportations by railway in Ukraine is about 550 km, which leads to considerable competition between the railway and motor transport. Thus, in order to increase the competitiveness of domestic grain in foreign markets, it is relevant the problem of adapting world experience and developing our own technical and technological solutions aimed at reducing the costs associated with its transportation from elevators to seaports.

3. Technical equipment of grain elevators for unit train loading

In order to minimize mileage of motor transport and reduce motor traffic volumes in cities, grain elevators should be placed in agricultural areas. However, in this case, elevators adjoin intermediate stations that do not have shunting locomotives. Therefore, provision of shunting operations at grain elevators is a key task that should be solved to form there unit trains. In order to organize the movement of grain unit trains according to the schedule it is accepted that duration of operations with the train at loading station should not exceed 20 hours. Other 4 hours of the day are used as a reserve.

The volumes of grain loading in Ukraine are insufficient to create grain elevators, which will ensure the daily formation of unit trains, as this will lead to an irrational increase in the distance of grain road transportation to them from the rural households. Studies performed in the work [4] show that the unit train loading once or twice a week is rational. In such conditions, in order to provide the railway employees with constant work at the private siding of one elevator, duration of the train loading should make it possible to organize 8 or 12-hour work shifts.

The work of railway transport is directly related to the work of continuous-handling systems of grain elevator. In this case, two technological schemes for the loading of cars with grain are possible – when grain from the storage is delivered directly to the car and when between storage and car the processing and storage bins are arranged. They provide an opportunity for a uniform loading of the elevator's transport facilities in the conditions where release devices have inconstant productivity in time. Estimation of the loading time of the unit train of 54 cars with different equipment of grain elevators is given in the Table 1.

Table 1. The time expenditures for car loading

Characteristics	Variants					
	1	2	3	4	5	6
Schema for the loading of cars	directly from a storage				from processing and storage bins	
Noria loading capacity, tons per hour	350	350	500	500	500	500
Number of loading tracks	1	2	1	2	1	2
Duration of loading, hour	29.7	17.1	26.0	13.4	21,0	10,5

Data analysis in the Table 1 shows that for the unit train loading a grain elevator should have two loading tracks. Taking into account the relatively low cost of loading bins and the possibility of reducing the duration of cargo operations by 19–21 %, it is advisable to equip the tracks with operational or storage bins. In this regard, it is advisable to take the variant 6 of the Table 1 as a basic version of continuous-handling system of grain elevator.

Possible variants of organizing the unit train loading are:

- placement of empty cars and picking up the loaded train at the connecting station with servicing by the locomotive of railway;
- creating at the elevator own track capacity for cars of unit train with supplying and picking up of cars by locomotive of railway and performing shunting operations at the elevator by own shunting facilities;
- creating at the elevator own track capacity for cars of unit train with supplying and picking up of cars, as well as performing shunting operations at the elevator by own shunting facilities.

Principal schemes of track development of elevators are shown in the Figure 2 *a–i*. Schemes at the Figure 2 *a, d, f* have one loading track and cannot provide unit train loading during the day. Therefore, these schemes are not considered in the future.

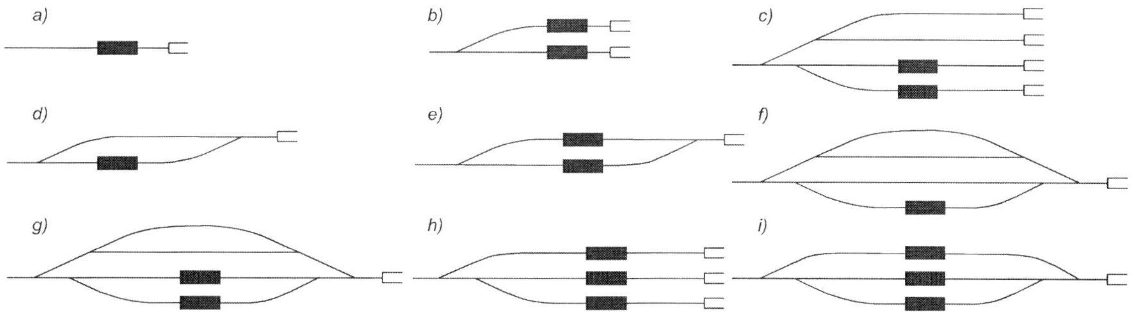

Fig. 2. Principal diagrams for track layout of elevators for unit trains loading

Schemes at the Figure 2 (*b, c, e, g*) have two loading tracks. It should be noted that when such railway locomotives as CHME3 or TEM2 are used for grain elevator servicing, the amount of car supply is limited by productivity and permissible time of continuous operation of locomotive compressor when pumping off cars air brakes. Based on this requirement, supply should consist of 20–21 cars. In these conditions, it is expedient to divide the unit train into 3 parts with 18 cars and load them at 2 tracks with 9 cars.

The useful length of loading tracks should provide placement of double length of car supply to move them during loading, taking into account the reserve. It is determined by the expression

$$L_{lt} = 2m_c l_c + 10\,; \qquad (2)$$

where m_c – is the number of cars placed at the loading track; l_c – is the length of grain carrier, m.

The schemes 2 *b* and *e* are characterized by the fact that they involve the construction of loading tracks only. So in this case connecting station should have two tracks to place empty and loaded cars of unit train. This can create problems for passing trains at the station and limits the possibility of performing parallel freight operations and the supply-removal of cars to the station. The schemes 2 *c* and *g* provide for creation of own track capacities to place the cars of unit train at grain elevator. It is possible to use the schemes 2 *b* and *c* in cases when car supply from the station is performed with the cars to the front. Application of the schemes 2 *e* and *g* makes it possible to perform car supply with the locomotive to the front.

Pulling the cars at loading fronts is carried out by shunting winches. Characteristics of winches that can be used to load grain unit trains are given in the Table 2.

Current Norms [6] set the requirements for loading of one car supply during 3.66 hours. In this regard, locomotive servicing of the unit train loading is efficient for the railway only at the marshaling yards and freight stations that have their own locomotives, which can be used for other operations too. When loading the unit trains at intermediate stations significant idle periods in the operation of shunting locomotives take place. This reduces the use of locomotives and locomotive crews of the railway. The presence of own traction rolling stock at grain elevator is one of the methods to make the elevator loading operation independent from servicing by the railway shunting locomotives.

Table 2.Characteristics of shunting winches

Parameters	Types of shunting winches			
	LEM-10	LEM-15	LEM-20	LEM-25
Number of moving cars	10	15	17	21
Weight of cargo in moving cars, tons	550	950	1070	1300
Speed of car movement, m/sec	0.060	0.060	0.064	0.064
Estimated cost, thousand USD	11,1	14,8	29,6	33,9

The elevators can use such shunting locomotives as TGM23 and TGM4 and their modifications as traction rolling stock. The estimated cost for locomotive TGM23 on the secondary market in Ukraine is 37–45 thousand USD, and for the locomotive TGM4-96-119 thousand USD. It is also possible to use locotractors, for example MMT3, TM1.175 or KRT-1, as shunting facilities at grain elevators. They have some advantages. First of all, they can be used to perform various technological and domestic tasks when there is no grain loading at the elevator. Secondly, the requirements for track development are simplified, because they can move both by railroad and the highway. Their main disadvantage is that they do not have permission to work at the mainline railway stations. Estimated cost of new loco tractor is 81–100 thousand USD.

Alternative ways to ensure unit train loading at the grain elevator allow the use of scheme 2 *h* and *i*. When using these schemes, the train loading is carried out at all tracks of grain elevator, and supply and picking up of cars can be carried out by a train locomotive at one time. In this case, the track capacity of connecting station can be used only for passing the train. Unit train making up can be performed at the connecting track; useful length of grain elevator tracks in this case is:

$$L_{el} = 2m_{\text{ut}}l_{\text{c}}; \qquad (3)$$

where m_{ut} – is the number of cars in the unit train set.

To evaluate infrastructure parameters and shunting facilities of grain elevators, the plan designing of their track development has been performed; the results of calculations are summarized in the Table 3.

Table 3. Parameters of infrastructure and shunting facilities of grain elevators

Indicator	Schemes of track layout					
	b	c	e	g	h	i
Number of tracks at the station	2	1	2	1	0	0
Number of switch points at grain elevator	1	3	2	6	2	4
Completed length of grain elevator siding, m.	670	1303	870	1731	1843	2120
Number of winches	2	2	2	2	3	3
Winch power, cars	10	10	10	10	21	21
Number of locomotives or locotractors	0	1	0	1	0	0
Number of loading tracks	2	2	2	2	3	3
Approximate capital investments, thousand USD	281	615	363	804	767	893

Taking into account the fact that technical supply of grain elevators according to the requirements [6] and the expression (1) should ensure the unit train loading at an annual volume of 66 thousand tons per year, the schemes *b* and *e* can be considered as basic variants. In these conditions, additional capital investments in the development of grain elevator for unit train loading at the intermediate station will be 330–530 thousand USD.

4. Operational evaluation of grain elevators for unit train loading

For technological calculations, it was assumed that grain elevator is at the distance of 3 km from the connecting station, the estimated slope of this section is 10‰, the curve radius is 180 m, and the track slope during shunting operations for car changing between the elevator tracks is 5‰. The time expenditures for placing and picking up of cars at the loading tracks are taken equal to 5 min. Additional time expenditures for preparatory and finishing operations related to the supply and picking up of cars at the station track (coupling-uncoupling of cars, fixing of cars and removing brake shoes, negotiations between employees, reversing the points, etc.) are taken equal to 5 min.

When using the railway locomotives for car loading, it is necessary to take into account the fact they can also perform other works at the station. Given that, according to the work [6], the interval between car supplies should be at least 2 hours, then the total duration of train loading when applying the schemes shown in the Figure 2 *b* and *e*, will be about 19 hours.

The car group weight, which can be moved by shunting locomotive, is determined as follows

$$Q = \frac{F_t - \left(w_0' + i + w_r'\right) \cdot P_{\text{ad}} \cdot g}{\left(w_0'' + i + w_r''\right) \cdot g}, \tag{4}$$

where F_p – is tractive power of traction facility, kN;
P_t – is adhesive weight of traction facility, tons;
w_0', w_0'' – the main specific movement resistance of traction facility and cars, respectively;
w_r', w_r'' – additional specific movement resistance of traction facility and cars in the curves, respectively;
i – is estimated slope of the track section;
g – is gravity acceleration.

Estimated number of cars in shunting trains, the displacement of which can be provided by elevator shunting locomotives and loco-tractors is shown in the Table 4.

Table 4. Number of cars in group during shunting operations

Slope, ‰	Locomotives			Locotractors
	TGM4	TGM23	MMT3	TM1.175, KRT-1
		Loaded cars		
5	21	12	10	7
10	14	7	-	-
		Empty cars		
5	85	49	40	28
10	21	14	-	-

Duration of brake testing in shunting train before shunting operations for supply and picking up of cars includes time for charging the brake line, checking its consistency and braking action of the cars. Time standard for complete air brake testing was determined using the following expression

$$t_{bt} = 8,5 + 0,64 m_{cg}; \tag{5}$$

where m_{cg} – is the number of cars in the group.

Time expenditures for shunting displacements were determined taking into account the number of cars in the train and permissible movement speeds based on the methodology presented in the work [12].

In order to evaluate the variants of technical equipment and operation technology of grain elevators sidings, there were constructed daily scheduled plans for the schemes presented in the Figure 2 *c* and *g*, as well as for Figure 2 *h* and *i*. Summarized evaluation results of technical and operational performance indicators of grain elevator sidings when loading a unit train are presented in the Table 5.

For track development schemes presented in the Figure 2 *c* and *g* the following variants are considered: when supply, picking up and shunting at grain elevator sidings are performed by the locomotive TGM4; and when car supply to grain elevator siding is performed by the railway shunting locomotives CHME3 or TEM2, or by the train locomotive 2TE116, and the shunting operations at grain elevator are performed by its own locomotive TGM23 or by the locotractor MMT-3.

Table 5. An example of a table

Parameters	Schemes				
	b, e	*c, g*			h, i
Traction rolling stock of grain elevator	-	TGM4	TGM23, MMT3		-
Supply-picking up of cars by locomotive	railway	elevator	railway	railway	railway
Type of locomotive performing supply-picking up	CHME3,TEM2	TGM4	CHME3, TEM2	2TE116	2TE116
Duration of unit train loading, h	19	16	17	16	11
Occupancy duration of elevator's locomotive, h	-	16	2×0.83	2×0.83+2×0.17	-
Occupancy duration of railway's locomotive, h	18	-	2×3.2	0.8+1.2	0.8+1.2

Analysis of the obtained results shows that the schemes presented in the Figure 2 *c* and *g* make it possible to load at grain elevator one unit train during a day. The advantage of using the TGM4 locomotive at grain elevator is that the railway can use both electric and diesel traction to deliver empty cars to the connecting station and pick up the loaded train. The disadvantage of this variant is that, because of the considerable amount of shunting operations at grain elevator, two locomotive drivers and two pickers must work when train loading during two eight-hour working changes.

Performing supply-picking up of cars at grain elevator by the railway locomotive makes it possible to use low-power locomotives TGM23 or locotractors. However, when performing supply and picking up of cars by shunting diesel locomotives, they will be occupied by train loading about 6.4 h. In addition, there are costs associated with the delivery and picking up of locomotive from intermediate station. Therefore, it is advisable to apply this variant only at the sections serviced by electric traction. At the sections serviced by diesel traction, it is expedient to perform supply and picking up of cars from grain elevator by a train locomotive. Such approach will make it possible to minimize the use of station infrastructure and, subject to changing the freight rate structure reduce transportation costs.

Using the schemes shown in the Figure 2 *h* and *i*, allows providing the train loading during 11 hours and, if it is necessary to load two trains per day. Disadvantage of these variants is that the supply and picking up of cars from loading station should perform only train diesel locomotives.

Conclusions

The performed researches make it possible to conclude the following. Improvement of transportation technologies of grain cargoes by rail in Ukraine due to implementation of unit train dispatching requires development of technical means of grain elevators private sidings. In order to ensure the unit train loading during the day, the elevator should have at least two loading tracks equipped with grain loading bins. In order to perform shunting operations at grain elevators it can be used own traction rolling stock – locomotives or locotractors. At the stations serviced by electric locomotive traction, it is expedient to use at grain elevator the TGM4 locomotives performing shunting operations for moving the cars at the elevator tracks, as well as for supplying and picking them up to the station. At the stations serviced by diesel traction shunting operations for cars moving at the elevator tracks can be performed by the locomotives TGM23 or locotractors with the supply and picking up of cars by the railway locomotive. One can also use the schemes of track development, which allow placing cars at the loading tracks for one supply by train diesel locomotive. The necessary additional capital investments in the development of grain elevator for unit train loading at intermediate station will be 330–530 thousand USD.

References

[1] Feidengold VB., Serebryany VL., Belecky SL. Technological design and efficient operation of the lines of acceptance and delivery of grain to the wagons at the elevators of the Rosrezerva. *Innovative technologies of production and storage of material values for state needs* 2015, 4: 225–242.

[2] Kozachenko DM., Korobyova RG., Rustamov RSh. Improvement of technical support and technologies of export transportation of grain cargoes in Ukraine. Bulletin Dnipropetrovsk State Agrarian University of Economics 2015, 4: 121–127.

[3] Berezovy MI. Transportation os steel products circular route schedule with private lokomotive. East European Journal of advanced technologies 2014; 3 (68): 51–55.

[1] Golubeva EV., Zubkov VN., Chebotaryova EA. Logistic technologies of formation of ship parties on the road range. Bulletin of Rostov State Transport University 2007, 4 (28): 56–61.

[5] Kozachenko DM., Ochkasov OB., Shepotenko AP., Sannycky NM. Prospects for the use of private locomotives for the transportation of goods in the direction of sea ports. Science and progress of transport 2017, 6 (72): 7–19. DOI : 10.15802/stp2017/118196.

[2] Kozachenko DM., Zhuravel IL., Levycky IYu. Normalization of the duration of maneuvering movements taking into account the limitation of the speed of movement on individual elements of following of trains. Railway transport of Ukraine 2014, 6: 30–36.

[7] Kolodiychuk VA. Efficiency of logistics of grain and products of its processing: monograph. Lviv: Ukrainian bestseller, 2015. 574 p.

[8] Bodnar B., Ochkasov A., Bobyr D. Improving Operation and Maintenance of Locomotives of Ukrainian Railways. Technology and Art 2016, 7: 109-114.

[3] Standards of technological design of grain reception enterprises and elevators: VNTP-05-88. Moscow: TsNIIPROMZENROPROEKT: Ministry of Bread Products of the USSR, 1988. 139 p.

[4] Rustamov RSh. Evaluation of the prospects for the development of grain logistics in Ukraine. Transportation systems and transportation technologies. Dnipropetrovsk, 2014, 8: 127–133. DOI : https://doi.org/10.15802/tstt2014/38101.

[11] Ndembe E. Hard Red Spring Wheat Marketing: Effects of Increased Shuttle Train Movements on Railroad Pricing in the Northern Plains. Journal of the Transportation Research Forum 2015, 54(2): 101–115.

[12] Vernyhora RV., Rustamov RSh. Analysis of the storage system of Ukrainian grain. Transportation systems and transportation technologies. Dnipro, 2017, 13: 10–18. DOI : 10.15802/tstt2017/110763.

ITMS'2018

The 12th International Scientific Conference Intelligent Technologies in Logistics and Mechatronics Systems (ITELMS'2018), 26–27 April 2018, Panevėžys, Lithuania

Remote Controlled Laboratory Station

Justyna Kras[a], Adam Bartnicki[a]

[a]*Military University of Technology, gen.W. Urbanowicza 2 str., 00-908 Warsaw, Poland*

Abstract

In this article a remotely controlled laboratory station is presented. Didactic problem related to the lack of laboratory re-sources is described. The usage of mechatronic station for online education is proposed (its construction and technological process) and the safety function for the station is suggested in order to ensure safety.

© 2018 J. Kras, A. Bartnicki
Peer-review under responsibility of the Kaunas University of Technology, Panevėžys Faculty of Technologies and Business

Keywords: laboratory resources, online education, safety function

---

1. Introduction

Development of driving systems in modern machines and devices causes rise in necessity of qualified engineers in exploitation and maintenance/operating those systems. This problem concerns both training of future engineering staff and developing their skills.

Discussions with representatives of educational centres indicate that due to lack of laboratory equipment in schools and universities there is a problem with high quality of education. Expensive training stations and money restriction on the development equipment laboratories cause that practical lessons are very difficult to implement. Furthermore, the virtual presentations are not enough to educate properly [5, 6].

This is more common that educational centres dynamically develop e-learning. They provide not only long-distance education, but they offer remote access to laboratory equipment which are part of so-called blended learning [1, 2, 3, 4].

This work presents innovative mechatronic laboratory of mobile robot's drives which was built in Institute of Machine Building at MUT the Mechatronic department supported by Ministry of National Defence [11].

© 2018 J. Kras, A. Bartnicki **B427H0017** . Ed*Learning*.it
Peer-review under responsibility of the Kaunas University of Technology, Panevėžys Faculty of Technologies and Business

2. Existing solutions for virtual laboratories for e-learning

The virtual laboratories existing in Institute of Technology and Operation – National Research Institute are technical and didactic laboratories. This virtual laboratory allows PLC programming in fields such a (Fig. 1):

- pressure regulation;
- liquid level regulation;
- technological transport;
- regulation of the hydraulic and pneumatic system;
- servo-motor control;
- photovoltaic cell control;
- controlling a smart home.

The laboratory is characterized by architecture of open, modular model in which the most important are measuring serves, managing server and computer applications integrated with didactic stations. This laboratory offers multi-select functions, remote access management, completing databases with research results and experiments, acquire and management knowledge up to teaching process management. The laboratory allows online lessons wherever and whenever (only condition is having computer or mobile device with net access).

The functioning of the station allows creating simulations of operation of control procedures on real object. Students offered remote programming of the PLC controller can observe on the screen practical implementation of their control algorithm entered into the system [7, 8].

Fig. 1. Laboratory station in Institute of Technology and Operation [7]

3. Automation laboratory for mobile robot drives in WAT

One of the biggest issue is elaboration of control systems for the system of mobile platforms. These systems are more often build with microprocessor systems and programmable controllers which communicate by CAN buses (Fig. 2).

Practical excercises are necessary in further desiging such control systems. Students have an opportunity programming three stations: the process of transport, assembly and storage of components which are made of different materials.

The stations present complete mechatronic transport and assembly system an consist of:

- magazine stations (Fig. 3);
- hydraulic press processing station (Fig. 4);
- storage station (Fig. 5).

Fig. 2. Structure of the modular Mechatronic System

The magazine station can work autonomously and consist of elements such a (Fig. 3):
- seperator magazine 1;
- separator magazine 2;
- conveyor belt long;
- testing unit;
- control panel;
- maintenance unit;
- programmable logic controller.

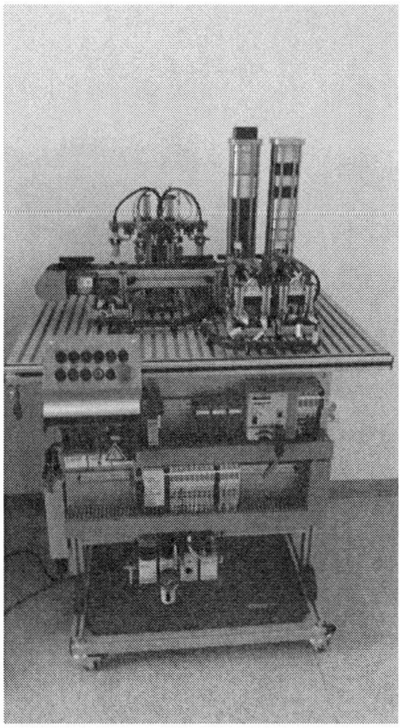

Fig. 3. Magazine station

Each of two separate magazines ensure for minimum 8 workpieces. The measurement of material properties takes places through 4 sensors: optical, capacitive, inductive and pneumatic cylinder. In this station, students program the process of downloading items from a magazine with specific properties.

The processing station can work autonomously and consist of elements such a (Fig. 4):

- pins unit;
- control panel;
- conveyor belt short;
- turning unit;
- handling device;
- portal;
- hydraulic press;
- maintenance unit;
- programmable logic controller.

Fig. 4. Hydraulic press – processing station

Selected elements are transported to processing station where hydraulic press assembles two elements together (Fig. 4), which were prepared with connecting elements previously.

Hydraulic press is central point in second station because it ensures maximum pressure during working in hydraulic system – 45 bars. Minimal flow rate of the working medium in hydraulic system is equal to 0,8 dm^3/min. The press is controlled by microprocessor controller and process take place behind the cover made of plastic which ensures safety and opportunity to observe the assembly process.

The storage magazine can work autonomously and consist of elements such a (Fig. 5):

- high bay racking;
- Cartesian robot;
- handling device;
- control panel;
- maintenance unit;
- programmable logic controller.

This station is equipped in high bay racking which can storage 25 elements. The Cartesian robot moves assembled products to their storage places. Pneumatic arm of handling device places it on demanded position in magazine.

Fig. 5. Storage station

Station is controlled by properly configured connection between computer with IndraWorks software and *L20* controller. At the beginning the user specify output and input signals. IndraWorks program interface is presented on Figure 6.

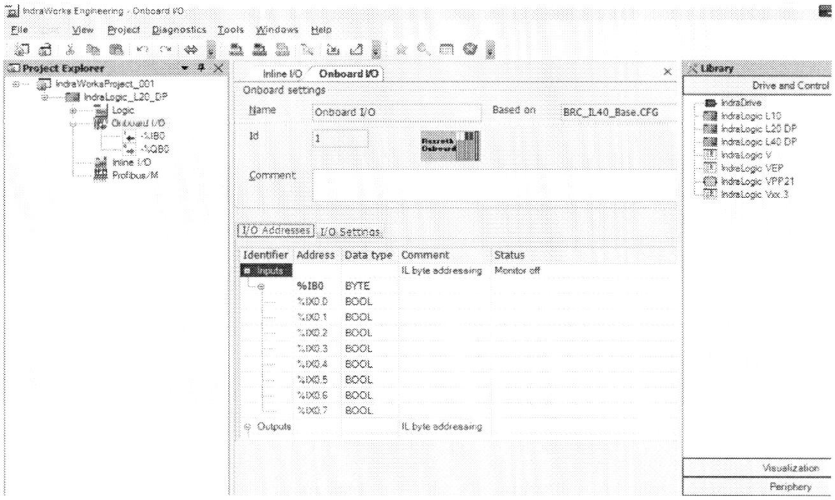

Fig. 6. Configuration window for I/O in IndraWorks

After defining the type of modules Inline I/O and Profibus/M, the user can declare relevant global variables. It allows to create a configuration program code:

- operation sensor which are responsible for choosing an element with specific properties;
- operation of the E-Stop safety button;
- operation of control diode;
- the proper work of the tray.

The structure of the control system presented in laboratory station is based on programmable microprocessor controllers. It allows easy implementation for the purposes of e-learning. Apart from hardware and software modernization it is necessary to provide safety for remote station in both technical and IT aspects. It is excellent solution because it protects against the implementation of incorrect control procedures and interference by not admitted people [9, 10].

4. Laboratory modification for the needs of remote education

At present in the institute of Mechanical Engineering works are being carried out to adapt laboratory architecture to the needs of remote control (Fig. 7), which allows starting the station, modification of work parameters and program code from anywhere and whenever.

The most important element of stations, which will be responsible for safety, will be safety controller. Its most important task will be controlling of created programs and blocking unwanted scenarios. This safety controller will be invisible to users who correctly wrote theirs program.

The use of the vision system will allow user to observe processes realized by integrated stations. Regarding data archived on the measurement server and analysing records from the vision system allow trainer to evaluate student's results.

Crucial aspect of modernized laboratory will be providing safety of processing and authorized access which will prevent interfere in the system by not admitted people, realized by data encryption.

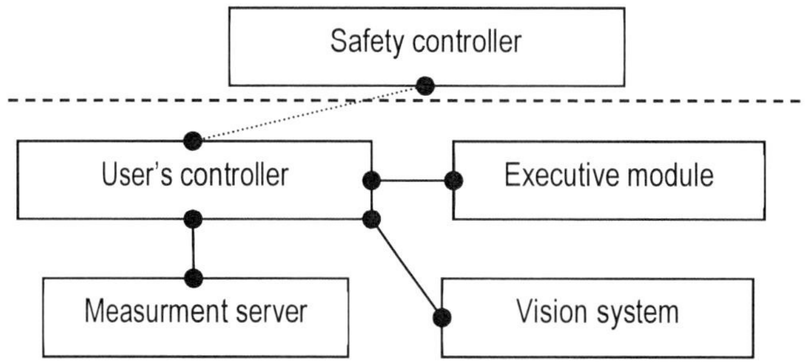

Fig. 7. Simplified architecture of a remote controlled station

The possibility of remote control and supervision over the correctly functioning of the laboratory will not eliminate the need for interference by personnel servicing the laboratory. It is related to dangerous element of the integrated station as hydraulic press, which generates high pressure forces.

To minimize this risk, it is proposed to install a safety system which uses light curtain C2C-SA03030A10000 deTec2 Core made by SICK company. This light curtain cooperates with Safety controller FX3-CPU000000 which is connected with additional modules I/O FX3-XTDI80002 and FX3-XTIO84002 (Fig. 8). In the event of accidental human interference or a threatening event, based on the signal from the light curtain, immediately the emergency procedure is started by the controller, which turns off the power of the hydraulic press.

The Figure 9 presents logic diagram program of safety system which uses light curtain. If the light curtain detects object (I1 will get "high" state) or E-stop button (I5) is turned on due to danger (it will get "high" state, too), function block "Reset" on the output signals Q3 will inform about danger by means of light impulses from lamp Q3 which means that the machine is not working. To turn the device back on, it is necessary to press button "Reset" I3 that will unlock block "Restart" and it will be signalled by the lamp Q4. The Restart function is used for additional safety.

After restarting the station by press button "Start" I4, device starts working again – inclusion "Motor contactor".

Sensor	Processing unit	Executive element
	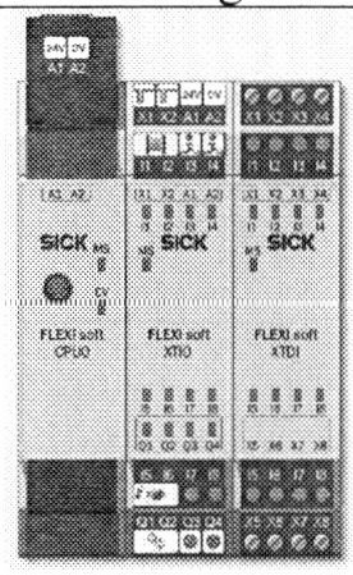	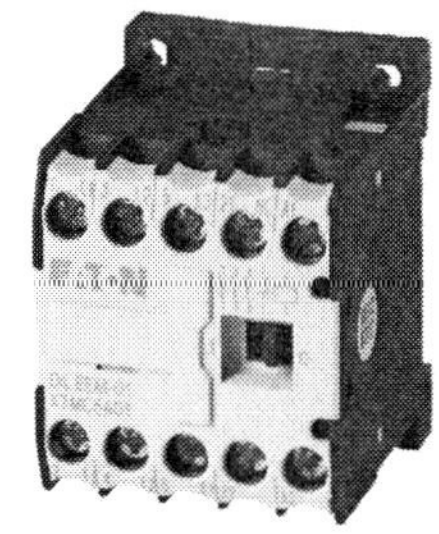
a) Light curtain	b) Safety controller	c) Contactor

Fig. 8. Elements of the laboratory station safety system: a) light curtain; b) safety controlled; c) contactor

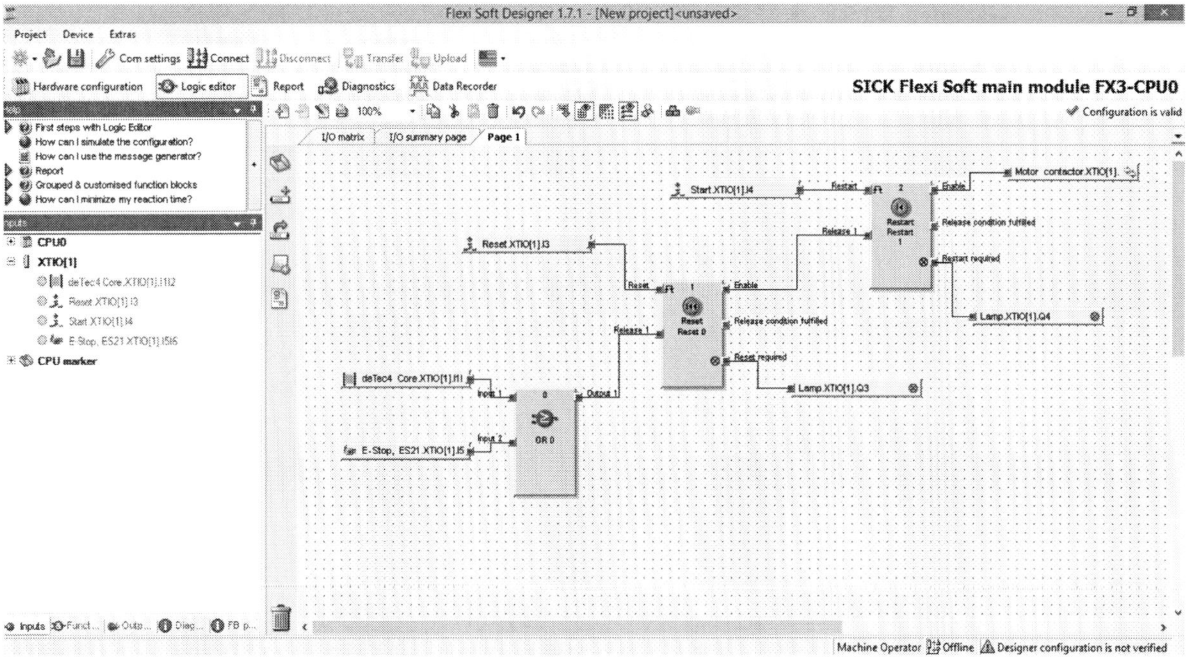

Fig. 9. Diagram of the logic of operation of the security system program using the light curtain

The Safety system used minimizes the risk of danger at the hydraulic press workplace.

Elements at the safety system have been purchased for founds as part of the activity of the Scientific Club. At the present works on assembly and commissioning light curtain, controller and executive elements of system on laboratory station are made. In the future, it is planned to design covers for manipulators with locking. Manipulators are another dangerous elements in the station.

Conclusions

Transfer of practical knowledge for students requires proper laboratory facilities. Experience allows to combine theory with practise.

Automation of laboratory for mobile robot drives enable training of future engineering staff.

Adaptation of the laboratory to the innovation method of education while the remote training will allow implementation of practical work regardless of the student's place of residence, requiring him access to the Internet.

The remote control will allow to use the station not only during opening hours.

Remote access system to station will facilitate service by administrator, who can interfere in the system without physical presence in the laboratory. The proposed safety system will allow safe service of the station by technical personnel.

This type of remotely programmable laboratory stations may be the basis for organizing a network of laboratories at different didactic centres.

References

[1] Agarwal N., Srivastava E. E-learning: New trend in Education and Training. International Journal of Advanced Research (2013), Volume 1, Issue 8, 797–810.

[2] Dąbrowski M. E learning w szkolnictwie wyższym. Studia BAS 3(35) 2013, pp. 203–212.

[3] Kowalik D., Wojutyński J., Siczek M. E-learningowe stanowiska dydaktyczne w nowoczesnym kształceniu zawodowym. Międzynarodowa Konferencja Naukowa – Rozwój zawodowy nauczycieli, 20–21 listopada 2014, Jedlnia-Letnisko. Edukacja Ustawiczna Dorosłych, 4/2014, pp. 199.

[4] Łabędzka J. Współdzielenie zasobów cyfrowych w edukacji – platformy informatyczne, Edukacja Ustawiczna Dorosłych, 3/2014, p. 132.

[5] Rekomendowane wyposażenie pracowni i warsztatów szkolnych. Access via internet: http://new.koweziu.edu.pl/wyposazenie-pracowni [Accessed: 2017-10-26].

[6] Standardy kompetencji zawodowych. Access via internet: https://www.mpips.gov.pl/praca/bazy-danych-mpips-dla-potrzeb-edukacji-i-pracy/rejestr-instytucji-szkoleniowych/ [Accessed: 2017-10-26].

[7] Wojutyński J., Dobrodziej J., Siczek M., Kaczyński J. Opracowanie modelowego stanowiska programowania systemów pomiarowo-sterujących w konwencji wirtualnego laboratorium. Prace ITeE-PIB, Radom 2011.

[8] Wojutyński J., Siczek *M.* Wirtualne laboratorium kształcenia zawodowego. Edukacja Ustawiczna Dorosłych, 2/2015, pp. 148–156.

[9] Norma PN-EN 61508-1 Bezpieczeństwo funkcjonalne elektrycznych/ elektronicznych/ programowalnych elektronicznych systemów związanych z bezpieczeństwem.

[10] Norma PN-EN 61508-2 Bezpieczeństwo funkcjonalne elektrycznych/ elektronicznych/ programowalnych elektronicznych systemów związanych z bezpieczeństwem.

[11] Rozporządzenia Ministra Edukacji Narodowej z dnia 7 lutego 2012 w sprawie podstawy programowej kształcenia.

ITMS'2018

ktu 1922 panevėžys faculty of technologies and business

The 12th International Scientific Conference Intelligent Technologies in Logistics and Mechatronics Systems (ITELMS'2018), 26–27 April 2018, Panevėžys, Lithuania

Identification Research of a Hydrostatic Drive System for Multi Degree of Freedom Manipulator

Piotr Krogul[a], Marian J. Łopatka[a], Rafał Typiak[a*]

[a]*Military University of Technology, gen. W. Urbanowicza street 2, Warsaw 00-908, Poland*

Abstract

The increase in functionality of unmanned land platforms is accompanied by the development of new manipulation equipment. In many cases, this development complicates their cinematic structure, increasing their degrees of freedom (DOF) to 6 and beyond. Based on extensive research carried out at the Institute of Machine Building of the Military University of Technology, using such hydraulically powered structures creates a need to operate two different actuators simultaneously. In such scenarios, the total volumetric flow of the hydraulic fluid within the power system reaches op to several liters per minute, which allows for use of small diameter hydraulic hoses. Based on the above-mentioned assumptions, a new power system has been developed for these types of manipulators, which uses a spool valve bank and thermoplastic hydraulic hoses. Since these types of manipulators are dedicated to handling high loads, there's a problem of vibrations being transferred to the work tool from the mechanical structure of the unit. Because of this, the following paper focuses on assessing the possibilities of the newly developed power system regarding limiting the vibrations of the work tool during precise control.

© 2018 P. Krogul, M. J. Łopatka, R. Typiak
Peer-review under responsibility of the Kaunas University of Technology, Panevėžys Faculty of Technologies and Business

Keywords: manipulator, control, hydraulics

1. Introduction

The manipulating equipment is being used in a wide range of industrial applications. These units used are being sold as both stationary and mobile solutions. The stationary solutions are mostly used at the production lines, e. g for

* Corresponding author. Tel.: +48 261 83 71 07
E-mail address: piotr.krogul@wat.edu.pl

© 2018 P. Krogul, M. J. Łopatka, R. Typiak

Peer-review under responsibility of the Kaunas University of Technology, Panevėžys Faculty of Technologies and Business

handling welding equipment or during assembly. On the other hand, the mobile versions are widely used as work machine's and robot's equipment, e. g for picking up and handling complex mechanisms, handling dangerous materials, inspection of hard to reach areas or handling objects specific to urban environments [1, 2]. Control of stationary manipulators is mostly done using automated processes and is being conducted in a constant (regarding movement trajectory) environments. On the other hand, mobile solutions are mostly remotely controlled, sometimes using teleoperation. They have a dynamically changing and diverse environmental working conditions. Further study focuses on these manipulators as they pose a more complex control challenge. The analyzed manipulators are expected to have lifting capabilities of several dozens of kilograms, a working range of several meters and a functionality like that of a human being [3, 4]. These requirements force the developed solutions to have complex kinematic structures, a high number of DOFs and a hydrostatic drive system, which is basically the most commonly used type of power generation for these manipulators. This is because of a wide range of benefits from using these solutions [5, 6]. The most common solution for a manipulator's hydrostatic drive system is a multi-section spool valve bank, typical double shielded (e. g 2SN) elastic hydraulic hoses and hydraulic actuators. Many of these systems are single circuit solutions, that's why there are other systems used alongside them (e. g Load Sensing) which allow for simultaneous operation of several receivers. These applications are being characterized by high volumetric flows, reaching several dozen liters per minute [7]. Conducted analysis [3] show that in order to move a work tool of a multi-axis manipulator in an external reference coordinate system its crucial to provide the ability to move two actuators simultaneously at the minimum. Knowing this and keeping in mind that the required lifting capabilities of these manipulators are of several dozens of kilograms requires limited flow values. This means that it's possible to not only limit the number of power lines but also their diameters and decrease their bending radius. A good solution in such situations would be a usage of thermoplastic hydraulic hoses, which are characterized by similar operating pressures as the steel braided ones but with a lower weight and bending radius. A major aspect of precise controlling the discussed manipulators is maintaining a stabile trajectory of the work tool, which is accomplished by limiting its vibrations during different phases of the work process. It's especially apparent during the startup and braking phases, where it's possible to minimize the with an appropriate conditioning of the control signals. The analysis carried out has shown however that the literature concerning the matter is limited, especially regarding controlling hydraulically driven manipulators with two actuators working simultaneously. That is why a series of test have been conducted to gain knowledge on the mentioned subject.

2. Test object

To carry out necessary research, a special test object has been developed, which is based on a 3 DOF hydrostatically drive manipulator (Fig. 1). It consists of a boom's actuator (1), boom (2), arm's actuator (3), arm (4), work tool's rotation actuator (5), work tool (6) and load (7). The manipulator has been equipped with a hydrostatic power system consisting of a Hawe 3 section spool valve bank (Fig. 3, *c*), Hansa-Flex NY 104 DN05 thermoplastic hydraulic hoses for connecting the spool valve bank with the actuators and the spool valve bank with a hydraulic power generator with a constant flow of 10 dm^3/min (Fig. 3, *a*). Additionally, the spool valves were equipped with asymmetrical spools which allowed for a nominal flow of 10 dm^3/min in one direction and 6 dm^3/min in the opposite direction. The power system has been equipped with a data acquisition system consisting of hydraulic pressure sensors connected to the boom's and arm's actuators (Fig. 3, *f, g*) and displacement sensors (Fig. 3, *d, e*) for measuring extension length of the boom's and arm's actuators. Data has been registered using a Q.birxx data registering system from Gantner Instruments (Fig. 3, *b*).

The manipulator's control system has been based on Danfoss microcontroller with a control application developed using PLUS+1 GUIDE software. It's possible to have independent control of every spool valve in the bank. The list of control parameters includes (Fig. 2):

- control signal's amplitude – A_i;
- signal's total duration time – t_c;
- ramp up time – t_r;
- ramp down time – t_h.

a)

b)

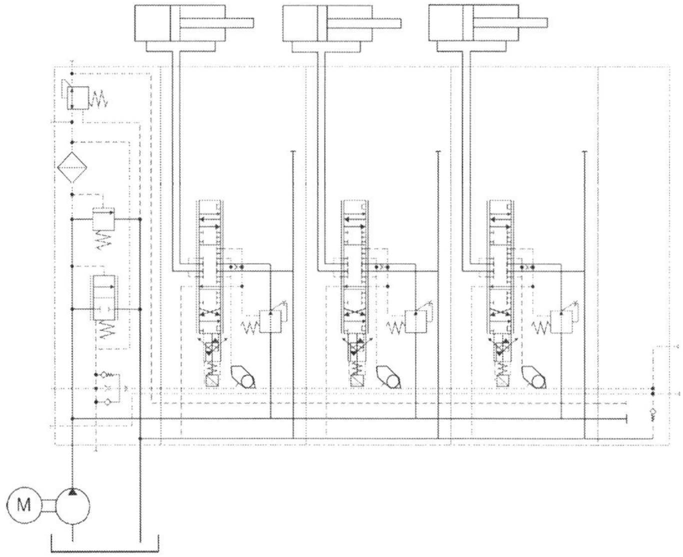

Fig. 1. a) Tested manipulator; b) a diagram of the propulsion system for the tested manipulator

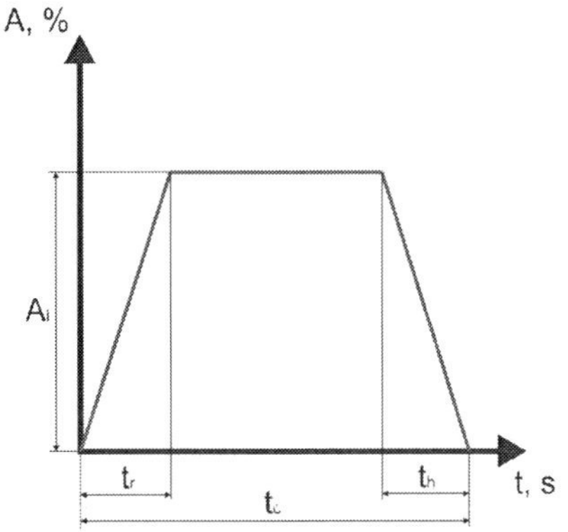

Fig. 2. A diagram of the parameters defining the control signal's ramp conditioning

a)

b)

c)

d)

e)

f)

g)

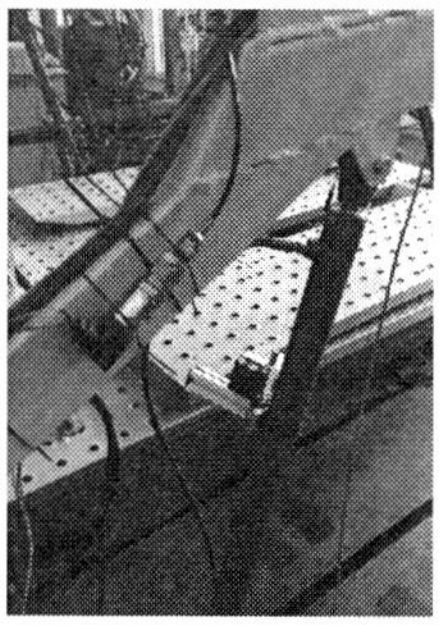

Fig. 3. a) UH.9.20 Pro-Hydros' hydraulic generator [8]; b) Gantner technologies' data acquisition system; c) Hawe PSL hydraulic spool valve bank [9]; d) Kubler's displacement sensor for the boom actuator[11]; e) Kubler's displacement sensor for the arm' actuator [11]; f) Kobold pressure sensor for the boom actuator [12]; g) Kobold pressure sensor for the arm actuator [12]

3. Research methodology

Conducted tests assumed 50 kg and 80 kg loads to be attached to the work tool. The premise behind this was to enhance negative phenomena which may occur during the work process, especially during the startup and braking phases. Conducted test have been carried out based on the values presented in Table 1.

Table 1. List of the tested parameter configurations used during testing

Pos.	Signal's total duration time t_c, s	Ramp up time t_r, s	Ramp down time t_h, s	Signal's amplitude A	Load's weight, kg	Retract(Re)/ Extend(Ex) Boom's actuator	Arm's actuator
1	5	0	0	$0.5Q_n$	50	Ex	Re
2	5	0	0	$0.5Q_n$	50	Re	Ex
3	5	0	0	$0.5Q_n$	50	Ex	Ex
4	5	0	0	$0.5Q_n$	50	Re	Re
5	5	0	0	$0.5Q_n$	80	Ex	Re
6	5	0	0	$0.5Q_n$	80	Re	Ex
7	5	0	0.5	$0.5Q_n$	80	Ex	Re
8	5	0	1	$0.5Q_n$	80	Ex	Re
9	5	0	0.5	$0.5Q_n$	80	Re	Ex
10	5	0	1	$0.5Q_n$	80	Re	Ex
11	5	0	1.5	$0.5Q_n$	80	Re	Ex
12	5	0	2	$0.5Q_n$	80	Re	Ex

The presented research plan acknowledges different work configurations which are the effect of different ramp down times – T_h. This has allowed to thoroughly test the power system's work parameters. The system evaluation has been carried out based on damping ratio and the maximum displacement of the actuator during vibrations values for the boom's actuator, as it was the source of the most noticeable vibrations. The signal's amplitude has been set to half of the maximum flow value of the hydraulic pump Q_n, which corresponds to the speed of the work tool at 0.1 m/s. The damping ratio has been calculated based on the expression presented below (1) [13, 14].

$$\xi = \frac{\lambda}{\sqrt{4\pi^2+\lambda^2}}; \tag{1}$$

where λ is a logarithmic damping decrement calculated using the following expression (2).

$$\lambda = \ln\frac{A_1}{A_2}; \tag{2}$$

where A1, A2 – are the amplitude values of following pressure pulses of the boom's actuator (Fig. 7).

4. Test results

First tests were conducted using a rectangle shaped control signal and saw bidirectional, up-down movement of the tip of the manipulator with a 50 kg load attached. The test saw the actuators move in the same and opposite directions. Based on the hydraulic pressure values in the actuators it has been determined that the highest oscillation amplitudes were present during the process of lowering the load while in the braking phase when the actuators were operating in the same direction (Fig. 4). While analyzing the actuator movement in the opposite directions (Fig. 5), during the braking phase, the pressure oscillations were self-quenching and had minimal impact on the movement of the tip of the manipulator. Because of the observed behavior, this movement configuration has not been further analyzed as it wasn't negatively impacting the movement of the manipulator's tip.

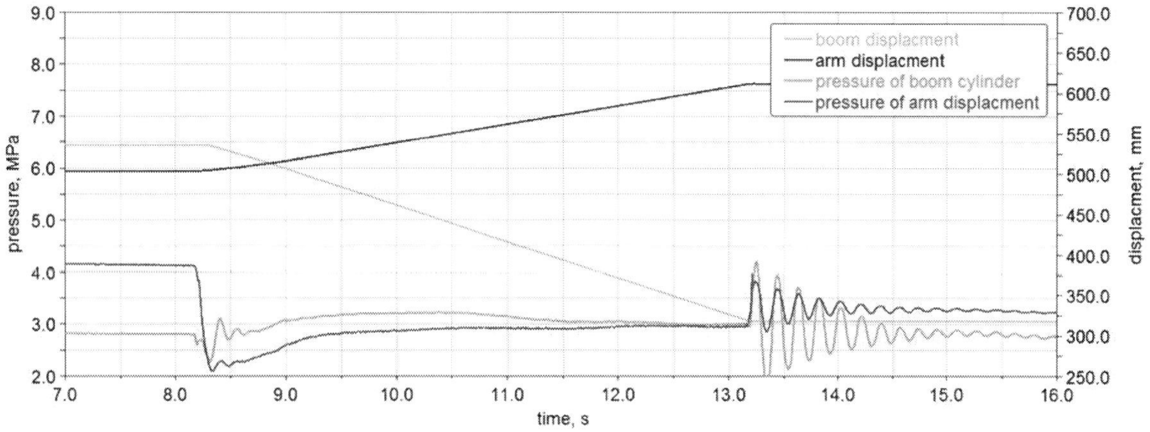

Fig. 4. Work parameters chart with a load of 50 kg during the retraction of the boom's actuator and the extension of the arm's actuator

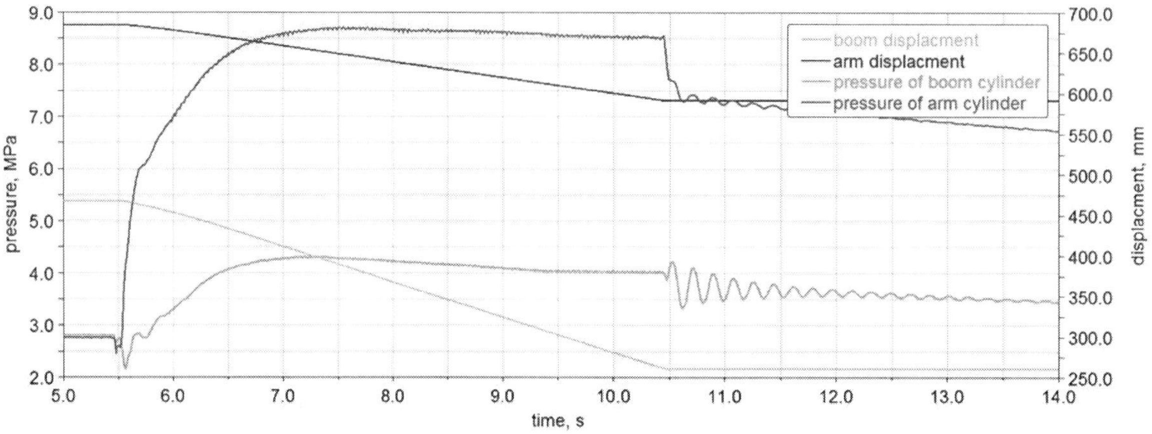

Fig. 5. Work parameters chart with a load of 50 kg during the retraction of both the boom's and the arm's actuators

Conducted tests have shown that the major impact on the value of the vibration amplitude of the manipulator's work have the vibrations of the boom's actuator. However, said vibrations do not correspond with the pressure value pulsations inside the actuator (the pressure pulsations are present even after the actuator has stopped vibrating). Because of this, further studies were carried out based on the signals from the displacement sensors of the boom's actuator. Additionally, the load attached to the tip of the manipulator has been increased from 50 kg to 80 kg, to highlight occurring vibrations. Figure 6 and 7 show the displacement values of the boom's actuator during the final stage of the braking phase for both an instant halt of the control command and a linear decrease of the command value with a ramp lasting 1.5 s.

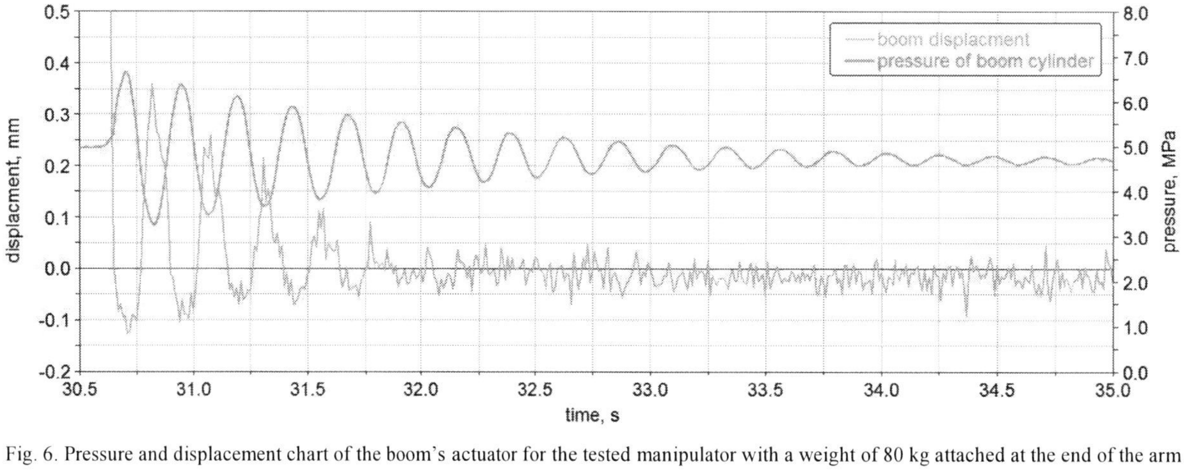

Fig. 6. Pressure and displacement chart of the boom's actuator for the tested manipulator with a weight of 80 kg attached at the end of the arm during the last stage of lowering the tip and without any ramp conditioning of the command signal

It's noticeable that a steady decrease of the command value significantly limits the vibration amplitude and its duration. Table 2 shows test results for different ramp times during the braking phase. Based on acquired pressure value data, a damping ratio for the braking phase of the boom's actuator has been calculated. It's noticeable that its value increases with the increase of ramp's length. On the basis of pressure values alongside the displacement data from the sensors, a natural period of a system equipped with thermoplastic hydraulic hoses has been determined at 0.2 s. It's worth noting that during the process of lifting the work tool, the braking phase's ramp time needed to minimize vibrations is shorter than when lowering the work tool.

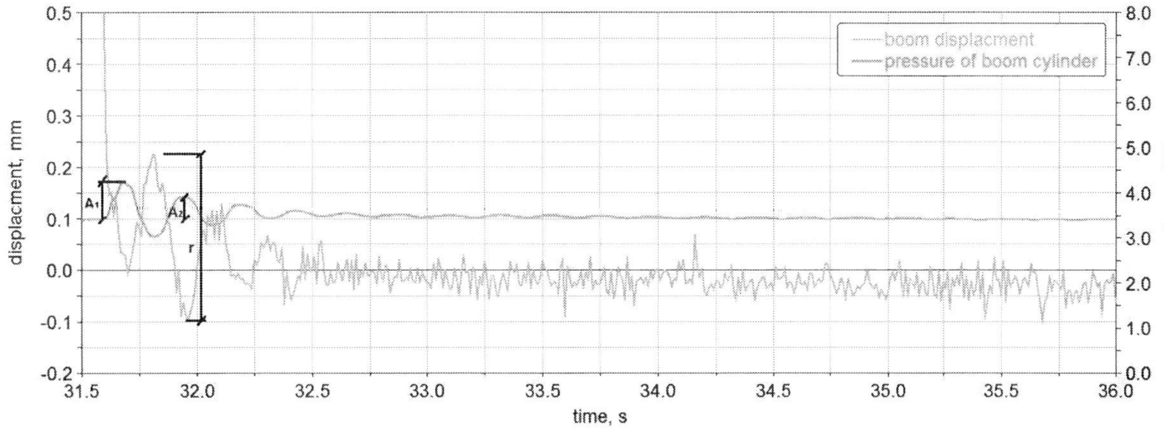

Fig. 7. Pressure and displacement chart of the boom's actuator for the tested manipulator with a weight of 80 kg attached at the end of the arm during the last stage of lowering the tip and with a ramp conditioning with a decrease time of 1.5 s

Table 2. Test results

t_h, s	A_1, MPa	A_2, MPa	λ	ξ	Vibration displacement r, mm	Retract(Re)/ Extend(Ex) Boom's actuator	Arm's actuator
0	1.29	0.98	0.267	0.042	0.55	Ex	Re
0.5	0.47	0.28	0.521	0.083	0.35	Ex	Re
0	1.97	1.68	0.154	0.024	0.56	Re	Ex
0.5	1.22	0.85	0.356	0.056	0.52	Re	Ex
1	0.84	0.52	0.472	0.075	0.36	Re	Ex
1.5	0.78	0.48	0.489	0.078	0.32	Re	Ex

Conclusions

Based on conducted research, a solution for minimizing the vibration amplitude of a work tool mounted on a hydrostatically driven, two-actuator-configuration manipulator, has been presented. Highest amplitude values have been registered during the process of stopping movement in downward direction. During other movement phases, the pressure pulsations were self-quenching, thus negating the need of using extended slow-down and breaking periods. Additionally, the startup phase was lacking any of the boom's and arm's actuator vibrations present during the braking phase. To minimize the vibrations of objects placed inside the work tool, during the braking phase, it was necessary to decrease the command signal linearly to zero during a period of at least 2 seconds. On the other had some use scenarios won't allow for such signal conditioning. That is why a two-stage braking phase has been proposed, with each phase having a different signal conditioning algorithm. An example of such conditioning would be a first phase of the braking process having a steep decrease characteristic, as was the case with the t_h=0.1 second, while the second one having a flatter shaped characteristic, as was the case with the t_h=2 s.

Acknowledgements

The work presented in this article has been supported by the Polish National Center for Research and Development – Grant No. DOBR-BIO4/083/13431/2013.

References

[1] Łopatka M. J., Muszyński T., Rubiec A. *Simulation identification of Fire Rescue Robot suspension loads.* 18th International Conference on Methods and Models in Automation and Robotics MMAR 2013. *IEEE Conference Publications.* Międzyzdroje 2013, pp. 408–413.

[2] Dabrowska A., Przybysz M., Rubiec A., Spadlo K. *Hydropneumatic suspension efficiency in terms of the teleoperated unmanned ground vehicle tests,* Intelligent Technologies In Logistics And Mechatronics Systems – ITELMS'2015, pp. 110–116, Panevezys, Lithuania 2015.
[3] Cieślik K., Konopka S. *Concept and preliminary research of anthropomorphic manipulator with hydrostatic drive system for mobile robot,* 21st International Conference on Methods and Models in Automation and Robotics, pp. 1154–1159, Miedzyzdroje 2016.
[4] Fujita S., Sudoh D., Baba K. *Rescue robot "T-52 Enryu"*, Proceedings o 23rd International Symposium on Automation and Robotics in Construction, pp. 237–240, 3–5 październik, Tokio 2006.
[5] Pizoń A. *Elektrohydrauliczne analogowe i cyfrowe układy automatyki.* Wydawnictwo Naukowo-Techniczne. Warszawa 1995.
[6] Galal Rabie M. *Fluid power engineering.* Wydawnictwo McGraw-Hill. 2009.
[7] Research project Nr. 0012/R/T00/2008/06: Bezzałogowy pojazd do wykonywania zadań w strefach zagrożenia. Wojskowa Akademia Techniczna. Warszawa 2011.
[8] Technical data: Opis techniczny zasilacza hydraulicznego UH.9.20. Pro-Hydro. 2017.
[9] Technical data: *Proportional directional spool valve type PSL and PSV according to the Load-Sensing principle size 2,* Hawe hydraulic. Sierpień 2011.
[11] Technical data of Kubler, *Linear measuring technology.*
[12] Technical data of Kobold. Przetwornik ciśnienia z celą ceramiczną.
[13] Gutowski R., Swietlicki W. A. *Dynamika i drgania układów mechanicznych.* PWN. Warszawa 1986.
[14] Szulej J. *Wyznaczenie ekwiwalentnego wiskotycznego tłumienia drgań w konstrukcjach wielomateriałowych.* Rozprawa doktorska. Wydawnictwo Politechniki Lubelskiej. Lublin 2010.

The 12th International Scientific Conference Intelligent Technologies in Logistics and Mechatronics Systems (ITELMS'2018), 26–27 April 2018, Panevėžys, Lithuania

Effect of Micrometakaolin Addition on Mechanical Properties of Concrete

Marcin Małek[a*], Mateusz Jackowski[a], Wojciech Życiński[a], Waldemar Łasica[a], Marcin Wachowski[b]

[a] *Military University of Technology in Warsaw, Faculty of Civil Engineering and Geodesy, gen. Witolda Urbanowicza 2 Street, 01-476 Warsaw, Poland*

[b] *Military University of Technology in Warsaw, Faculty of Mechanical Engineering, gen. Witolda Urbanowicza 2 Street, 01-476 Warsaw, Poland*

Abstract

Micrometakaolin is a pozzolan concrete additive. Generally, it is used as a partial replacement for Portland cement. To modify concrete micrometakaolin as a filler in 5, 10 and 15 wt.% were used. The reference sample was prepared using three parts of aggregates: 0.125–0.250; 0.250–0.500 and 0.500–1.000 mm. In this article, the results of characteristic properties such as chemical composition SEM micrographs, mechanical bending test and compressive strength are presented. After 1, 7, 14 and 28 days of curing mechanical properties were tested. Obtained results shows increasing effect of micrometakaolin modification as a filler. Research was proven that 10 wt.% of micrometakaolin exhibit the most increase effect on final mechanical properties of studied concrete and are very perspective for future application in modern building technologies.

© 2018 M. Małek, M. Jackowski, W. Życiński, W. Łasica, M. Wachowski
Peer-review under responsibility of the Kaunas University of Technology, Panevėžys Faculty of Technologies and Business

Keywords: micrometakaolin, concrete mixtures, modern building technologies, mechanical properties, chemical modification

* Corresponding author. Tel.: +48 261 83 96 55
E-mail address: marcin.malek@wat.edu.pl

© 2018 M. Małek, M. Jackowski, W. Życiński, W. Łasica, M. Wachowski B427H0019 . EdLearning.it
Peer-review under responsibility of the Kaunas University of Technology, Panevėžys Faculty of Technologies and Business

1. Introduction

One of the basic materials used in modern building technologies is concrete, therefore its strength and durability play an important role. Based on the available additives, it is possible to achieve the required concrete properties [1–2].

The construction market offers a wide range of concrete admixtures. Among natural pozzolanic materials such as zeolite and fly ash is micrometakaolin. There is very little knowledge of its use in concrete production and its influence on the concrete mixture. However, based on specialist literature, it can be concluded that micrometakaolin should give similar results to metakaolinite, which has larger grains than first one. There are proofs that smaller grains can give more compacted structure what results in higher mechanical strength and etc. This statement allows to hypothesize that micrometakaolin gives many benefits in [3–6]:

- strength increase;
- reduction of water absorption;
- increased frost resistance;
- increased thermal resistance;
- porosity reduction;
- increased chemical resistance;
- reduction of concrete efflorescence forming;
- increased salt resistance;
- reduced alkali-silica reaction.

Micrometakaolin is a natural pozzolan produced by heating kaolin-containing clays to temperature about 600–900 °C which it recrystallizes, rendering it to mullite. Depending on the clays, three different chemical composition can be found: mullite ($Al_6Si_2O_{13}$), spinel ($MgAl_2O_4$) and amorphous silica [7–8]. Metakaolinite can be activated by sodium/potassium silicate or hydroxide. In calcium hydroxide environment, which originates during the hydration of Portland cement process, the micrometakaolin undergoes pozzolanic reaction and calcium aluminum hydrates and aluminum silicon hydrates as the main products are created. Thus, obtained hardened product contains lower amount of portlandite and the final material shows higher compressive strength as well as durability in different environments [9–10].

This study includes the results of own experimental research of that topic. In this work investigation of zeolite applied to the concrete mixtures were characterized.

2. Materials and methodology

The subject of this research was micrometakaolin. To investigation of this substance SEM JEOLJSM-6610 was used with SE mode and 5 kV voltage. In addition to characterize of micrometakaolin chemical composition was done using EDS technique. Laser diffraction method was used to obtain grain size of filler. Furthermore, Portland cement, quartz sand with aggregates of 0.125–0.250; 0.250–0.500 and 0.500–1.000 mm, water and deflocculant based on polycarboxylates were added to the concrete mixtures. For concrete mixtures fabrication mechanical mixer and aluminum reactor were used. Apart from research of micrometakaolin powder, concrete mixtures were prepared and tested for technological properties characterization.

After 1, 7, 14 and 28 days seasoning process from the day of concreting, the strength parameters of samples were studied. Dividing the value of the maximum sample load across the cross-sectional area of the sample, the bending and compression strength of each sample was determined. Based on the test results, the average bending and compressive strengths were calculated for samples from all concrete series.

3. Results and discussion

In this study to preparation of concrete mixtures micrometakaolin was used. The maximum particle size of studied micrometakaolin was 4.2 μm as 67.3 % of amount is present in Figure 1.

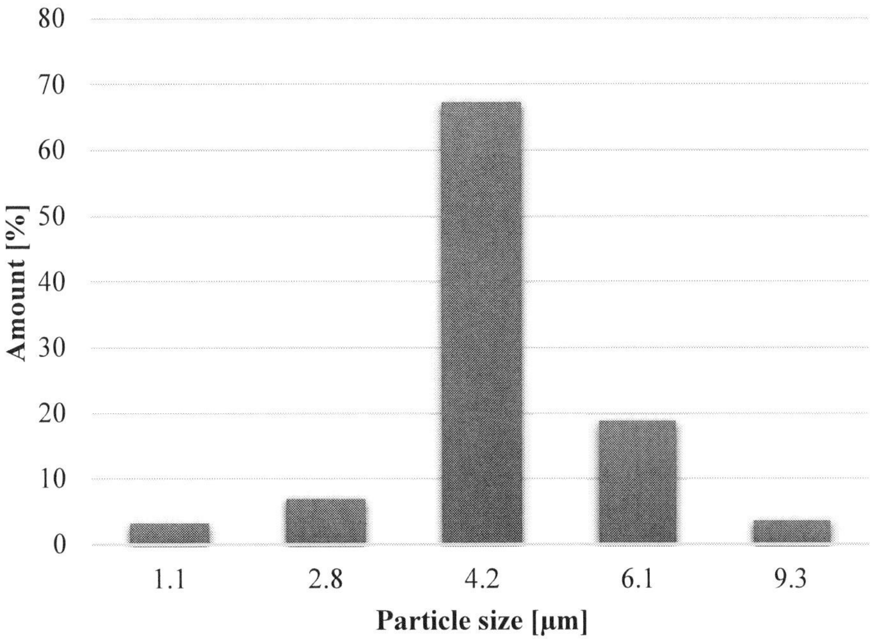

Fig. 1. Particle size distribution of micrometakaolin powder

Figure 2 shows the original scanning electron microscopy (SEM) images of tested powder. Researched images shown that micrometakaolin powder exhibits typically sharp-edged structure and create large agglomerates.

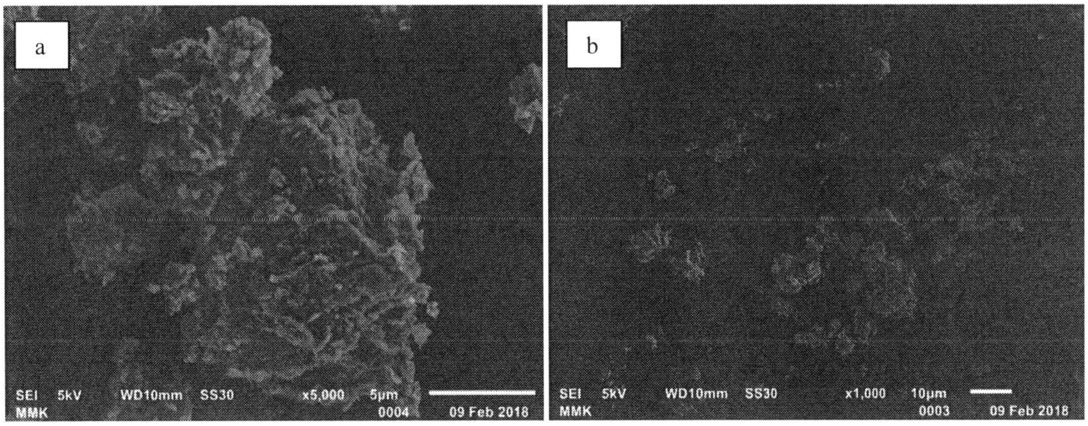

Fig. 2. SEM images of tested micro metakaolinite powder:(a) magnification 5000x; (b) magnification 1000x

Figure 3 presents the chemical composition of tested micrometakaolin powder, which has shown that micrometakaolin based most of silicon and aluminum oxides. Rest of compounds from of the natural environment and depend of kaolinite base clay.

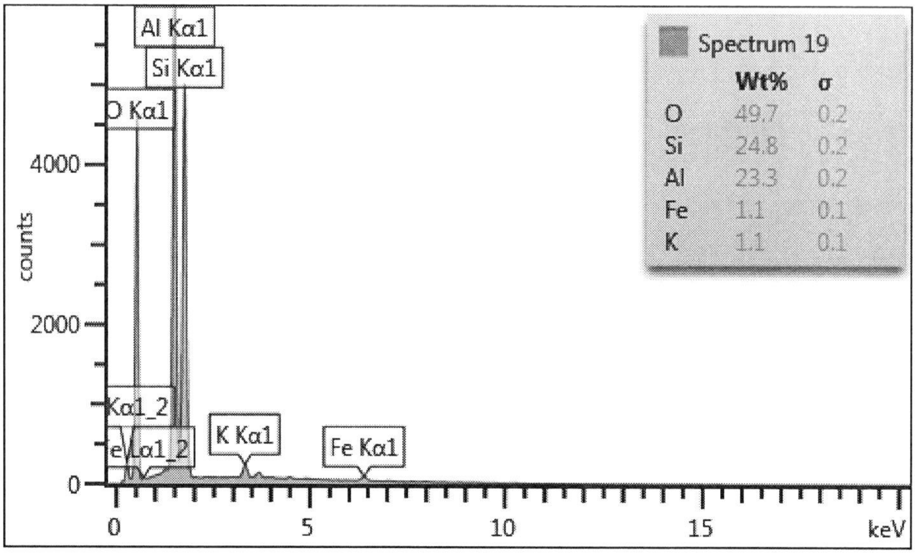

Fig. 3. EDS distribution of tested micro metakaolinite powder

Measurements of the mechanical strength of the concrete samples tested are shown in Figures 4–5, considering the curing time. Last day of experiment exhibit the highest values of mechanical properties, which is dependent on type of cement addition and time of curing. Comparing with reference sample, samples with micrometakaolin addition revealed higher values of bending force and compressive strength. All studied amounts (5 wt.%, 10 wt.% and 15 wt.%) of micrometakaolin addition has shown increase effect on mechanical properties. Sample with 10 wt.% of micrometakaolin addition revealed the highest values after bending and compressive tests.

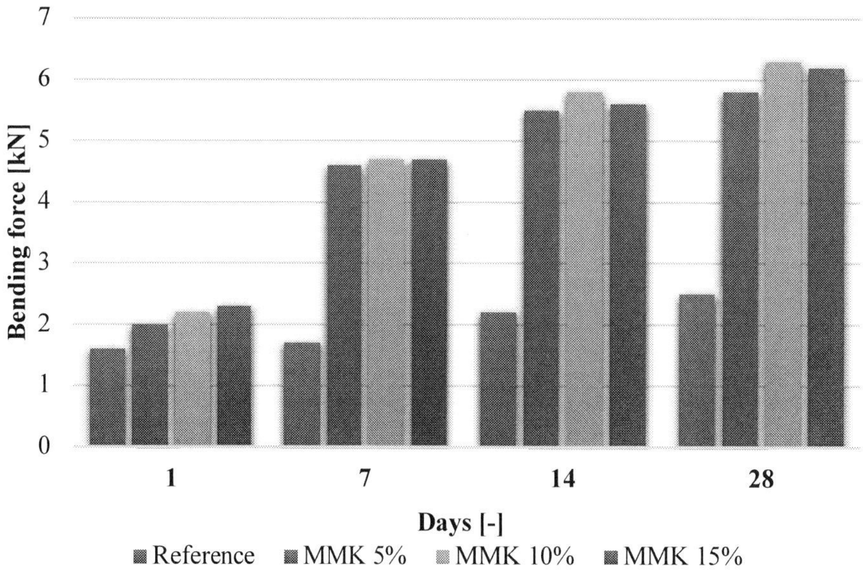

Fig. 4. Bending force distribution of examined concrete mixtures

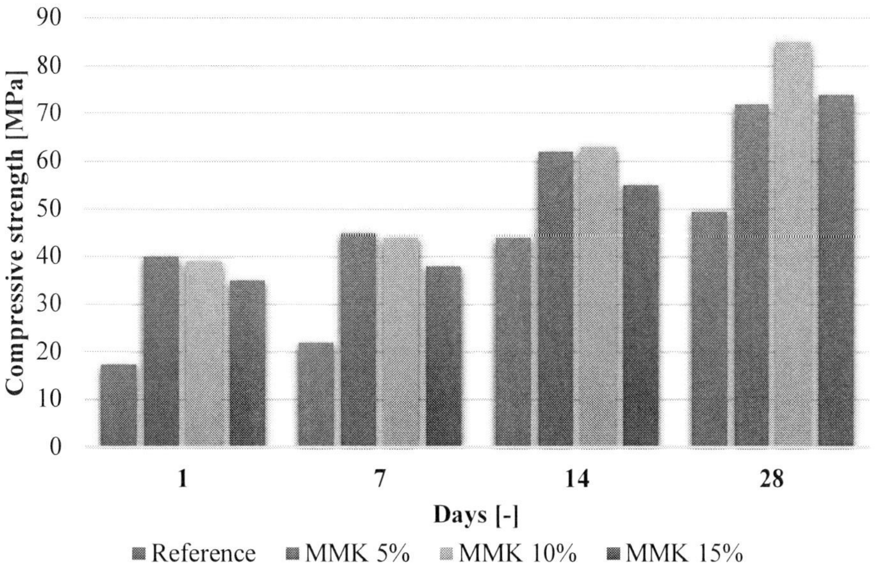

Fig. 5. Compressive strength distribution of examined concrete mixtures

Conclusions

In connection with the growing environmental standards, a new composition of the concrete mix has been developed. By substituting part of Portland cement with metakaolinite, which is obtained by baking naturally occurring kaolinite clays, the principle of sustainable development has been implemented, thus limiting the harmful effect of cermet production on the natural environment (the cement production process is a high energy process). The aim of the conducted research was to determine the influence of applied micrometakaolin to concrete mixtures. Test results show that all properties have been increased by adding micrometakaolin to mixture. According to the works [1–10], many authors focus on properties of concrete mixtures. However, it is difficult to compare test results without knowledge of concrete recipe,

The presented tests and test results are one of the first in the world in the field of using micrometakaolin as an admixture for concrete.

References

[1] Badogiannis E., Tsivilis S. Exploitation of poor Greek kaolins: Durability of metakaolin concrete. *Cement and Concrete Composites 2009*; 31: 128–133.

[2] García R., de la Villa R. V., Rodríguez O., Frías M. Study of hydrated phases present in calcined paper sludge (metakaolinite)/saturated CaO dissolution system cured at 40°C and 28 days of reaction. *Materials Science and Engineering: A 2010*; 527: 3936–3941.

[3] Heller-Kallai L., Lapides I. Reactions of kaolinites and metakaolinites with NaOH—comparison of different samples (Part 1). *Applied Clay Science* 2007; 35: 99–107.

[4] Konkol J., Prokopski G. Fracture toughness and fracture surfaces morphology of metakaolinite-modified concrete. *Construction and Building Materials 2016*; 123: 638–648.

[5] Lotfy A., Karahan O., Ozbay E, Hossain K. M. A, Lachemi M. Effect of kaolin waste content on the properties of normal-weight concretes. *Construction and Building Materials 2015*; 83: 102–107.

[6] Matějka V., Matějková P., Kovář P., Vlček J., Přikryl J., Červenka P., Lacný Z., Kukutschová J. Metakaolinite/TiO_2 composite: Photoactive admixture for building materials based on Portland cement binder. *Construction and Building Materials 2012*; 35: 38–44.

[7] Rashad A. M. Metakaolin as cementitious material: History, scours, production and composition – A comprehensive overview. *Construction and Building Materials 2013*; 41: 303–318.

[8] Shafiq N., Nuruddin MF., Khan SU., Ayub T. Calcined kaolin as cement replacing material and its use in high strength concrete. *Construction and Building Materials 2015*; 81: 313–323.

[9] Teklay A., Bøjer M., Adelsward A., Yin C., Rosendahl L. Simulation of Flash Dehydroxylation of Clay Particle Using gPROMS: A Move Towards Green Concrete. *Energy Procedia 2014*; 61: 556–559.

[10] Wu HC., Sun P. New building materials from fly ash-based lightweight inorganic polymer. *Construction and Building Materials 2014*; 21: 211–217.

ITMS'2018

The 12th International Scientific Conference Intelligent Technologies in Logistics and Mechatronics Systems (ITELMS'2018), 26–27 April 2018, Panevėžys, Lithuania

Characterization on New Filler Addition for Concrete Fabrication Based on Portland cement

Marcin Małek[a*], Mateusz Jackowski[a], Wojciech Życiński[a], Waldemar Łasica[a], Marcin Wachowski[b]

[a] *Military University of Technology in Warsaw, Faculty of Civil Engineering and Geodesy, gen. Witolda Urbanowicza 2 Street, 01-476 Warsaw, Poland*

[b] *Military University of Technology in Warsaw, Faculty of Mechanical Engineering, gen. Witolda Urbanowicza 2 Street, 01-476 Warsaw, Poland*

Abstract

In this work obtained results of chemical modification of concrete based of portland cement was summarized. Study was focused on new fillers addition such as: 10 wt.% of micrometakaolin and 10 wt.% of zeolite. Additionally, 10 wt.% of steel fibers of recycling origin of tires were also used. The main goal of this research was to identify effect on new compound addition on final properties of concrete. Microstructure of fillers and steel fiber was characterized by SEM and LM technique. Bending force, compression strength, chemical composition, grain size and thickness of the fibers have been estimated. Final samples of concrete have been characterized after 1, 7, 14, 28 days of curing. Obtained results shows that application of 10 wt.% of steel fibers obtain the highest compression strength and bending force.

© 2018 M. Małek, M. Jackowski, W. Życiński, W. Łasica, M. Wachowski
Peer-review under responsibility of the Kaunas University of Technology, Panevėžys Faculty of Technologies and Business

Keywords: zeolite, micrometakaolin, recycled fibers, concrete, mechanical properties

* Corresponding author. Tel.: +48 261 83 96 55
E-mail address: marcin.malek@wat.edu.pl

© 2018 M. Małek, M. Jackowski, W. Życiński, W. Łasica, M. Wachowski
Peer-review under responsibility of the Kaunas University of Technology, Panevėžys Faculty of Technologies and Business

B427H0021 . EdLearning.it

1. Introduction

The concrete modification may contain organic and inorganic substances. The effectiveness of chemical admixtures should be considered in three aspects [1–4]:

- technical – as the quotient of the required basic effect, i.e. properties change of the concrete mixtures and the minimum amount of admixture necessary to fabricate concrete;
- technological – assessing the ease and safety process of admixture and the sensitivity of its effects to changes in technological conditions;
- economic – as the unit cost of the basic effect of the admixture modification.

In the practice the choice of compound is mainly influenced by economic and technological aspects, the fulfillment of which means obtaining the lowest possible cost of the intended effect of using the new compound. Therefore, to achieve the expected result of modifying the properties of a concrete mixtures and concrete with admixtures it is necessary to properly design the modified concrete composition and the correct technology for its implementation. Failure to meet these conditions guarantees the ineffectiveness of the mixtures. Among the technological factors on which the effectiveness of admixtures depends, the most important are [5–9]:

- its properties;
- chemical and mineral composition;
- application size;
- presence of other chemical admixtures;
- type and properties of mineral additives;
- addition of the admixture in relation to mixing of the cement with water;
- mixing time and temperature of the mixture.

Each of these factors influences of the effectiveness of admixtures, especially dependence effect of the admixtures addition on the properties of cement, and compatibility of the admixture with cement. The purpose of additives, which are in the form of dusts, crumbs or fibers, is to improve selected concrete properties, save cement and supplement silty aggregate fractions. Additives require weight dosing, very careful distribution in a concrete mixture and often sharpened care conditions [10].

The effectiveness of chemical modification was defined as the criterion and characteristics of the quality of their performance due to the function performed and the associated basic effect. When assessing the effectiveness of the admixture and its use, secondary effects must be also considered, because there is a possibility of an adverse effect of the admixture implementation on the important features of the concrete mixtures and / or hardened concrete [11].

In this study process of compering the most typical commercial additives for concrete production was presented.

2. Materials and methodology

Concrete mixtures were prepared by mechanical mixing of portland cement, quartz sand with aggregates of 0.125–0.250; 0.250–0.500 and 0.500–1.000 mm, water and deflocculant addition. Next, after mixing process, 10 wt.% of modifiers were added. Time of mixing was equal 5 min I lab condition of 21 °C and 50 % of humidity. After that, concrete mixtures have been transported to production forms. The morphology, thickness of fibers and the grain size was determined by SEM (JEOLJSM-6610 at voltage equal to 5 kV), LM (Olympus LEXT OLS4100) technique and laser diffraction method. Additionally, chemical composition of used fillers were widely done by EDS detector. The measurements of mechanical properties such as bending test and compressive strength was determined by Zwick machine. The results are average value of 6 samples after 1, 7, 14, and 28 days of curing.

3. Results and discussion

Figure 1 presents the powders and fibers used in production of concrete based on portland cement. Presented images of fillers exhibit very similar structure of surface. It is characterized by sharp edges and non-regular structure. Agglomerates have been observed. Steel fibers are characterized by regular structure. On LM images rust have been observed. All fibers used in process of preparation concrete mixtures revealed different length due to the recycling process of tires cord. The grain size of modifier, thickness and circumference of fibers have been presented in Table 1. Obtained results shows that are very different admixtures were used to fabrication concrete mixtures. Micrometakaolin exhibits the smallest particles size in this study. Zeolite revealed 10 times higher average particle size compare with micrometakaolin ~ 45 μm.

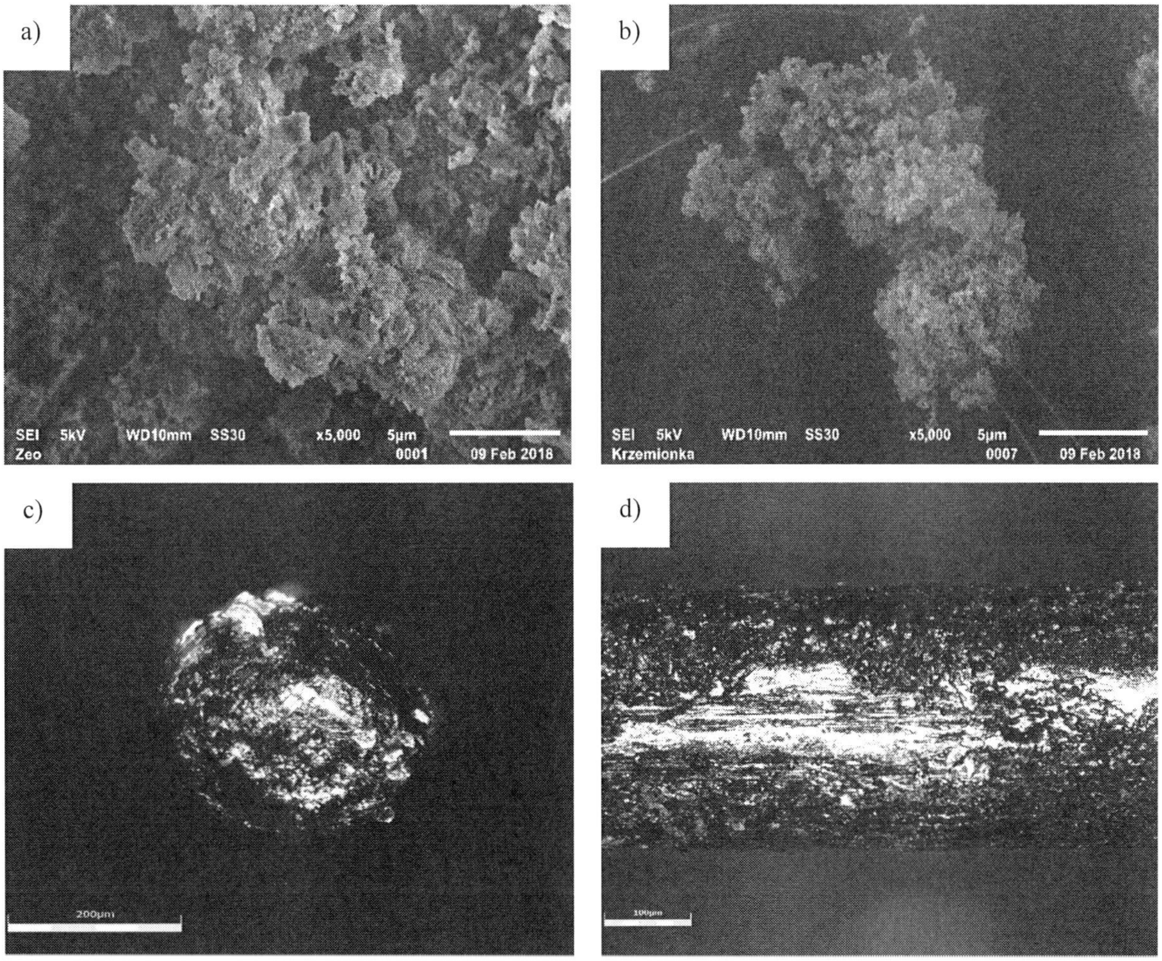

Fig. 1. SEM and LM images of fillers and fibers, where: a) zeolite; b) micrometakaolin – MMK; c, d) steel fibers

Table 1. Grain size of admixtures; thickness and circumference of steel fibers

Type of modifier			
Zeolite	Micrometakaolin	Steel fibers	
Mean grain size		Thickness	Circumference
45.8 μm	4.7 μm	356.7 μm	142.9 μm

Chemical composition of investigated admixtures was presented on Figure 2. As seen in Figure 2 concrete admixtures revealed similar chemical composition based on silicon and aluminum oxides with a small amount of Fe, Ca and K compounds. According to steel fiber, its based of iron with the addition of alloying elements. However, rest of the chemical compound comes from process of recycling cord of tires and was cut. Negative effect of recycled fibers addition have not been observed.

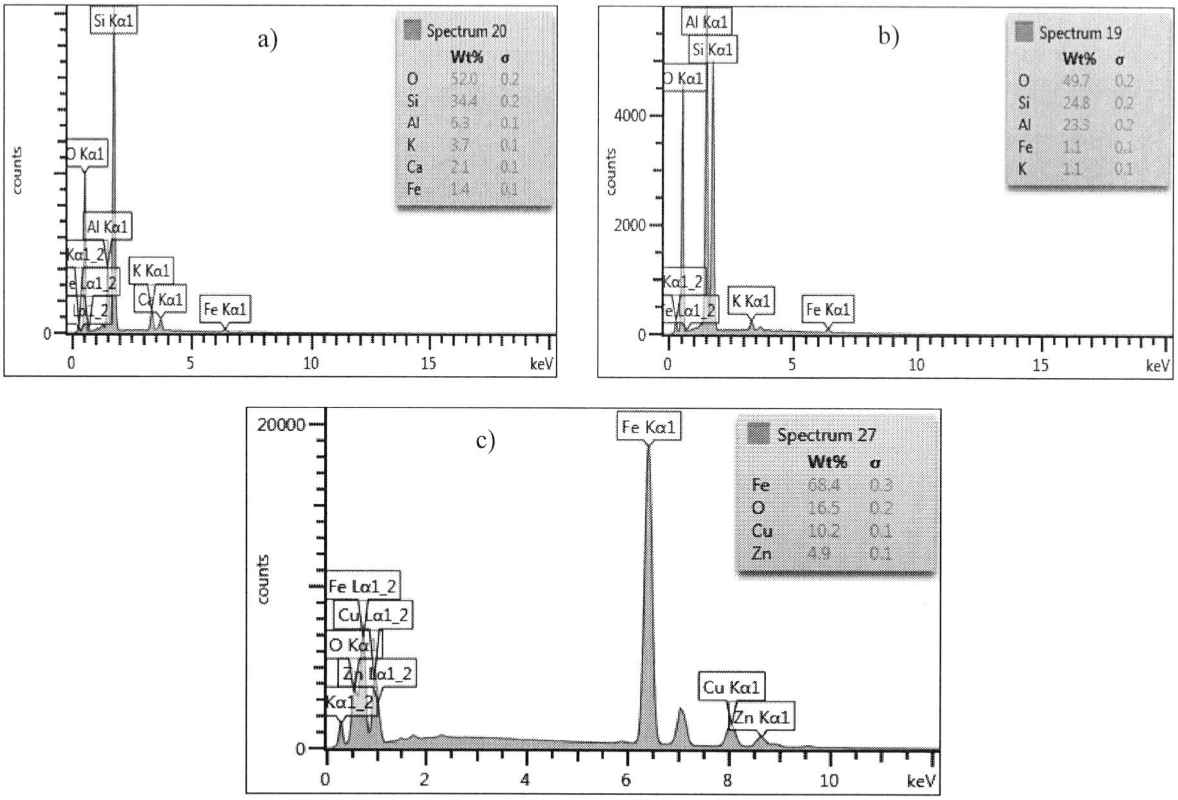

Fig. 2. Chemical composition of fillers and fibers, where: a) zeolite; b) micrometakaolin – MMK; c) steel fibers

Figure 3 presents results of distribution of bending force of researched samples. The highest effect of this mechanical properties was noted for 10 wt.% of MMK addition, after 7, 14 and 28 days of curing equal ~ 6.3 kN. On first day of measuring all admixtures represent similar value of bending force nearby 2.5 kN. Increase effect of bending force was noted for all used additives.

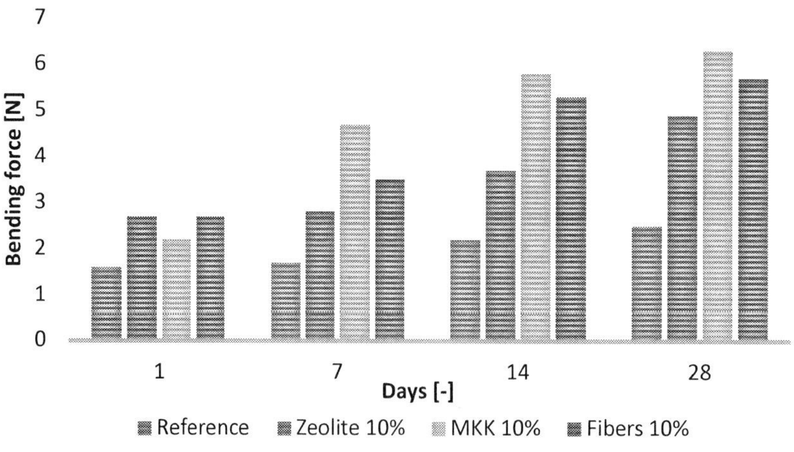

Fig.3. Distribution of bending force of studied concrete mixtures with 10 wt.% addition of zeolite, micrometakaolin and steel fibers

Figure 4 shows distribution effect of compressive strength of modified concrete mixtures by 10 wt.% of zeolite, MMK and steel fibers. The increasing effect of all used additives were observed. The results show that addition of steel fibers exhibit the highest values of compression strength ~ 120 MPa after 28 days of measuring. On first day 67 MPa was noted for the same samples. From the below observations of compressive strength, other modifiers revealed similar values of compressive strength to 14 days of curing ~ 60 MPa. On 28 day the MMK sample revealed the higher value, equal 85 MPa in compare with zeolite sample ~ 71 MPa.

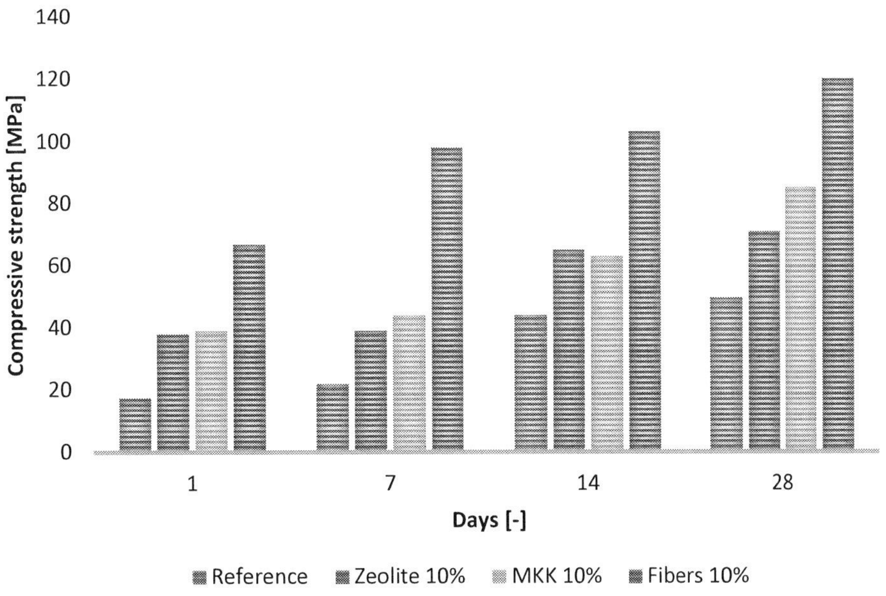

Fig.4. Distribution of compressive strength of studied concrete mixtures with 10 wt.% addition of zeolite, micrometakaolin and steel fibers

Conclusions

In this research obtained result of chemical modification of concrete mixtures based on quartz sand and Portland cement was presented. Based on the conducted research, the following conclusions were made:

1. All tested modifiers exhibit increase effect on mechanical properties of concrete mixtures.
2. Addition of 10 wt.% of steel fiber has the highest value of compressive strength equal 120 MPa.
3. The relatively low values of compressive strength measured for zeolite and MMK samples can be improved in new recipe including small particles of quartz sands and dust.
4. Addition of each components provide very easy way to increase the final mechanical properties of these concrete samples.
5. It is possible to fabricate new concrete based on portland cement and recycled steel fibers in commercial conditions without changes in the production line.
6. The research gives a new insight into unconventional concrete modification in laboratory condition and might use in future study on this purpose.

Reference

[1] Markiv T., Sobol K., Franus M., Franus W. Mechanical and durability properties of concretes incorporating natural zeolite. *Archives of Civil and Mechanical Engineering 2016*; 4: 554–562.

[2] Malek M., Wisniewski P., Szymańska J., Mizera J., Kurzydlowski K. J. Technological properties of ceramic slurries based on silicon carbide with poly(vinyl alcohol) addition for shell moluds fabrication in precision casting process. *Acta Physica Polonica A 2016*; 129 (4): 528–530.

[3] Guneyisi E., Gesoglu M., Mermerdas K. Improving strength, drying shrinkage and pore structure of concrete using metakaolin. *Materials and Structures 2008; 41:* 937–949.

[4] Abbas R., Abo-El-Enein SA., Ezzat ES. Properties and durability of metakaolin blended cements: Mortar and concrete. *Materiales De Construccion 2010;* 60*:* 33–49.

[5] Valipour M., Yekkalar M., Shekarchi M., Panahi S. Environmental assessment of green concrete containing natural zeolite on the global warming index in marine environments. *Journal of Cleaner Production 2014;* 65: 418–423.

[6] Vejmelková E., Konˇáková D., Kulovaná T., Keppert M., Zˇumár J., Rovnaníková P., Keršner Z., Sedlmajer M., Cˇerny´ R. Engineering properties of concrete containing natural zeolite as supplementary cementitious material: Strength, toughness, durability, and hygrothermal performance. *Cement and Concrete Composites 2015*; 55: 259–267.

[7] Małek M., Wiśniewski P., Matysiak H., Zagórska M., Kurzydłowski KJ. Technological properties of SiC-based ceramic slurries for manufacturing investment casting shell moulds. *Archives of Metallurgy and Materials* 2014; 59 (3): 1059–1062.

[8] AzariJafari H., Tajadini A., Rahimi M., Berenjian J. Reducing variations in the test results of self-consolidating lightweight concrete by incorporating pozzolanic materials. *Construction and Building Materials 2018*; 166: 889–897.

[9] Li JJ., Niu JG., Wan CJ, Liu XQ., Jin ZY. Comparison of flexural property between high performance polypropylene fiber reinforced lightweight aggregate concrete and steel fiber reinforced lightweight aggregate concrete. *Construction And Building Materials 2017*; 157: 729–736.

[10] Othman H., Sabrah T., Marzouk H. Feasibility of using ultra-high performance fiber reinforced concrete for radioactive waste containers: Drop test simulation. *Nuclear Engineering And Design 2017*; 325: 113–123.

[11] Wisniewski P, Szymanska J, Malek M, Mizera J. Optimizing the Lightweight Ceramic Proppants Properties. *Acta Physica Polonica A 2016;* 129(4): 501–503.

ITMS'2018

ktu
1922
panevėžys faculty of technologies and business

The 12th International Scientific Conference Intelligent Technologies in Logistics and Mechatronics Systems (ITELMS'2018), 26–27 April 2018, Panevėžys, Lithuania

Mechanical Strength Characterization of New Manufactured Concrete With Zeolite addition

Marcin Małek[a*], Mateusz Jackowski[a], Wojciech Życiński[a], Waldemar Łasica[a], Marcin Wachowski[b]

[a]*Military University of Technology in Warsaw, Faculty of Civil Engineering and Geodesy, gen. Witolda Urbanowicza 2 Street, 01-476 Warsaw, Poland*
[b]*Military University of Technology in Warsaw, Faculty of Mechanical Engineering, gen. Witolda Urbanowicza 2 Street, 01-476 Warsaw, Poland*

Abstract

In this work chemical modification of concrete was done. To modify, zeolite as a filler in 5, 10 and 15 wt.% were used. The reference sample was prepare using three parts of aggregates: 0.125–0.250; 0.250–0.500 and 0.500–1.000 mm. To obtain characteristic properties of concrete components chemical composition and grain size test were widely done. SEM micrographs, mechanical compressive strength and bending test was analyzed. After 1, 7, 14 and 28 days of curing mechanical properties were tested. Obtained results show increasing effect of zeolite modification as a filler. Research was proven that 5 wt.% of zeolite exhibit the most increase effect on final mechanical properties of studied concrete and are very perspective for future application in modern building technologies.

© 2018 M. Małek, M. Jackowski, W. Życiński, W. Łasica, M. Wachowski
Peer-review under responsibility of the Kaunas University of Technology, Panevėžys Faculty of Technologies and Business

Keywords: modern building technologies, zeolite, concrete mixtures, chemical modification, mechanical properties

* Corresponding author. Tel.: +48 261 83 96 55
E-mail address: marcin.malek@wat.edu.pl

© 2018 M. Małek, M. Jackowski, W. Życiński, W. Łasica, M. Wachowski **B427H0021** . Ed*Learning*.it
Peer-review under responsibility of the Kaunas University of Technology, Panevėžys Faculty of Technologies and Business

1. Introduction

Nowadays concrete is one of the main building material used in construction. Depending on the purpose, possibilities to increase its properties are existing, including durability, thermal insulation, frost resistance, etc [1–3].

There are many differentcompounds used in process of concrete production. On the other hand, there is not too much knowledge of zeolite effect on concrete mixture [3–5].

Natural zeolites are formed in hydrothermal processes of rock transformation in sedimentary environments. The most common zeolites occurring in the nature are: clinoptilolite, laumontite, filipse, stibnite and mordenite. Depending on the type of rocks containing zeolites, chemical composition is different. The structure of zeolites is buildusing tetra oxides of silicon and aluminum (SiO_4 and AlO_4), which are create the basic structure. By combining with oxygen and each other form a 3D structure.

The three-dimensional structure of zeolites ensures unique properties of these minerals. They are characterized by negative charge of the crystal lattice, easy exchange of non-network cations, uniform micropore size, thermal and hydrothermal stability. In modern building technology zeolites are used in the process of cement production, the production of ready-mixed concrete and cobblestones, concrete constructions, cellular concrete and reinforced concrete. They exhibit directly effect of the strength and quality of concrete, e.g.: increase durability to changing weather conditions, including frost, increase mechanical properties (resistance to bending and compression), and increase the resistance to cracking. In addition, zeolites improve resistance to harmful chemicals present in the air [5–10].

This study includes the results of own experimental research of that topic. In this work investigation of zeolite applied to the concrete mixtures were characterized.

2. Materials and methodology

The subject of this research was natural zeolite. To investigation of this powder SEM JEOLJSM-6610 was used with SE mode and 5 kV voltage. In addition to characterize of zeolite chemical composition was done using EDS technique. To examine the grain size of zeolite laser diffraction method was used. Moreover, for concrete production, Portland cement, quartz sand with aggregates of 0.125–0.250; 0.250–0.500 and 0.500–1.000 mm, water and deflocculant based on polycarboxylates were added to the concrete mixtures. For concrete mixtures fabrication mechanical mixer and aluminum reactor were used. Apart from research of zeolite powder, concrete mixtures were prepared and investigated for technological properties characterization.

After seasoning process for 1, 7, 14 and 28 days from the day of concreting, the samples were studied on strength machine. Then, the bending and compression strength of each sample was determined, dividing the value of the maximum sample load across the cross-sectional area of the sample. Based on the strengths determined, the average bending and compressive strengths were calculated for samples from all concrete series.

3. Results and discussion

In this study to preparation of concrete mixtures zeolite 50 was used. The maximum particle size of studied zeolite was 50.9 μm as 59.8 % of amount is present in Figure 1.

Figure 2 shows the representative scanning electron microscopy (SEM) images of characterized powder. Researched images shown that Zeolite powder exhibits typically sharp-edged structure and create large agglomerates.

Figure 3 presents the chemical composition of tested zeolite powder. Studied diagram shown that zeolite based most of silicon and aluminum oxides. Rest of compounds from of the natural environment and depend of mining region.

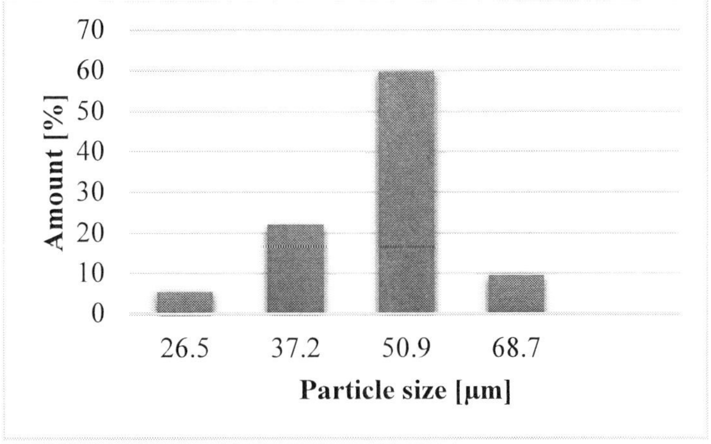

Fig. 1. Particle size distribution of zeolite 50 powder

Fig. 2. SEM images of tested zeolite powder: a) magnification 5000x; b) magnification 1000x

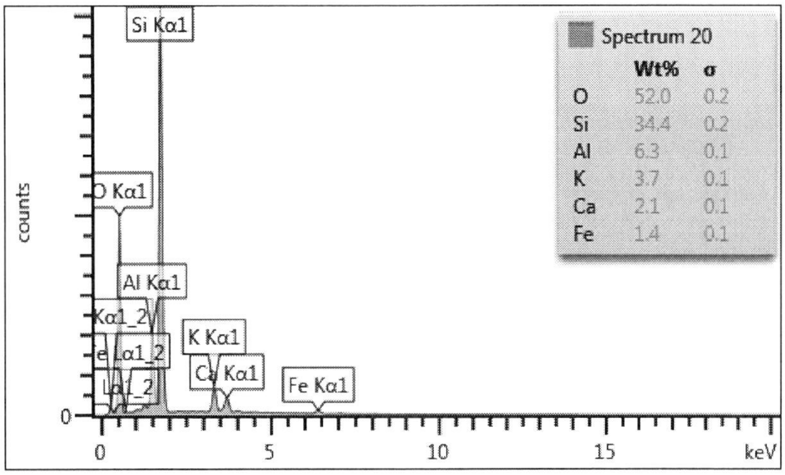

Fig. 3. EDS distribution of tested zeolite powder

Figures 4–5 show mechanical strength measurements of tested concrete sample after 1, 7, 14 and 28 days of curing. As seen in all figures, last day of experiment exhibit the biggest values of mechanical properties. It is dependent on type of cement addition and time of curing. Comparing with reference sample, samples with zeolite addition revealed higher values of bending force and compressive strength. All studied amounts (5 wt.%, 10 wt.% and 15 wt.%) of zeolite addition exhibits increase effect on mechanical properties. Sample with 5 wt.% of Zeolite addition revealed the highest values after bending and compressive tests.

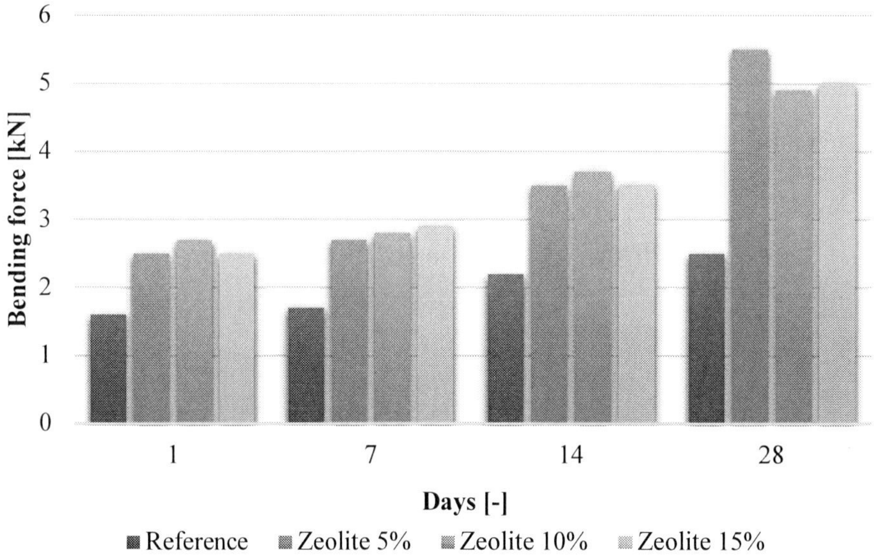

Fig. 4. Bending force distribution of examined concrete mixtures

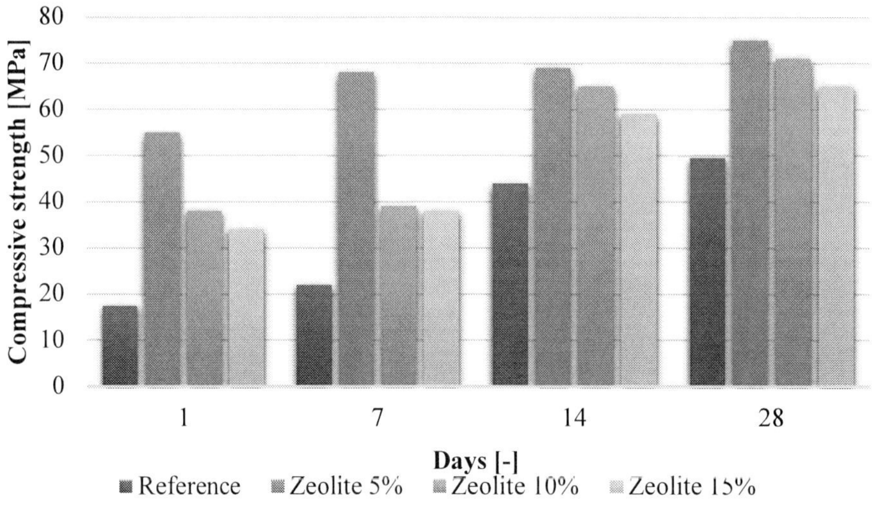

Fig. 5. Compressive strength distribution of examined concrete mixtures

Conclusions

According to study assumptions, as well as in connection with industrial standards, these studies were focused to examine concrete mixtures, which are characterized by low cost, relatively high mechanical test results and the biggest than usually materials for building construction value of filler addition. The main purpose was to use new zeolite for fabrication concrete mixtures. It has been proved that all suitable mechanical parameters are revealed by concrete mixture with 5 wt.% amount of Zeolite. According to the works [1–10], many authors focus on properties of concrete mixtures. First of all, it is difficult to compare obtained technological properties with other authors, because authors prepare concrete mixtures using different types of cement and fillers. Second technological properties and original recipes are strongly protected by companies engaged in the fabrication of concrete.

References

[1] Girskas G., Skripkiūnas G. The effect of synthetic zeolite on hardened cement paste microstructure and freeze-thaw durability of concrete. *Construction and Building Materials 2017*; 142: 117–127.

[2] Girskas G., Skripkiunas G., Šahmenko G., Korjakins A. Durability of concrete containing synthetic zeolite from aluminum fluoride production waste as a supplementary cementitious material. *Construction and Building Materials 2016*; 117: 99–106.

[3] Ikotun BD., Ekolu S. Strength and durability effect of modified zeolite additive on concrete properties. *Construction and Building Materials 2010*; 24: 749–757.

[4] Markiv T., Sobol K., Franus M, Franus W. Mechanical and durability properties of concretes incorporating natural zeolite. *Archives of Civil and Mechanical Engineering 2016*; 16: 554–562.

[5] Nagrockiene D, Girskas G. Research into the properties of concrete modified with natural zeolite addition. *Construction and Building Materials 2016*; 113: 964–969.

[6] Ramezanianpour AA., Mousavi R., Kalhori M., Sobhani J., Najimi M. Micro and macro level properties of natural zeolite contained concretes. *Construction and Building Materials 2015*; 101: 347–358.

[7] Samimi K., Kamali-Bernard S., Maghsoudi AA, Maghsoudi M., Siad H. Influence of pumice and zeolite on compressive strength, transport properties and resistance to chloride penetration of high strength self-compacting concretes. *Construction and Building Materials 2017*; 151: 292–311.

[8] Valipour M., Yekkalar M., Shekarchi M., Panahi S. Environmental assessment of green concrete containing natural zeolite on the global warming index in marine environments. *Journal of Cleaner Production 2014*; 65: 418–423.

[9] Vejmelková E., Konˇáková D., Kulovaná T., Keppert M., Zˇumár J., Rovnaníková P., Keršner Z., Sedlmajer M., Cˇernyˊ R. Engineering properties of concrete containing natural zeolite as supplementary cementitious material: Strength, toughness, durability, and hygrothermal performance. *Cement and Concrete Composites 2015*; 55: 259–267.

[10] Zhang J., Ding X., Wang Q., Zheng X. Effective solution for low shrinkage and low permeability of normal strength concrete using calcined zeolite particles. *Construction and Building Materials 2018*; 160: 57–65.

ITMS'2018

panevėžys faculty
of technologies
and business

The 12th International Scientific Conference Intelligent Technologies in Logistics and Mechatronics Systems (ITELMS'2018), 26–27 April 2018, Panevėžys, Lithuania

Effect of Recycled Fibers Addition on Mechanical Properties of Concrete

Marcin Małek[a*], Wojciech Życiński[a], Mateusz Jackowski[a], Waldemar Łasica[a], Marcin Wachowski[b]

[a] *Military University of Technology in Warsaw, Faculty of Civil Engineering and Geodesy, gen. Witolda Urbanowicza 2 Street, 01-476 Warsaw, Poland*

[b] *Military University of Technology in Warsaw, Faculty of Mechanical Engineering, gen. Witolda Urbanowicza 2 Street, 01-476 Warsaw, Poland*

Abstract

In this work results of concrete modification based on portland cement (CEM 1–42.5 R) by large amount of steel and polymer recycled fibers into the matrix was done. The main purpose of this study was to obtain directly influence addition of fibers on concrete mechanical properties as eg. tensile strength. The recipe of concrete was prepared use three parts of aggregates: 0.125–0.250; 0.250–0.500 and 0.500–1.000 mm. SEM and LM observations, chemical composition was widely done. Samples of concrete was identified by compressive strength and bending tests after 1, 7, 14 and 28 days of curing. Summarized results were compare with reference concrete. This study was proven that all chosen modifiers revealed increase effect on final mechanical properties of studied concrete and are very perspective for future application in civil engineering.

© 2018 M. Małek, W. Życiński, M. Jackowski, W. Łasica, M. Wachowski
Peer-review under responsibility of the Kaunas University of Technology, Panevėžys Faculty of Technologies and Business

Keywords: concrete, steel fibers, polymer fibers, polypropylene, mechanical strength

* Corresponding author. Tel.: +48 261 83 96 55
E-mail address: marcin.malek@wat.edu.pl

© 2018 K. Malon, J. Lopatka, M. Lopatka **B427H0025** . Ed*Learning*.it
Peer-review under responsibility of the Kaunas University of Technology, Panevėžys Faculty of Technologies and Business

1. Introduction

In the field of concrete technology, the procedure of reinforcing structures dispersed throughout the entire material with fibers has been observed for several decades. Today's state of the art and the advancement of FRC (Fiber Reinforced Concrete) technology indicates that it is one of the basic directions of modifying ordinary concrete. Dynamic development is mainly observed for polymer macrofibres.

The main area of interest is the modification of mechanical parameters of concrete. Increasing the compressive, bending or impact strength and reduction of shrinkage allow limiting the cross-section or eliminating reinforcing meshes in elements [1]. All industrial construction: floor and wall slabs, floors, columns, foundations could be modifying.

Currently, the most popular materials of which fibers are made are steel and polymers such as polypropylene (PP) or polyethylene (PE). Mainly, because of their high strength and in the case of polymers also high chemical resistance.

First type of fibers (steel) was obtained from the recycling car tires. Grinding tires, fibers with a length of 4 - 50 mm are formed. The rubber and steel are separated from each other, fall on sieves and given the appropriate fractions and lengths.

Polypropylene fibers can be obtained by recycling polypropylene products for example: plastic packaging boxes. In the production process, a continuous wire is extruded, then pre-cooled, shaped, cooled again and cut to the desired length [2–4].

This material was used in everyday life and just as tires clutter the natural environment to a great extent. As a result, it is ideally suited to produce fibers with structural application, reducing the amount of steel used in bars [5–8].

2. Materials and methodology

The subject of the research were steel fibers obtained from the recycling tires and two types polypropylene fibers. The steel fibers were made from high carboned steel with a strength 2850 MPa. Shape waved, hooked with a diameter 0.15 mm and a length 25 mm. In 15 wt.% addition.

Polymer fibers made from polypropylene (PP) with waved shape. Length 39.0 mm and diameter 0.78 mm. Tensile strength 520 MPa. In 1,8 wt.% addition both (white and green with different surface structure). The concrete mixture has been made on quartz sand aggregates 0.125, 0.250 and 0.500 mm, Portland cement CEM I 42.5 R, deffloculant for liquefaction based on polycarboxylates and water.

Samples were seasoned for 1, 7, 14 and 28 days after transfer from forms in water. Subsequently concrete samples were tested on a strength machine. The average compressive and bending strength were determined for each series additive and compared with the reference mixtures (without additives). Surface morphology of the additives, thickness and circumference has been analyzed by LM microscope (Olympus). For steel fibers chemical composition was widely done with the EDS technique.

3. Results

Figure and Table 1 show LM images and basic properties such as thickness and circumference of researched fibers. The higher thickness and circumference revealed PP white fiber. The assessing of the steel fiber the smallest basic properties were noted. Presented fibers were characterized by different morphology of surface: steel and PP green fibers – homogeneous surface structure with sharp edges, PP white fibers – heterogeneous structure of surface with oval edges.

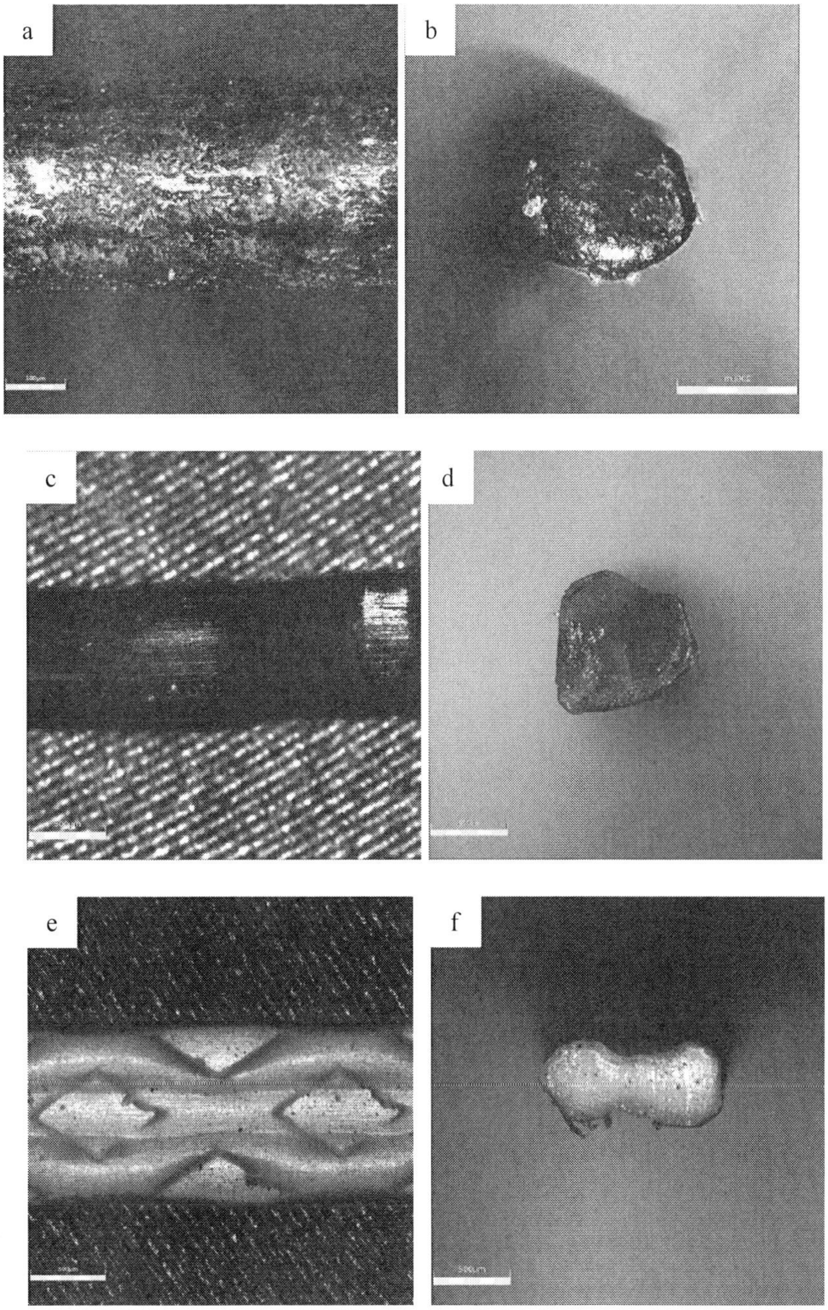

Fig. 1. LM images of studied fibers: (a – surface, b – cross-section) – steel fibers, (c – surface, d – cross-section) PP-1 green fibers, (e – surface, f – cross-section) PP-2 white fibers

Table 1. Basic properties of studied fibers

	Steel fiber [μm]	PP-1 white [μm]	PP-2 green [μm]
Thickness	344.8	1109.3	929.5
Circumference	156.9	546.5	471.6

Figure 2 presents the bending force distribution of researched fibers.

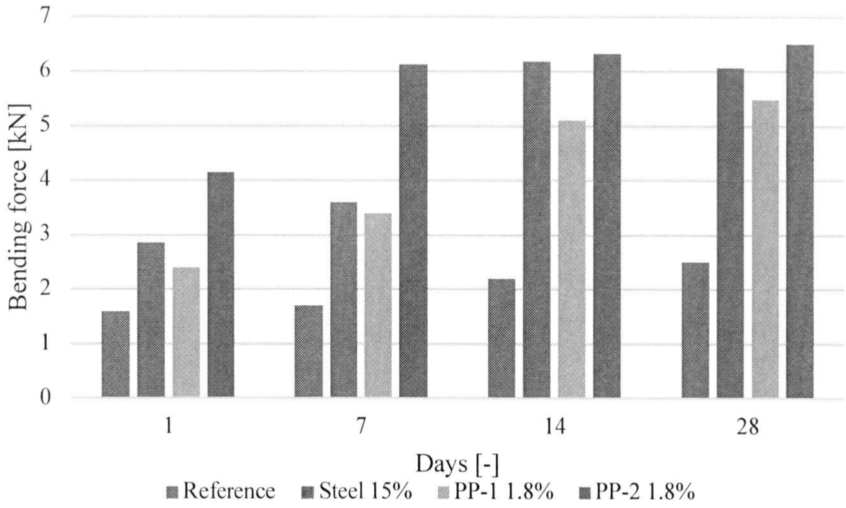

Fig. 2. Bending force distribution of researched samples

The obtained results shown increase effect of bending force to the 6.5 kN noted for PP white fibers after 28 days of curing. The lowest increase of the bending force was observed for PP green fibers and the maximum of these force was 5.5 kN in the same day of measuring. All samples exhibit the increase influence of admixtures addition.

Figure 3 shows the results of compressive strength distribution of prepared samples. The largest strength was observed for a steel fiber added with amount of 15 wt.%. Other samples present the very similar characteristic of compressive strength. Results after 14 and 28 days for these sample were almost identical. The reason of this phenomenon was to high fiber content into concrete mixture. Results of this process was formed of large air voids and decomposition of the concrete.

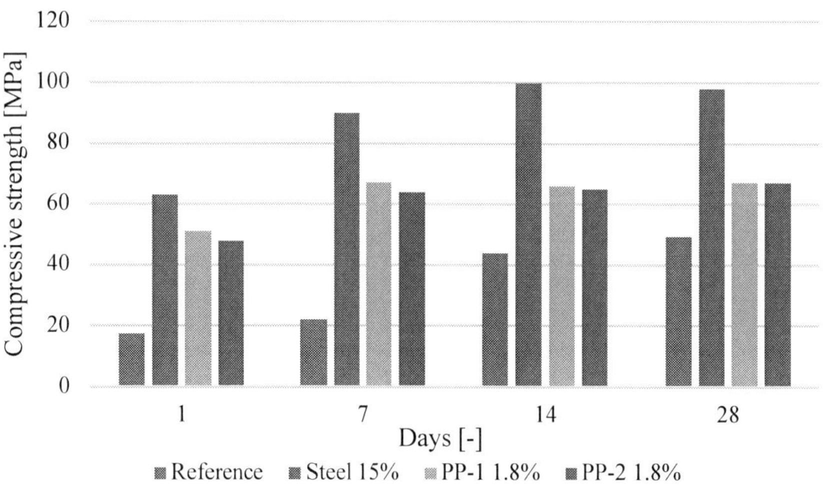

Fig. 3. Compressive strength distribution of researched fibers

Conclusions

According to the work assumptions increase effect of the addition of fibers on the strength and bending force of concrete has been proved. All chosen modifiers in the forms of fibers revealed the increase effect of both mechanical properties. There is a large development potential in the use of recycled fibers, from waste materials that is difficult to dispose. In nowadays there is a lot of polypropylene comes from recycling, used in the construction industry and this is a new direction to use its. The concrete in which are modification by using a fiber recycling origin can be successfully used for increasing the mechanical properties of final product. Environment friendly concrete mixtures are very promising for future fabrication of civil engineering and new building technology.

References

[1] Smrkić M. F., Damjanović D., Baričević A. Application of recycled steel fibres in concrete elements subjected to fatigue loading. *Gradevinar* 10/2007: 893–905.

[2] Glid M., Sobrados I., Ben Rhaiem H., Sanz J., Amara AB. Alkaline activation of metakaolinite-silica mixtures: Role of dissolved silica concentration on the formation of geopolymers. *Ceramics International* 2018; 43: 12641–12650.

[3] Malek M., Wisniewski P., Szymańska J., Mizera J., Kurzydlowski KJ. Technological properties of ceramic slurries based on silicon carbide with poly(vinyl alcohol) addition for shell moluds fabrication in precision casting process, *Acta Physica Polonica* A 2016, 129 (4): 528–530.

[4] Centonze G., Leone M., Micelli F., Aiello M., Petito G. Concrete reinforced with recycled steel fibres from scrap tires: a case study, *8th International Conference Fibre Concrete 2015*; 3, 121–125.

[5] Zhang J., Ding XP., Wang Q., Zheng X. Effective solution for low shrinkage and low permeability of normal strength concrete using calcined zeolite particles. *Construction and Building Materials* 2018; 160: 57–65.

[6] Abbas R., Abo-El-Enein SA., Ezzat ES. Properties and durability of metakaolin blended cements: Mortar and concrete. *Materiales De Construccion* 2010; 60: 33–49.

[7] Patil BB., Kumbhar PD. Strength and durability properties of high performance concrete incorporating high reactivity metakaolin. *International Journal of Modern Engineering Research* 2012; 2: 1099–1104.

[8] Dinakar P., Sahoo Pradosh K., Sriram G. Effect of Metakaolin Content on the Properties of High Strength Concrete. International Journal of Concrete Structures and Materials 2013; 7: 215–223.

ITMS'2018

The 12th International Scientific Conference Intelligent Technologies in Logistics and Mechatronics Systems (ITELMS'2018), 26–27 April 2018, Panevėžys, Lithuania

Influence of New Admixture Addition on Concrete Mechanical Properties

Marcin Małek[a*], Wojciech Życiński[a], Mateusz Jackowski[a], Waldemar Łasica[a], Marcin Wachowski[b]

[a] *Military University of Technology in Warsaw, Faculty of Civil Engineering and Geodesy, gen. Witolda Urbanowicza 2 Street, 01-476 Warsaw, Poland*

[b] *Military University of Technology in Warsaw, Faculty of Mechanical Engineering, gen. Witolda Urbanowicza 2 Street, 01-476 Warsaw, Poland*

Abstract

The work shows the influence of tree different admixture addition such as: steel fibers, zeolite and micrometakaolin on increase effect on mechanical properties of concrete mixtures. The concrete mixtures were modified with the addition of 5 wt.% of steel fibers, 5 wt.% of micrometakaolin and 5 wt.% of zeolite. The mixture was made of aggregate 0.125, 0.250 and 0.500 mm and Portland cement CEMI 42,5 R. After 1, 7, 14 and 28 days of cure bending and compression strength was done. To characterize the compounds SEM and LM observations, chemical composition and grain size test was widely done. All used additives have significantly improved the mechanical strength of studied concrete, which in the future gives the opportunity for further development in construction.

© 2018 M. Małek, W. Życiński, M. Jackowski, W. Łasica, M. Wachowski
Peer-review under responsibility of the Kaunas University of Technology, Panevėžys Faculty of Technologies and Business

Keywords: admixtures, concrete, zeolite, metakaolin, steel fibers, mechanical strength

* Corresponding author. Tel.: +48 261 83 96 55
E-mail address: marcin.malek@wat.edu.pl

© 2018 K. Malon, J. Lopatka, M. Lopatka
Peer-review under responsibility of the Kaunas University of Technology, Panevėžys Faculty of Technologies and Business

B427H0025 . Ed*Learning*.it

1. Introduction

Zeolites consist of a three-dimensional network of silicon-oxygen and aluminum – oxygen bonds [1]. The rock is hydrated aluminosilicate minerals of alkali and alkaline earth metals. Crystals of natural zeolite have a honeycomb structure with small channels and pores. Natural zeolites also have ability to adsorb and lose water up to 30% of their dry weight and exchange extra framework cations, without any change of crystal structure. As well as Portland cement Zeolite as pozzolan material is widely used in China in the production of cement increasing the strength of the concrete matrix [2].

Kaolinite can be filler and bonding material as well as cement. Is is widespread in paper and ceramic industries. Kaolinite converts to a pozzolan material named metakaolin after suitable thermal treatment [3–5]. It can be used in mortar and concrete to improve their properties. In addition, can be used as a source of cementing materials. Micrometakaolin is additionally ground to obtain a larger surface area and increase the binding properties [6].

Steel fibers were obtained from the recycling car tires. Grinding tires, fibers with a length of 4–50 mm are formed. The rubber and steel are separated from each other, fall on sieves and given the appropriate fractions and lengths. Due to the growing problem of used car tires, their partial management carries positive effects [7].

All of these additives have a great potential in binding concrete and improving its mechanical properties [8]. Due to the growing awareness of the scale of environmental pollution by the cement and construction industry.

2. Materials and methodology

The subject of the research was impact of tree additives on concrete. The steel fibers were made from high carboned steel with a strength 2850 MPa. Shape waved, sharp edges with a diameter 0.15 mm and a length 25 mm. added in amount of 5 wt.% of cement.

Micrometakaolin (MMK) and zeolite in 5 wt.% addition was mixed with cement. The concrete mixture has been made on quartz send aggregate 0.125, 0.250 and 0.500 mm, Portland cement CEM I 42.5 R, deflocculant for liquefaction based on polycarboxylates and water.

Morphology of the fillers has been analyzed with the use of scanning electron microscope JEOLJSM-6610 at voltage equal to 5 kV. Chemical composite determination was estimated with the EDS technique on analyzed surface.

Samples were seasoned for 1, 7, 14 and 28 days after transfer from forms in water. Subsequently concrete samples were tested on a strength machine. The average compressive and bending strength were determined for each series additive and compared with the reference mix (without additives). Surface morphology of the additives, thickness and circumference has been analyzed by LM microscope (Olympus). For steel fibers chemical composition was widely done with the EDS technique.

3. Results

The fillers were characterized by similar clusters of agglomerates and sharp edges as cement. Steel fibers - homogeneous surface structure with sharp edges. Figure 1 shows fillers and fibers images obtained with the use of scanning electron microscope. Chemical composition has been showed in Table 1.

Table 1. Chemical composition of researched fillers

Chemical Composition					
Steel fibers		Zeolite		Micrometakaolin	
Amount [%]	Element [wt.%]	Amount [%]	Element [wt.%]	Amount [%]	Element [wt.%]
Fe	68.4	O	49.2	O	50.0
O	16.5	Si	22.6	Si	25.5
Cu	10.2	Al	9.0	Al	23.7
Zn	4.9	Ta	8.2	K	0.8
		K	5.2		
		Ca	5.2		

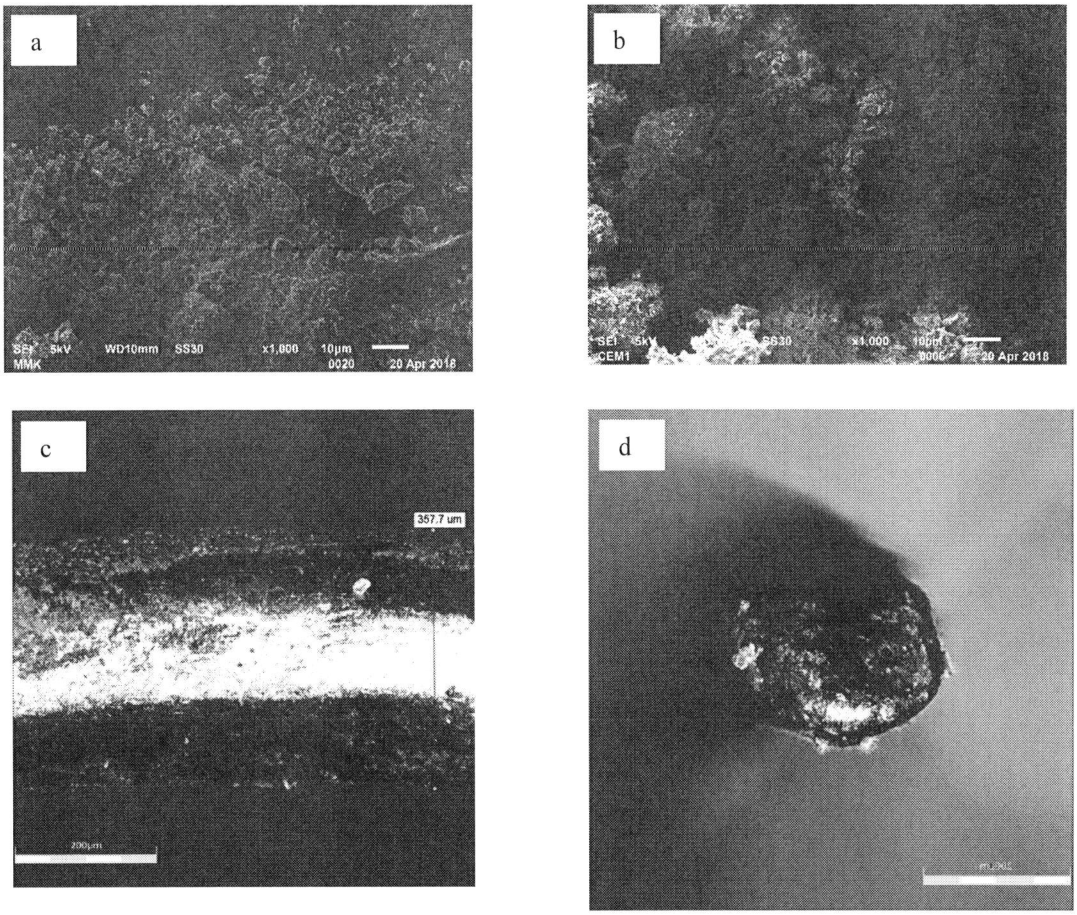

Fig. 1. SEM and LM images of studied fibers – (a) – micrometakaolin, (b) zeolite, (c, d) steel fibers

Figure 2 presents the bending force distribution of researched fibers.

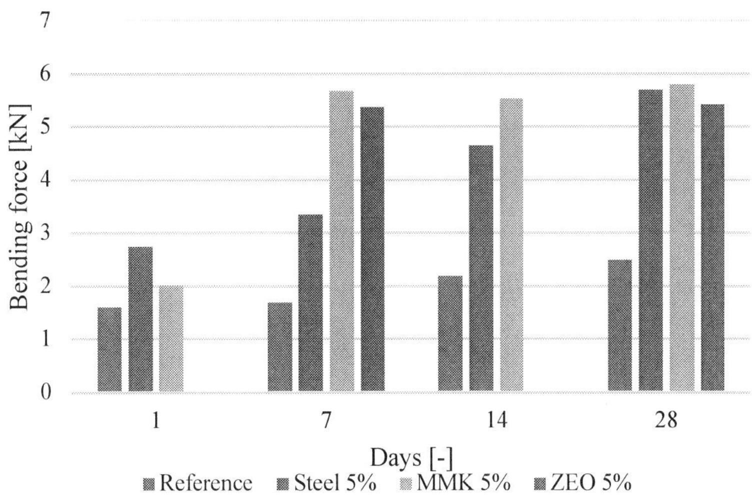

Fig. 2. Bending force distribution of researched samples

Micrometakaolin powder was based most of silicon and aluminum oxides. Rest of compounds from of the natural environment and depend of kaolinite base clay.

All tested samples shown a tendency to increase strength and reached its highest value on the last day of care – 28th (Figures 2–3).

The bending test shown that the increase in strength is already from 7th day for pozzolan additives (zeolite, micrometakaolin) and almost the same for 28th day for compressive and bending strength.

In the case of bending strength after 28 days, all additives significantly improved strength. Steel fibers improve compressive strength from 1 day in the same way as the other additives, but in the case of 28 days, the increase is significant and for fibers is 48%, for mineral additives 31% (MMK) and 30% (zeolite) compared with references sample (Fig. 3).

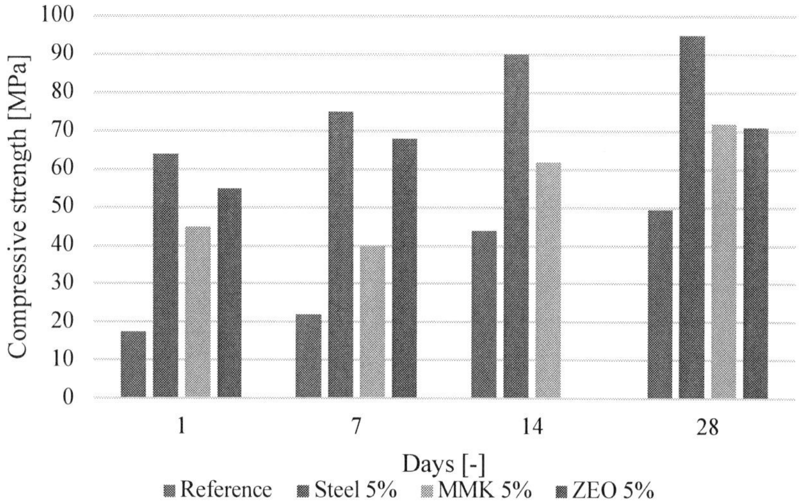

Fig. 3. Compressive strength distribution of researched fibers

Conclusions

According to the assumptions, effect of the addition of fibers on the strength of concrete has been proved. The mineral additive increased the cross-linking of the matrix, meanwhile increasing the strength. Correlation between the zeolite and metakaolin addition on final mechanical properties has been proved. There is a large development potential in the use of new puzzolans materials due to the significant emission of carbon dioxide in the production of cement, the use of additives such as zeolite and micrometakaolin may significantly reduce its volume. The concrete in which are modification by using a fiber recycling origin can be successfully used for increasing the mechanical properties of final product. Environment friendly concrete mixtures are very promising for future fabrication of civil engineering and new building technology.

References

[1] Glid M., Sobrados I., Rhaiem H., Sanz J, Amara AB. Alkaline activation of metakaolin-silica mixtures: Role of dissolved silica concentration on the formation of geopolymers. *Ceramics International* 2018; 43: 12641–12650.

[2] Malek M., Wisniewski P., Szymańska J., Mizera J., Kurzydlowski KJ., Technological properties of ceramic slurries based on silicon carbide with poly(vinyl alcohol) addition for shell moluds fabrication in precision casting process, *Acta Physica Polonica* A 2016, 129 (4): 528–530.

[3] Zhang J., Ding XP., Wang Q., Zheng X. Effective solution for low shrinkage and low permeability of normal strength concrete using calcined zeolite particles. *Construction and Building Materials* 2018; 160: 57–65.

[4] Abbas R., Abo-El-Enein SA., Ezzat ES. Properties and durability of metakaolin blended cements: Mortar and concrete. *Materiales De Construccion* 2010; 60: 33–49.

[5] Patil BB., Kumbhar PD. Strength and durability properties of high performance concrete incorporating high reactivity metakaolin. *International Journal of Modern Engineering Research* 2012; 2: 1099–1104.
[6] Dinakar P., Sahoo Pradosh K., Sriram G. Effect of Metakaolin Content on the Properties of High Strength Concrete. International Journal of Concrete Structures and Materials 2013; 7: 215–223.
[7] Smrkić M. F., Damjanović D., Baričević A. Application of recycled steel fibres in concrete elements subjected to fatigue loading. *Gradevinar* 10/2007: 893–905.
[8] Centonze G., Leone M., Micelli F., Aiello M., Petito G. Concrete reinforced with recycled steel fibres from scrap tires: a case study, *8th International Conference Fibre Concrete 2015*; 3, 121–125.

The 12th International Scientific Conference Intelligent Technologies in Logistics and Mechatronics Systems (ITELMS'2018), 26–27 April 2018, Panevėžys, Lithuania

Effect of new Filler Addition on the Mechanical Properties of Concrete

Marcin Małek[a*], Wojciech Życiński[a], Mateusz Jackowski[a], Waldemar Łasica[a], Marcin Wachowski[b]

[a] *Military University of Technology in Warsaw, Faculty of Civil Engineering and Geodesy, gen. Witolda Urbanowicza 2 Street, 01-476 Warsaw, Poland*

[b] *Military University of Technology in Warsaw, Faculty of Mechanical Engineering, gen. Witolda Urbanowicza 2 Street, 01-476 Warsaw, Poland*

Abstract

This work summarizes the results of chemical modification of concrete based on portland cement by zeolite, metakaolin and micrometakaolin addition in 5 wt.% of concrete amount into the mixture as a filler. The main purpose of this study was to characterize directly influence addition of new commercial fillers on concrete mechanical properties. The recipe of concrete was basic and contained three parts of aggregates: 0.125–0.250; 0.250–0.500 and 0.500–1.000 mm. To characterize the compounds SEM and LM observations, chemical composition and grain size test was widely done. Samples of concrete were characterized by compressive strength and bending tests after 1, 7, 14 and 28 days of curing. Obtained results were compare with reference concrete samples without chemical modification. Research was proven that all chosen modifiers exhibit increase effect on the final mechanical properties of studied concrete samples and are very perspective for future application in civil engineering.

© 2018 M. Małek, W. Życiński, M. Jackowski, W. Łasica, M. Wachowski
Peer-review under responsibility of the Kaunas University of Technology, Panevėžys Faculty of Technologies and Business

Keywords: modern building technologies, zeolite, metakaolin, micrometakaolin, chemical modification, mechanical properties.

---

* Corresponding author. Tel.: +48 261 83 96 55
E-mail address: marcin.malek@wat.edu.pl

© 2018 K. Malon, J. Lopatka, M. Lopatka **B427H0025** . Ed*Learning*.it
Peer-review under responsibility of the Kaunas University of Technology, Panevėžys Faculty of Technologies and Business

1. Introduction

The basic principles of concrete production have not changed over a century, but the production and implementation of a concrete elements is not so simple. To obtain the assume concrete mix, precision and error-free is required. Nevertheless, special attention should be paid to the selection of the mixture component and time. Process of preparing concrete mixture is focused on choosing a suitable recipe, which can be completely unsuccessful or far from expected. Preparation the appropriate recipe is the most important, because the consequences of the error can be tragic [1–4].

Currently, the largest amount of concrete production exhibits fly ash and silica dust, referred to as pozzolana. Pozzolans are material rich in silica and alumina, which combine with lime in the presence of water. Most significantly, they form identical compounds as in the process of cement hydration. Metakaolin and Zeolite are another environmental friendly binder. This kind of components are produced by roasting kaolin at a temperature of 600–700°C. Depending on the origin, it may contain a different amount of iron, titanium and potassium oxides. The partial replacement of cement by metakaolin or zeolite affects not only for the hydration of cement, but also of the properties of ordinary concrete, its strength, creep, porosity and durability [5–11].

The aim of this study was to characterize addition effect of new commercial fillers on concrete mechanical and physico-chemical properties.

2. Materials and methodology

The following study concerns relation between zeolite, metakaolin and micrometakaolin. Obtained results of three fillers were comparable quantitative and qualitative content. The concrete mixtures were prepared from Portland cement, water and deflocculant based on polycarboxylates. Additionally, 5 wt.% of fillers were added to the concrete mixtures reducing amount of cement. Base of the concrete mixtures were quartz sand with aggregates of 0.125–0.250, 0.250–0.500 and 0.500–1.000 mm. Furthermore, samples have been prepared in the aluminum mixer in lab condition 21°C and 50% humidity. Study of grain size have been carried out in laser diffraction method. Morphology of the fillers has been analyzed with the use of scanning electron microscope JEOLJSM-6610 at voltage equal to 5 kV. Chemical composite determination was estimated with the EDS technique on analyzed surface.

The results of the mechanical properties were measured after seasoning process of 1, 7, 14, 28 days. They are involved compression strength and bending tests on Zwick machine on few samples. The average value was calculated and presented in this study.

3. Results and discussion

Figure 1 shows fillers images obtained with the use of scanning electron microscope. Presented fillers were characterized by similar morphology and typical shape for materials used in concrete production – sharp edges and agglomerates. Results of fillers chemical composition have been presented in Table 1.

Table 1. Chemical composition of researched fillers

Chemical Composition					
Zeolite		Metakaolin		Micrometakaolin	
Amount [%]	Element [wt.%]	Amount [%]	Element [wt.%]	Amount [%]	Element [wt.%]
O	52.0	O	55.4	O	49.7
Si	34.4	Si	22.2	Si	24.8
Al	6.3	Al	20.7	Al	23.3
K	3.7	Fe	0.7	Fe	1.1
Ca	2.1	Ti	0.6	K	1.1
Fe	1.4	Ca	0.4	-	-

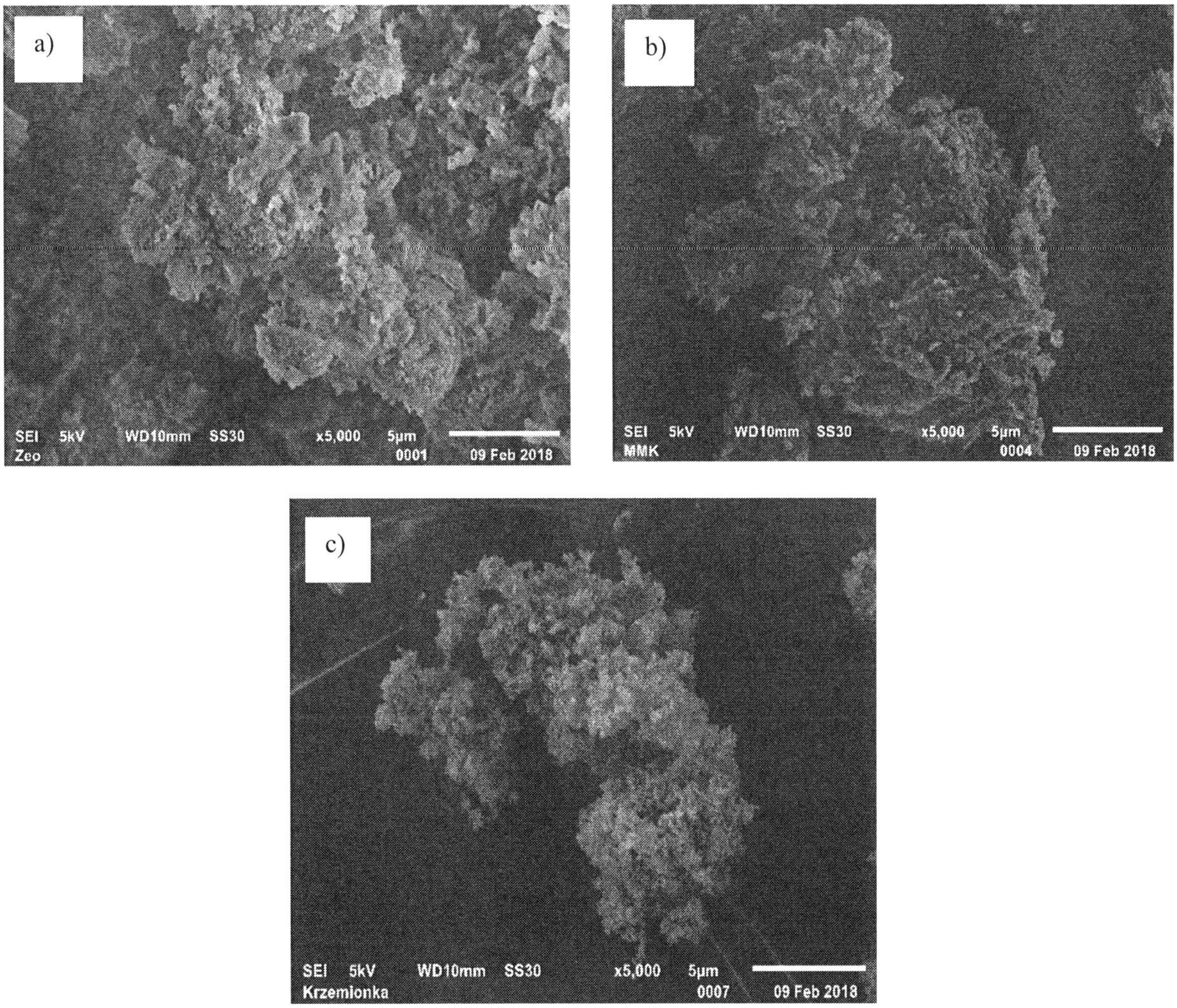

Fig. 1. SEM images of fillers, where: a) zeolite, b) metakaolin – MK, c) micrometakaolin – MMK

Chemical composition of studied fillers it different in a narrow range of oxides and silicon content. Rest of occurring elements are residual substances. Each of studied fillers based on natural composition of kaolin and zeolite. However, for commercial use studied fillers were chemically purified from heavy elements found naturally in these compounds by the company supplies the final product.

Figure 2 presents the average particle size of studied fillers. Analysis of obtained fillers indicate similar particle size for zeolite and metakaolin fillers equal ~ 45μm. Micrometakaolin exhibit almost 10 times less average particle size equal ~ 5μm.

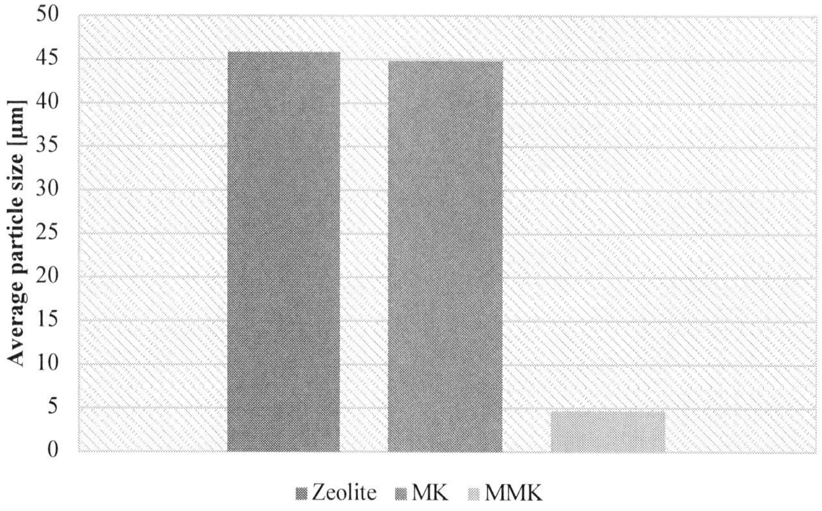

Fig. 2. Average particle size distribution of studied fillers

Figure 3 shows distribution of examined concrete mixtures after bending tests. Obtained results shown very similar values of bending force for MK and MMK samples after 7, 14 and 28 days of maturation process of concrete. On 28 days final bending force was ~ 6kN for all modified samples. Compare with the reference samples, 5 wt.% of adding new fillers exhibit increase effect on the bending force. The largest increase in bending force was noted for the sample with 5 wt.% of micrometakaolin addition.

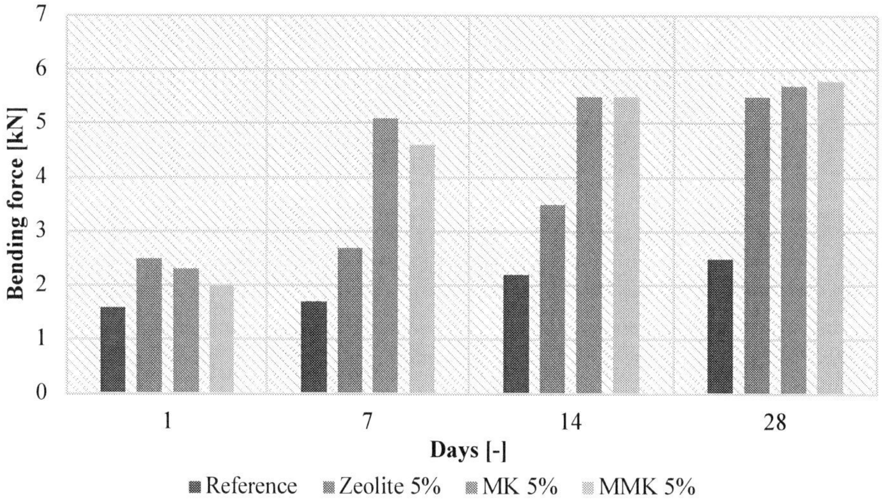

Fig. 3. Distribution of bending force of investigated concrete mixtures with 5 wt.% addition of zeolite, metakaolin and micrometakaolin

Figure 4 presents distribution of compressive strength of modified concrete mixtures by 5 wt.% addition of zeolite, metakaolin and micrometakaolin. The final step of comparison new used fillers with references samples, in concrete mixtures, were characterization of mechanical compressive strength. Obtained results shown that addition of all modifiers revealed increase effect on compressive strength. After one day of maturation process value of compressive increase to 55 MPa for Zeolite, 38 MPa for MK and 40 MPa for MMK. Further research confirmed the upward trend for each of the samples. The biggest increase effect was noted for 5 wt.% of zeolite addition equal 75 MPa after 28 days of maturation process of concrete.

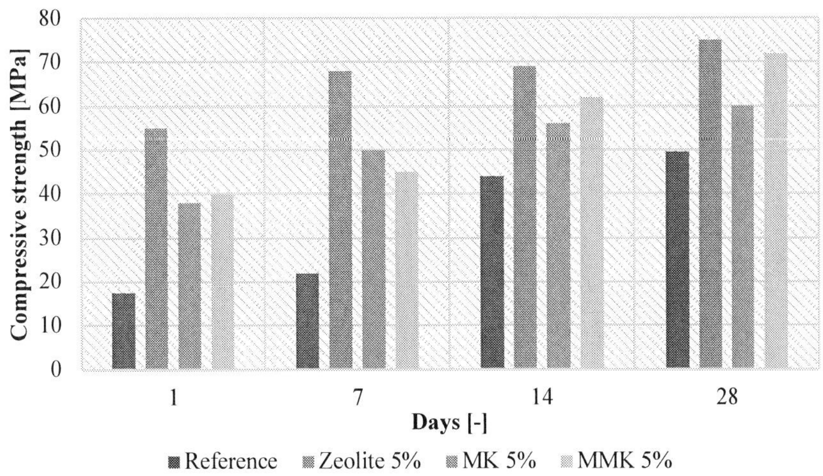

Fig. 4. Distribution of compressive strength of investigated concrete mixtures with 5 wt.% addition of zeolite, metakaolin and micrometakaolin

Conclusions

The results obtained in this study have revealed that the concrete mixtures properties are correlated with the type of modifiers use as well. Regulation of parameters such as content of modifiers, time of curing or kind of substance in some extent might contribute to better design of concrete mixtures. Increase effect of bending force by 5 wt.% of metakaolin addition and compressive strength by adding 5 wt.% of zeolites confirm this hypothesis. The measurement proved correlation between the zeolite and metakaolin addition on final mechanical properties. All chosen modifiers exhibit increase influence on final properties of concrete mixtures. Addition of each compound provide there is very easy way to increase of concrete class even by two classes. The result reported in this study give a new insight into unconventional concrete modification in laboratory condition and might use in future research on this purpose.

References

[1] Abbas R., Abo-El-Enein SA., Ezzat ES. Properties and durability of metakaolin blended cements: Mortar and concrete. *Materiales De Construccion 2010; 60*: 33–49.

[2] Guneyisi E., Gesoglu M., Mermerdas K. Improving strength, drying shrinkage, and pore structure of concrete using metakaolin. *Materials and Structures 2008*; 41: 937–949.

[3] Patil BB., Kumbhar PD. Strength and durability properties of high performance concrete incorporating high reactivity metakaolin. *International Journal of Modern Engineering Research* 2012; *2*: 1099–1104.

[4] Dinakar P., Sahoo Pradosh K., Sriram G. Effect of Metakaolin Content on the Properties of High Strength Concrete. *International Journal of Concrete Structures and Materials 2013;* 7: 215–223.

[5] Małek M., Wiśniewski P., Matysiak H., Zagórska M., Kurzydłowski KJ. Technological properties of SiC-based ceramic slurries for manufacturing investment casting shell moulds. *Archives of Metallurgy and Materials 2014*; 59: 1059–1062.

[6] AzariJafari H., Tajadini A., Rahimi M., Berenjian J. Reducing variations in the test results of self-consolidating lightweight concrete by incorporating pozzolanic materials. *Construction and Building Materials 2018*; 166: 889–897.

[7] Zhang J., Ding XP., Wang Q., Zheng X. Effective solution for low shrinkage and low permeability of normal strength concrete using calcined zeolite particles. *Construction and Building Materials 2018*; 160: 57–65.

[8] Wisniewski P., Szymanska J., Malek M., Mizera J. Optimizing the Lightweight Ceramic Proppants Properties. *Acta Physica Polonica A 2016;* 129(4): 501–503.

[9] Glid M., Sobrados I., Ben Rhaiem H., Sanz J., Amara AB. Alkaline activation of metakaolin-silica mixtures: Role of dissolved silica concentration on the formation of geopolymers. *Ceramics International 2018*; 43: 12641–12650.
[10] Hamidi S., Marandi SM. Clay concrete and effect of clay minerals types on stabilized soft clay soils by epoxy resin. *Applied Clay Science 2018*; 151: 92–101.
[11] Malek M., Wisniewski P., Szymańska J., Mizera J., Kurzydlowski KJ. Technological properties of ceramic slurries based on silicon carbide with poly(vinyl alcohol) addition for shell moluds fabrication in precision casting process. *Acta Physica Polonica A 2016*; 129 (4): 528–530.

ITMS'2018

panevėžys faculty of technologies and business

The 12th International Scientific Conference Intelligent Technologies in Logistics and Mechatronics Systems (ITELMS'2018), 26–27 April 2018, Panevėžys, Lithuania

UWB Based Follow-Me System for Unmanned Ground Vehicle Operator

Krzysztof Malon[a], Jerzy Lopatka[a], Marian Lopatka[a*]

[a]Military University of Technology, Urbanowicza Str.2, 00-908 Warsaw, Poland

Abstract

Man-machine cooperation is one of the most important aspects of unmanned platforms use. The follow me system significantly improves ergonomy of UGV control, because the platform is autonomously following the operator, copying his path and keeping safe distance, without necessary actions from the operator side. The proposed system is based on wireless ultra wide band (UWB) range meters. UWB modules are mounted both on the UGV and on the operator. Based on measurements, system estimates relative operator's position. Next, knowing platform position and orientation, it calculates control signals for platform driving system. This paper contains the analysis of possible accuracy of operator localization with UWB system and description of calculation method.

© 2018 K. Malon, J. Lopatka, M. Lopatka
Peer-review under responsibility of the Kaunas University of Technology, Panevėžys Faculty of Technologies and Business

Keywords: Unmaned Ground Vehicle, Follow-me System, Ultra Wide Band system, guide location

1. Introduction

Construction of a lightweight unmanned platform intended for direct support of operators working in a very difficult terrain is the research goal of many teams [1, 2, 9, 12, 14, 15, 16, 18]. To increase the operation ergonomics, the platform should move at a safe distance behind the guide, copying the trajectory of the guide movement (Fig. 1) because the guide is responsible for choosing the route and avoiding obstacles. Deviation from the trajectory determined by the guide may lead to collisions, so the deviation value is one of the most important

* Corresponding author. Tel.: +48 261 839 616; fax: +48 261 839 616.
E-mail address: marian.lopatka@wat.edu.pl

© 2018 K. Malon, J. Lopatka, M. Lopatka
Peer-review under responsibility of the Kaunas University of Technology, Panevėžys Faculty of Technologies and Business

B427H0025 . Ed*Learning*.it

parameters for the assessment of the follow-me system. It depends on a number of factors, including kinematics and dynamic properties and the adherence of the steering system and used control procedures, but one of the most important parameter is the accuracy of the location of the guide relative to the platform. In order to increase the accuracy, diverse systems are used – optical stereoscopic systems, differential GPS systems, laser scanners, thermal imaging cameras, etc. are used [3, 10, 13, 17, 19]. The costs of the most complex systems exceed the cost of the controlled platform [12].

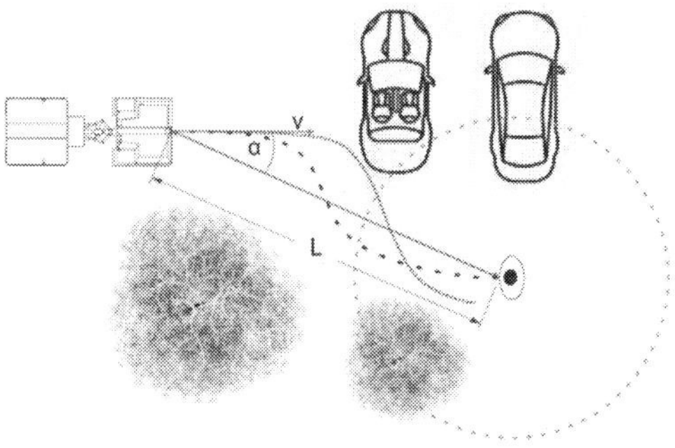

Fig. 1. The guide is responsible for setting a safe route (red line – platform route, dotted line – guide's route).

The aim of the current research is to determine the possibility of precise locating the guide using a relatively cheap UWB (Ultra Wide Band) system [8, 11]. It is anticipated that the speed of platform movement should be during the walk, in favourable terrain conditions at the level of 3–5 km/h. If fast transport is needed, it can reach 7–8 km/h (fast march), and even 12–15 km/h (running). Depending on the speed of movement, the safe distance should be between 2.5 and 7.5 m, so the guide should be ahead of the platform about 2 seconds. The maximum deviation from the trajectory should not exceed 0.5 m (it is desirable not to exceed 0.1 m). This will enable safe navigation of the system on roads with a width of 2–3 m.

2. Follow-me system concept

The tested commercial available UWB system [5, 6, 7] allows measuring the distance between a mobile transmitter called a label and a stationary transmitter called an anchor using radio signals from 3 to 8 GHz frequency range. The developed system is based on 4 distance measurements (Fig. 2 and 3), using triangulation, determines the position of the guide (carrying the label) relative to the platform.

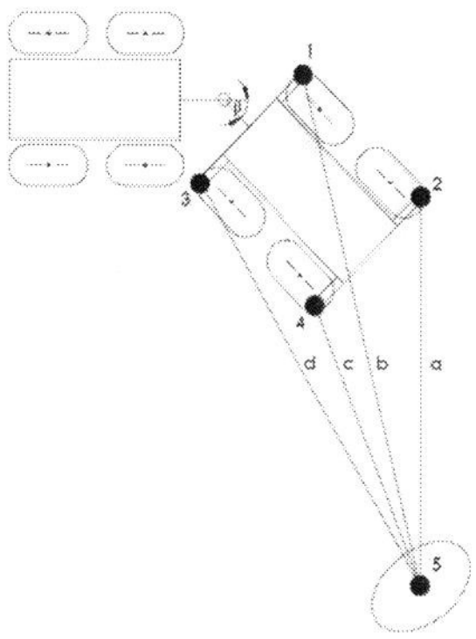

Fig. 2. The concept of the conductor position measurement system: a, b, c, d – measured distances using the UWB system; 1, 2, 3, 4 – anchors of the UWB system; 5 – UWB system label

On this basis, the speed and direction control signals are generated by the on-board computer. To determine the position of the guide, two anchors are necessary (located at a known distance from each other) however, the concept adopted the use of 4 anchors to increase the accuracy of measurements and reduce possible interferences.

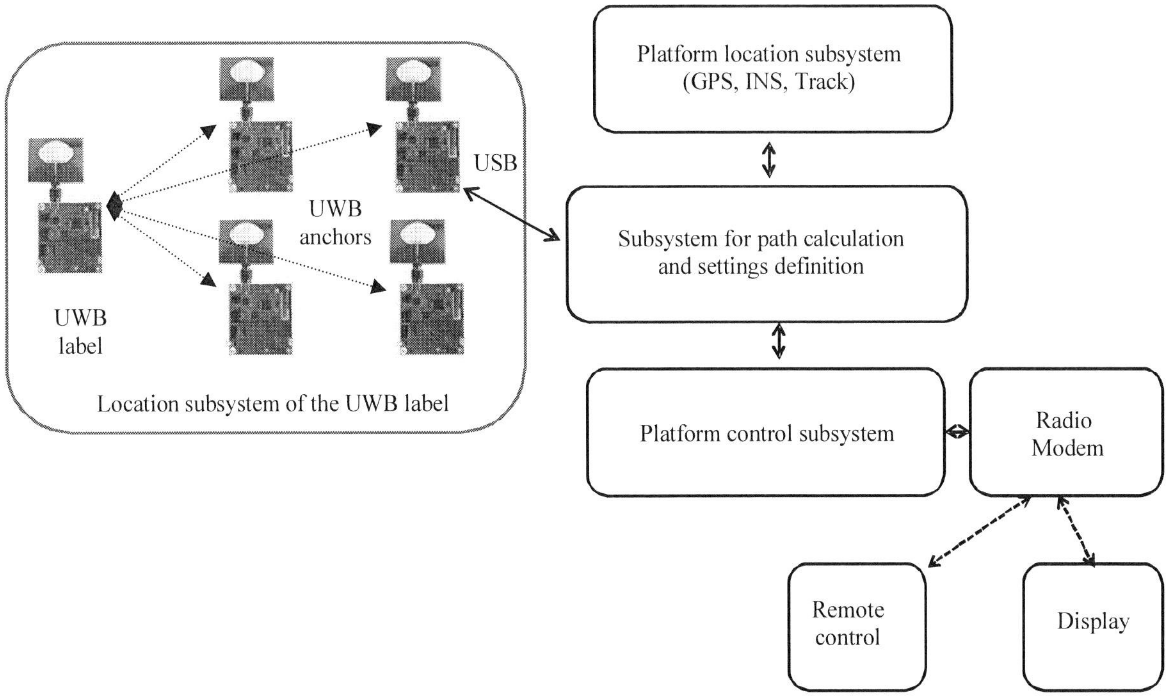

Fig. 3. System architecture

The method of determining the location of the operator using two sensors is presented in Figure 4. Assuming the distance between sensors (s_1 and s_2) of 1 m, which corresponds to the typical width of a lightweight unmanned platform, it is possible to determine the operator's location according to the following formulas.

$$s_1 = 0{,}0\,; \tag{1}$$

$$s_2 = 1{,}0\,; \tag{2}$$

$$\begin{cases}(x-0)^2+(y-0)^2=r_1^2\\(x-1)^2+(y-0)^2=r_2^2\end{cases}; \tag{3}$$

$$\begin{matrix}x^2+y^2=r_1^2\\x^2-2x+1+y^2=r_2^2\end{matrix}; \tag{4}$$

$$2x-1=r_1^2-r_2^2\,; \tag{5}$$

$$x=\frac{r_1^2-r_2^2+1}{2}; \tag{6}$$

$$y = \sqrt{r_1^2 - \frac{r_1^2 - r_2^2 + 1}{2}}; \tag{7}$$

$$y = \sqrt{\frac{r_1^2 + r_2^2 - 1}{2}}. \tag{8}$$

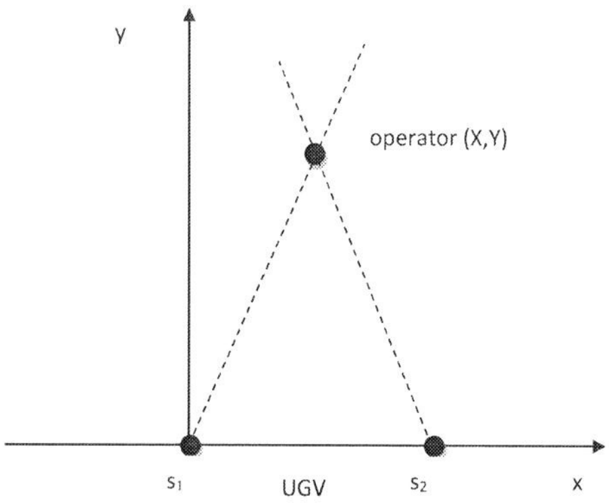

Fig. 4. Determination of the operator location using two sensors

3. Methodology of investigation

For distance measurements purposes, a TREK1000 evaluation set, consisting of four printed circuit boards EVB1000 (integrated circuit DW1000, ARM microprocessor, etc.), antennas as well as necessary cabling and software, was used. Each module can be used as an anchor or a tag, with an identification number. The real-time location system RTLS implemented by the manufacturer enables working on one of two frequency channels (3.993 GHz or 6.489 GHz) with the location rate equal to 3.57 Hz or 10 Hz. The testbed used for these measurements is presented in Figure 5.

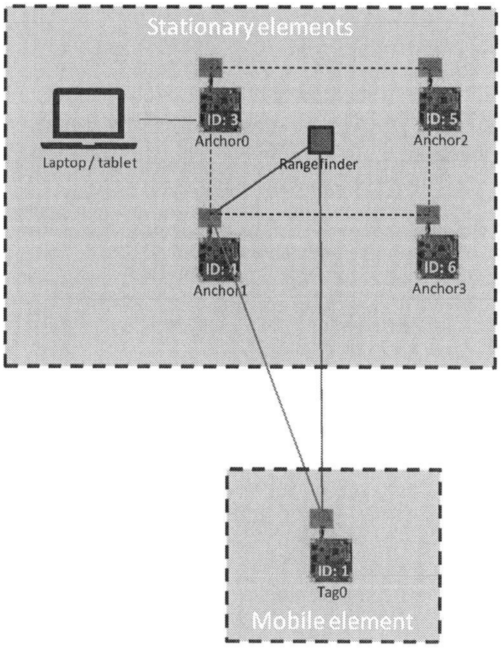

Fig. 5. Testbed for distances measurements accuracy of UWB modules

During the tests, five modules was used: four working as anchors and one acting as a tag. The real (reference) distance between the particular anchors and the tag was determined based on the Leica DISTO D510 rangefinder measurements. This laser distance meter is characterized by a certified typical measuring accuracy of +/- 1 mm and typical range up to 200 m (applies for 100 % target reflectivity, low background illumination, 25 °C) [4]. The method of determining the reference distance is presented in Figure 6.

$$d_{ref} = \sqrt{\left(d_{dal} - 2 \times d_{dys}\right)^2 + d_{kot}^{\ 2}}, \tag{9}$$

where d_{kot} = 30 cm;
d_{dyst} = 3 cm;
d_{dal} = distance measured by the rangefinder;
d_{ref} = reference distance – reference for a measurements made by UWB modules.

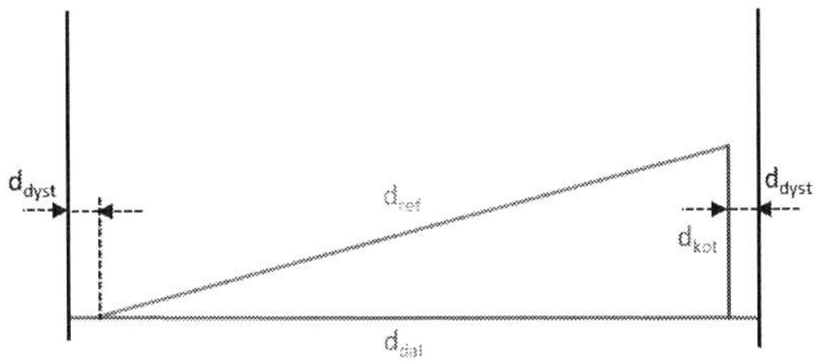

Fig. 1. Reference distance designation

Laptop with the own developed software, enabling the reception of measurement data from UWB modules and rangefinder, was connected to the Anchor0. In case of UWB sensors, bias corrected ranges were recorded. Detailed description of this data format is included in the manufacturer's documentation 0. Due to the possibility of using only specific frequency resources, channel 5 (6.489 GHz) was selected with the location rate of 10 Hz. Measurements were made at distances from 2 m up to 10 m with a step of approx. 25 cm and at distances from 10 m up to 18 m with a step of approx. 2 m. In each case, a series of 500 distance measurements were recorded for each link between the anchors and the tag.

4. Distance measurements results

Statistical data from measurements series for one anchor (Anchor0) are presented on Figure 7. On each box, the central mark indicates the median, and the bottom and top edges of the box indicate the 25th and 75th percentiles, respectively. The whiskers extend to the most extreme data points not considered outliers, and the outliers are plotted individually using the "+" symbol. Mean values error for different modules and reference distances are shown in Figure 8.

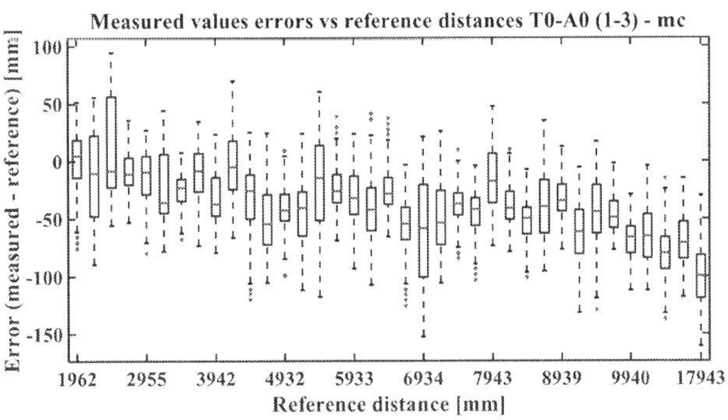

Fig. 7. Measured values errors vs reference distances

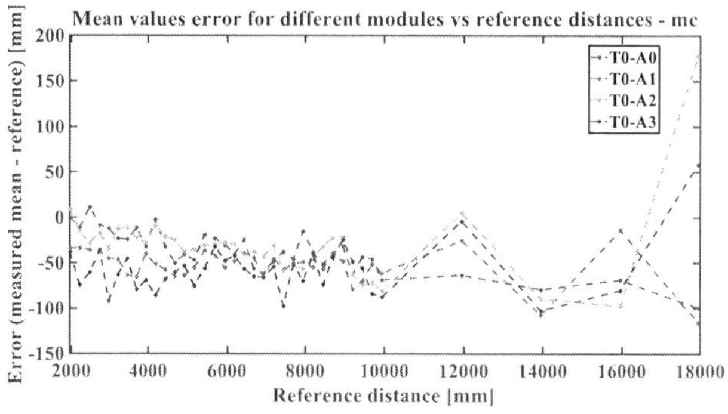

Fig. 8. Mean values error for different modules vs reference distances

Analyzing the presented research results, it can be concluded that the vast majority of distance measurements are in the tolerance range declared by the manufacturer. However, due to the large scattering of measured values (differences between the maximum and minimum values in a given series) and the occurrence of outliers, in the whole system, methods to minimize these effects, should be used. Proposed solutions may taking into account additional data, concerning the kinetic model, position and orientation of the platform or implementation of results averaging, etc. The vast majority of errors have a negative sign, which means that the distance values measured by UWB modules are shorter than the real distances.

5. Operator location measurements results

In order to estimate the accuracy of the localization, a simulation was performed in which the real operator location was assumed (x = 0.5 and y = 5). For the sensors location from and according to the equations (6) and (8), a series of simulated positions were generated. During the simulation, a standard error of distance measurement, sigma = 2 cm was used. For the presented assumptions, the calculated locations area, which is a common part of two rings with the widths given in (from 2 x sigma up to 8 x sigma), was also determined. The ratio of the simulated positions which are located in the calculated area, was designated and depicted as "in" values. In the next part of the results, the real and simulated operator's positions were determined based on the reference distance, derived from the measurements, according to the eq. (9). The operator's positions marked as "measurements" were calculated on the basis of data collected during the measurements presented in the previous point for selected reference distances.

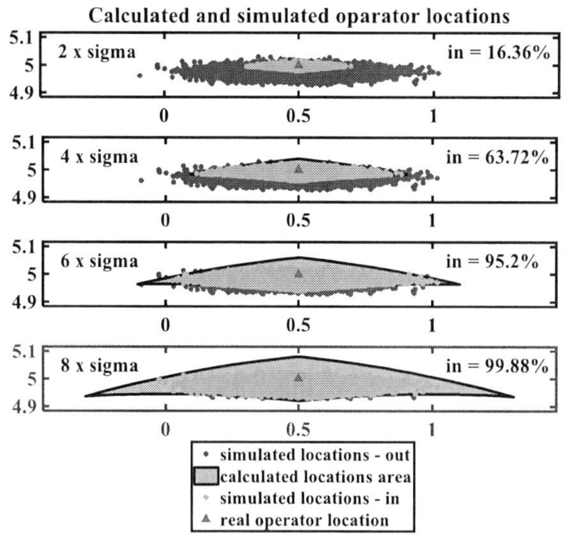

Fig. 9. Calculated and simulated operator locations

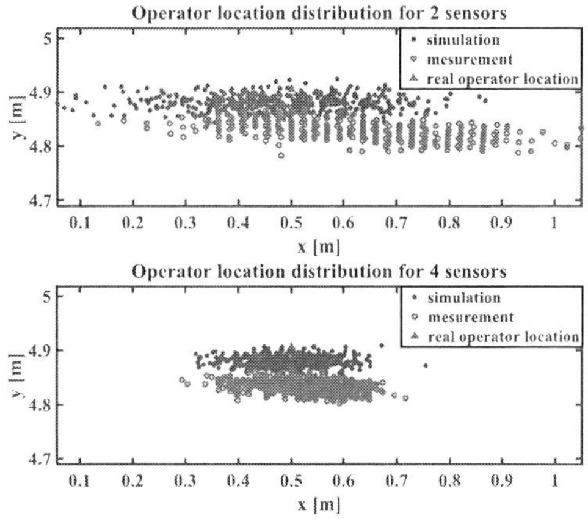

Fig. 10. Operator locations distribution for 2 and 4 sensors

In the case of four sensors, it was assumed to place two sensors in each of s_1 and s_2 points in. As a result, operator's positions for four possible combinations of sensor pairs were calculated, and then averaged. Fig. 12. Operator locations distribution in the Y plane for 2 and 4 sensors present the operator's locations distributions, which were determined based on the simulations and measurements for two and four sensors. In all cases, the improvement of location results accuracy can be observed for a larger number of sensors, especially in the X plane.

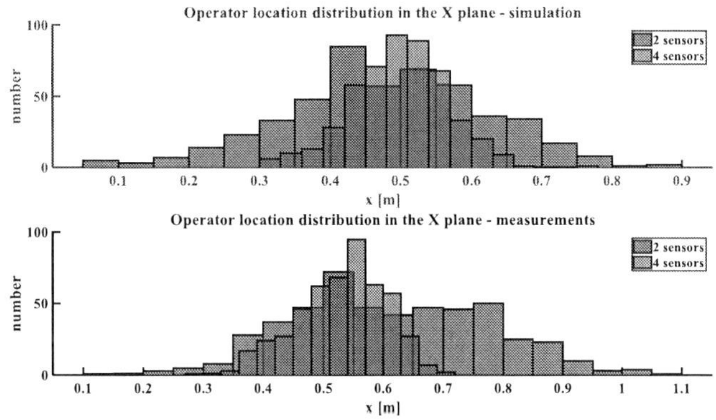

Fig. 11. Operator locations distribution in the X plane for 2 and 4 sensors

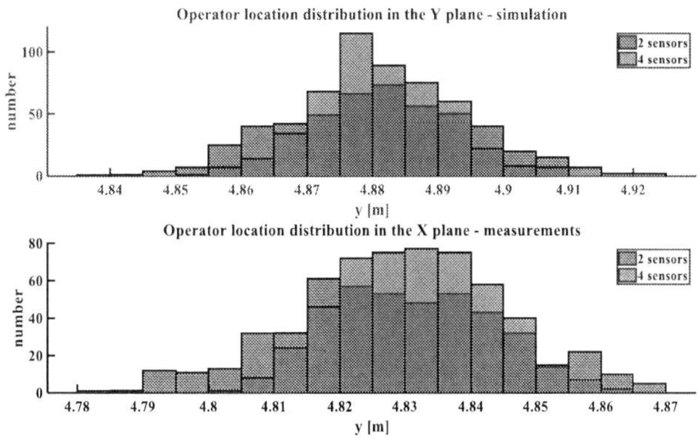

Fig. 12. Operator locations distribution in the Y plane for 2 and 4 sensors

Conclusions

Carried out measurements (Fig. 8) showed that the actual data obtained during distance measurements using UBW system and standard signal processing at a distance of up to 10m allow to get an accuracy of 50 mm, for each of tested anchors. However, they indicate a low value of the distance.as a result in case of use 2 anchors and distances between them about 1m, the distance of the guide from UGV can be determine with an accuracy of 0.05 m. Increasing the number of sensors up to 4 allows for a significant increase in the accuracy of the location of the guide – the error does not exceed 0.2 m.

Operator location distribution (Fig. 11) show that for big number of measurements with high probability, measurements error not exceed 0.1 m. It is promising results for using UWB system for calculating track of UGV operator. The next problem is working up of path calculation procedure on basic limited measurements during operator (guide) and UGV movements.

Acknowledgements

The work presented in this article has been supported by the Polis National Center for Research and Development – by grant DOBR-BIO4/083/13431/2013.

References

[1] Allyn D.B. The US Army Robotic and Autonomous System Strategy, US Army Training and Doctrine Command, March 2017.

[2] Armstrong N.C. Reducing the burden on the dismounted soldier. *Journal of Science and Medicine in Sport*. Volume 20, Supplement 2, November 2017, 147 p.

[3] Brandão A. S., Sarcinelli-Filho M., Carelli R. 2013. Leader-following control of a UAV-UGV formation, *16th International Conference on Advanced Robotics (ICAR)*, Montevideo, pp. 1–6.

[4] Leica Geosystems. Leica DISTO™ D510 User Manual.

[5] decaWave. DecaRangeRTLS ARM source code guide, Version 2.1, 2015.

[6] decaWave. DecaRangeRTLS ARM source code guide, 2015.

[7] decaWave. TREK1000 User Manual, Version 1.06, 2016.

[8] Djaja-Josko V. A new anchor nodes position determination method supporting UWB localization system deployment, *Signal Processing Symposium (SPSympo)*, Jachranka 2017, pp. 1–5.

[9] Holste S. T., Ciccimaro D. A. Increasing the Mobility of Dismounted Marines. Small Unit Mobility Enhancement Technology: Unmanned Ground Vehicles Market Survey. Technical Raport 1988. SPAWAR System Center PACIFIC October 2009.

[10] Kopfstedt T., Menges M., Bullmer S. Methods for control of UGV convoys in unexplored and partially unpaved terrain. 2nd IFAC Symposium on Telematics Applications Politehnica University, Timisoara, Romania October 5–8, 2010.

[11] Lazzari F., Buffi A., Nepa P., Lazzari S.: Numerical Investigation of an UWB Localization Technique for Unmanned Aerial Vehicles in Outdoor Scenarios. *IEEE Sensors Journal*, 2017, Vol. 17, No. 9, pp. 2896–2903.

[12] Massey K. Squad Mission Equipment Transport (SMET) Lessons Learned for Industry. NDIA Ground Robotics Capabilities Conference & Exhibition, 2–3 March 2016 Springfild VA, USA.

[13] Nishizumi Y. *et al.* FPGA implementation of object recognition processor for HDTV resolution video using sparse FIND feature, *IEEE International Workshop on Signal Processing Systems (SiPS)*, 2017, pp. 1–6.

[14] Reducing the Burden on the Dismounted Soldier Capability Vision Task 3 – The 'Energy-Efficient Soldier'. Centre for Defence Enterprise call for proposals & expressions of interest. United Kingdom Ministry of Defence. Release date: 24th March 2009 v1.0.

[15] Richardson C. JIEDDO's Robotics Programs. Military Robitics Summit. 29 August 2012 *Washington* DC, USA.

[16] Rogers P. Future Ground Vehicle Robotics. US Army Tank-Automotive Research Development and Engineering Center. 15 July 2015.

[17] Bo S., Weidong C., Yugeng X. Team-oriented formation control for multiple mobile robots. 16th Triennial World Congress, Prague, Czech Republic 4–8 July 2005.

[18] Unmanned Systems Integrated Roadmap 2013–2038. Washington, D.C.: Government Printing Office, 2007.

[19] Vyas M., Sarath H., Smitha K. and A. Bagubali Modern automotive embedded systems with special mention to radars," *2nd IEEE International Conference on Recent Trends in Electronics, Information & Communication Technology (RTEICT)*, 2017, pp. 1618–1625.

ITMS'2018

The 12th International Scientific Conference Intelligent Technologies in Logistics and Mechatronics Systems (ITELMS'2018), 26–27 April 2018, Panevėžys, Lithuania

Experiments on the Change of the Structure of the Sludge and the Complete Stabilization

Marius Mikolajūnas[a,b], Jovita Kaupienė[a,b]*, Danguolė Striukienė[a]

[a]*Kaunas Kaunas University of Technology, Nemuno g. 33, LT-37164 Panevėžys, Lithuania*
[b]*Panevėžys university of applied sciences, Laisvės a. 23, Panevėžys*

Abstract

Currently, the market offers a wide variety of sewage sludge and additives for reducing to distinguish odors, reducing the activity of the remaining heavy metals, phosphorus, nitrates impact on the environment, to reduce microbial, bacterial contamination, but they are not always cost-effective or does not meet the effects. Manufacturers are uncommonly testing the proposed substances in non-laboratory conditions, and in specific customer's production tanks or pools, due to environmental impacts and other conditions, a wide variety of sludge composition, biological contamination, temperature and other influential factors can occur.

© 2018 M. Mikolajūnas, J. Kaupienė, D. Striukienė
Peer-review under responsibility of the Kaunas University of Technology, Panevėžys Faculty of Technologies and Business

Keywords: construction, renovation, impact, environment, energy, savings, residential buildings

1. Introduction

Research object. Sewage sludge accumulated in *AB Aukstaitija Vandenys* in wastewater treatment tanks, smells, stabilization and smells of sludge in technological processes.

Purpose Carry out tests by adding the test dose of the stabilizer "Polonite" to the old sludge and evaluate:

– The changes to the structure, the result to be achieved – the sludge from a single limiting mass should turn into a crumbling, disintegrating structure.

* Corresponding author. Tel.: +370-652-14307
E-mail address: jovita.kaupiene@ktu.lt

© 2018 M. Mikolajūnas, J. Kaupienė, D. Striukienė **B427H0026** . Ed*Learning*.it
Peer-review under responsibility of the Kaunas University of Technology, Panevėžys Faculty of Technologies and Business

- Stabilization, the result to be achieved is the reduction of odors, gas emissions, elimination of microbiological activity.

Manufacturers' recommendations. There was no change in the manufacturers' suggestions for the structure of the drained sludge, so the pilot dose of "Polonite Stabilizer" was chosen. The dosage proposed by the manufacturers for the complete stabilization of sludge is about 20 %.

2. Literature review

Every year tones of phosphorus flow into the Baltic Sea. It's calculating that we can reverse the negative pollution trend in the Baltic. If we can recycle the phosphorus run-off, we can reduce our dependency on imported phosphorus [1]. Today are a lot of filter materials suitable for phosphorus (P) removal in wastewater treatment systems such as constructed wetlands or filter wells [2, 5, 6, 7, 8]. The mechanism of P removal involves sorption reactions. Efficiency of materials depends on the physical and chemical properties. Polonite is a commercial product used as reactive filter for wastewater treatment in Sweden [9]. Chemical treatment to reduce phosphorus in domestic wastewater is a very efficient method although uneconomic and unsustainable [10]. Polonite® is a compact natural filter commercialized by Ecofiltration AB. It is mainly used for on-site wastewater treatment and is produced from the calcium carbonate and silicate based rock Opoka. In studies on phosphorus removal with Polonite® have shown good results. There shown that phosphorus-saturated Polonite® is promising as a fertilizer in agriculture [6].

The P fertilizers today are becoming more expensive and less sustainable and phosphates are limited non-renewable resources [11]. Polonite and other materials after wastewater treatment may supply nutrients that could reduce the use of mineral P fertilizers. Studies on iron-rich sands have shown that P adsorbing to these substrates is as available [12]. The industrial product – blast furnace slag may efficiently improve wastewater treatment process [6]. Natural calcium-rich materials – Polonite, may help improving soil condition and fertility [13].

We can found many studies that use the special treatment of filter media to enhance P retention capacity [2, 3, 4]. Heating Ca-rich material at high temperatures is a common way to perform filter media improvement. During heating, CaO has a more reactive Ca-phase than commonly existing $CaCO_3$. The objectives of this paper are to summarize existing information on different filter media used for phosphorus removal from wastewater that could be applied to sludge, to clarify relationship between P retention capacity, pH level, Ca content and other parameters of the filter media.

3. Stage I. Keeping the specimens outdoors

The course of the research. The percentage of sludge in the material, using the old basin mud, is to choose: 15 percent stabilizer; 20 percent stabilizer; 25 percent stabilizer. Weighing and loading 5.0 kg of drained sludge, 750 g, 1000 g and 1250 g of Polonite stabilizer, corresponding to 15 %, 20 % and 25 % of the initial sludge weight, were weighed into 10 l plastic buckets. The sludge is evenly mixed with the stabilizer using a universal hand-held battery mixer.

After mixing using a pad, there was a change in structure. The stabilizer was inserted for several minutes, the water dampened, the sludge consistency became harder. It was possible to feel the consistency and hardness difference between samples with different amounts of stabilizer through the sludge.

One hour after the insertion of the stabilizer, it is **concluded** that such reagent volumes do not change the required structure. The required change would be fixed; if the sludge were no longer sticky, the structure would be close to the buffalo algae.

After one-day samples of sludge, compared with the sludge without stabilizer became more physical, but essentially has maintained its structural properties – still sticky, did not meet the criterion of flowability, felt this strong smell of ammonia, stronger, than in control samples (Fig. 1).

The samples were checked after three days and after 15 days after mixing. The samples do not expose sealed, only to cover up, the ambient temperature is above 15 C at night, at least 25 days C. Due to the evaporation of water from the inside of the sludge itself, the surface structure of the samples changed in the interior through the cracks created by it. Surface sludge became bulk, which was obviously seen with a 15% stabilizer in a blended sample. The sludge with a 25 % stabilizer was cemented and hardened, requiring little force when crushed. An obvious moisture and gravity gradient were observed when analyzing from the surface of the sample to the depth. 2–3 cm from the surface of the specimen (Fig. 2).

a) wet, not altering the properties of sludge. The specimens did not have a clear, distinct smell; they felt only after moving the specimen when the plastic deformations were carried out, and the smell from the lower, undamaged layers reached.

Fig. 1 Sediment samples with stabilizer after a day

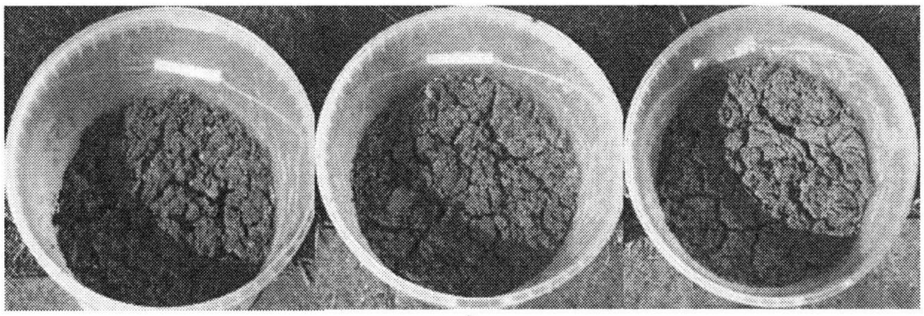

a)

b)

Fig. 2. Samples with sludge after fifteen days, 15 %, 20 %, 25 % stabilizer inserted from left to right: a) top view; b) side view

4. Stage II Storing specimens in the room

No exhaust emissions nor olfactometric measurements were performed in the field, as environmental conditions significantly influenced the results obtained. The experiment was repeated by holding specimens in the room. Also, the experiment includes materials that, by stabilizing the sludge, could reduce their odor by comparing their effects with the Polonite stabilizer. This is a natural grinded 2–3 mm fraction of dolomite and crushed lime. The amount of all substances in the sludge is selected by weighing and mixing well in separate 5 l containers the following: 10 percent – stabilizer "Polonite" / dolomite / extinguished lime; 20 percent – stabilizer "Polonite" / dolomite / extinguished lime; 30 percent – stabilizer "Polonite" / dolomite / extinguished lime.

Also, a sample of probiotics mixed with the test specimen without any additional substances is added to the entire volume, according to the ratio indicated by the manufacturer. All samples, 11 pieces were only numbered and placed in the room for proper olfactometric measurements (Fig. 3), kept loose to close – covered, the room was well ventilated, the temperature from the beginning of experiment 18–20 C, at the end of experiment 13–15 C.

Fig. 3. Sample sludge samples with stabilizer and other materials for gas emission and olfactometric measurements

Olphactometric measurements perform the measurement consists of keeping a minimum of 5–7 respondents were offered to evaluate the sample smell ugliness 5-point system, where 1 point – not filthy smell, and 5 points – very nasty smell. The first odor test was performed 1–3 hours after the preparation of the specimens, and the averages are shown in Figure 4.

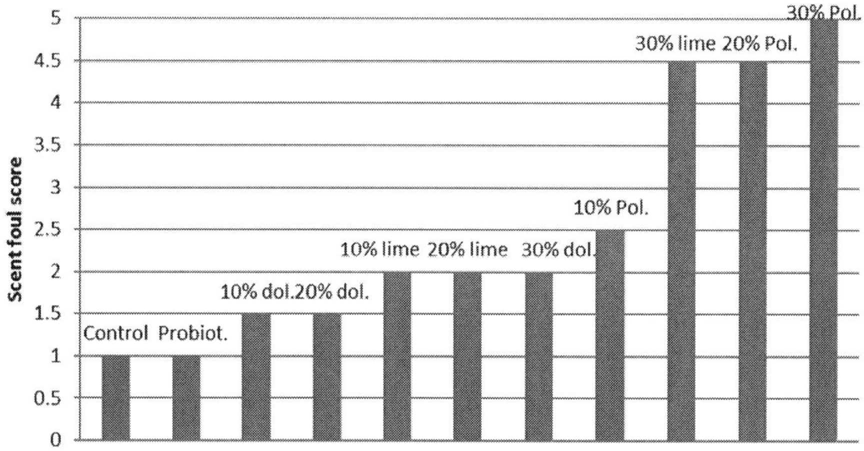

Fig. 4. Averages of test result evaluations 1 to 3 hours after mixing

All five respondents who evaluated the specimens found that the scent of the most abdominal smells felt a strong ammonia odor. Obviously, probiotics were not yet operational. Dolomite, which is a larger fraction than lime and the Polonite Stabilizer, was not active in the first hours.

The next dependency is given after two days. Olfactometric data have already been collected with six other respondents; their results are presented in Figure 5.

Samples are stored for 50 days. When comparing two materials – crushed dolomite and extinguished lime, both of them were embedded in the sludge, a nasty smell, according to respondents, was considered to be 8–10 days. Lime samples (20 % and 30 %) in the first 4–5 days deflected dolomite with bad smell intensity, and later the odor did not calm down. Meanwhile, samples of any concentration of dolomite not lift almost smell.

The smell of mud measurements with Polonite Stabilizer different doses of sludge and a control sample with probiotics are shown in Figure 6.

Polonite Stabilizer 20 % and 30 % of the mixing ratio among all the samples broke very high odor emissions, which after 8–10 days only declined, but not disappeared even before 40 to 50 days. And last olphactometric measurement time, 50 days, these two samples had a specific odor. The specimen with probiotics shortest and lowest intensity spread odors – already 7–8 day, as well as a control sample.

The experiment of the materials used and their effect on the percentage composition of odors formation comparison – Figure 7. It is the rejection of particularly strong impact and the lowest concentration variations.

Dolomite is the impact of odors were the least intensive, so the comparison is taken in the experiment used the maximum quantity.

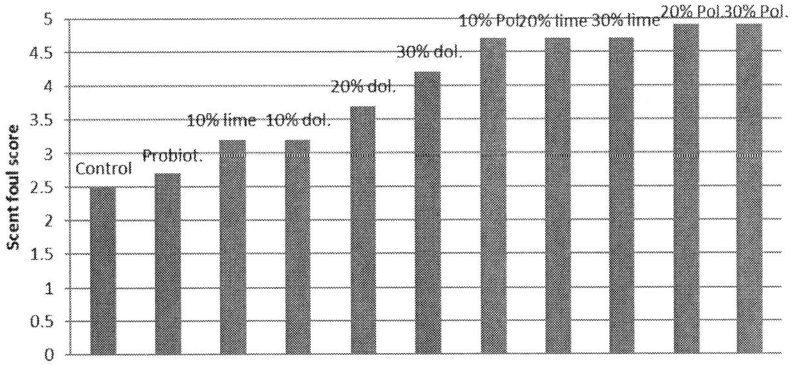

Fig. 5. Mean results of olfactometric evaluation of samples after two days after mixing of samples

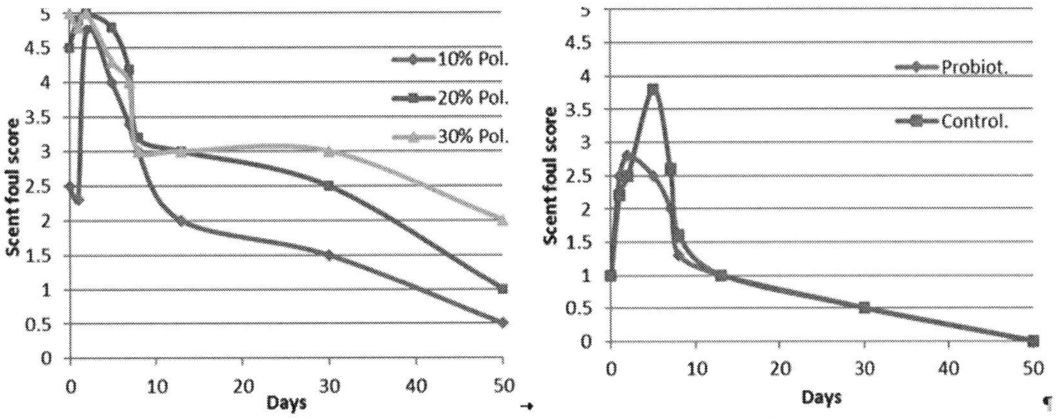

a) b)

Fig. 6. a) Sample olfactometric evaluation results for samples with averages Polonite Stabilizer; b) the samples with probiotics and the control sample

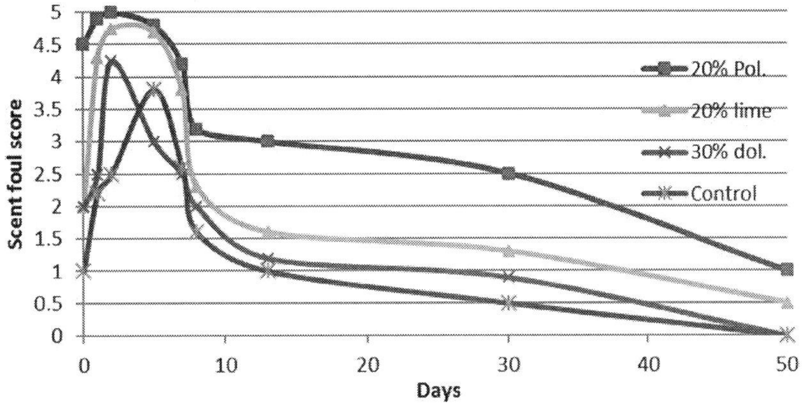

Fig. 7. Olphactometric subjective evaluation and comparison of three substances to the reference sample

Sludge structure change by adding the test substances was also documented visual assessment. Sludge samples for storage indoor and outdoor different. During the summer they had to keep under cover but exposed to sunlight and temperature sufficiently high noon. The samples were kept inside the lower ambient temperatures with a lower sun exposure. Specimens Polonite Stabilizer image after two days and thirty days (one after another) to the Figure 8.

a) b) c)

Fig. 8. a) Specimens Polonite Stabilizer image after two days after mixing and after thirty days 10 % of the ratio of; b) 20 % of the ratio of; c) 30 % of the ratio

A month later period of mixing relations with Polonite Stabilizer sludges surface darkened, contrary to the sun in summer standing. Water builds up on the surface, due to the reduced evaporation from the surface and improve their internal processes.

The sludge concentration in the measurement gas differing, used gas detector G460 (GFR Europe Ltd.) capable of measuring NH_3, CH_4, H_2S, SO_2, NO_2, CO and NO concentrations in the air. Gas evolved from the old sludge with probiotic substances, such as in room temperature tests have been conducted and reported previous to our reports [1]. The sludge mixed with stabilizing agents discovered bright time-varying NH_3 emissions and lowed SO_2 emissions. Sulfur dioxide after two hours after mixing and after two days, a similar 0.2–0.3 ppm concentration detected above four samples with 20 % and 30 % of slaked lime and 20 % and 30 % Polonite Stabilizer. The control sample and sample with probiotics for a few days at the beginning of the experiment fixed CH 4 and NO emissions. Their dynamics similar to previous experiments [1] was measured, only small concentrations.

Gas measurements were carried out at the open diffusion conditions, but using the gas tank is placed in the measuring device concentrating the cover. It tried to eliminate the uncertainty of the measurement result for the device installation space. NH_3 emissions by measuring the change in the samples with Polonite Stabilizer and probiotics and the control sample presented Figure 9.

The measurements performed by the mat shalt thou eat the NH_3 emissions immediately after mixing the sludge samples and the first days of the most significant in combination with lime. Stabilizer mixes with ammonia emissions only about 10 percent lower. At a later period of the experiment together for emissions. Gas, NH_3 emissions of samples and the control sample, dolomite and probiotics Figure 9 *b* was approximately twice lower.

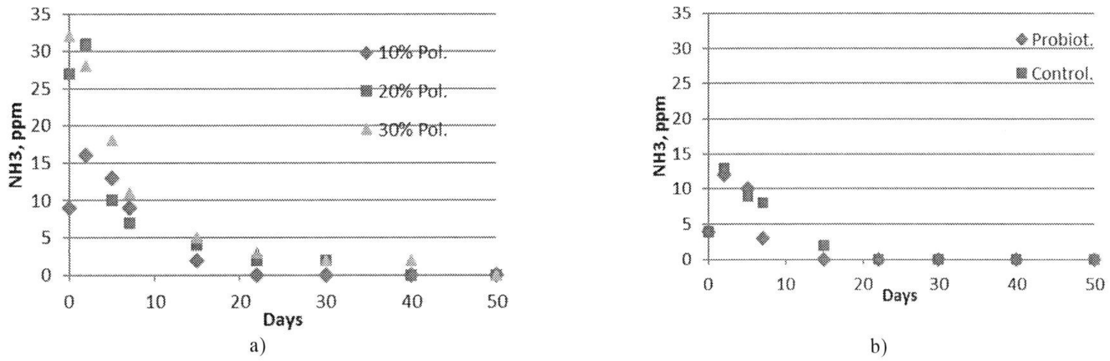

Fig. 9. a) Ammonia emission measurements during the period of the experiment sludge mixtures with Polonite Stabilizer; b) probiotics and the control sample

Perform general conclusion pH and sludge mass structure evaluations to provide 3 table. pH increases in proportion to the materials basics characteristics – mainly slaked lime, dolomite least. Material Polonite Stabilizer Interim their alkalinity, and its 20 percent of fully stabilized sludges, and making close to pH 10. The change in the structure of sludge depends on the percentage of the loaded amount of the reactants, but at ambient conditions, where the average temperature was 15 °C, there was an intense evaporation, and mass sludge structure is changed to the extent that the added material has absorbed residual moisture from the sludge mass. For the sample with 30 percent. Polonite liquid structure around the sludge volume, it can be assumed that the porous material is intensified evaporation of water from the sludge and its return back from the guise of the lid, drinking extra ambient moisture, so the mass became homogeneous liquid (Table 1).

Table 1. Sample pH and sludge mass structure evaluations

Specimen	pH	Sludge mass structure after 50 days
10 percent. stabilizer "Polonite"	8.5	Hard, adhesive on top of unchanged volume
20 percent. stabilizer "Polonite"	9.6	Hard, adhesive on top of a thick volume
30 percent. stabilizer "Polonite"	10.4	Fluid on top of the bottom liquid
10 percent. slaked lime	9.8	Unchanged top of the liquid, the liquid volume
20 percent. slaked lime	10.7	Thick top of unchanged volume
30 percent. slaked lime	11.4	Very tough, adhesive on top of a thick volume
10 percent. dolomite	8.3	Unchanged top of the liquid volume
20 percent. dolomite	8.6	Stiff, adhesive on top of a thick volume
30 percent. dolomite	9.3	Hard, adhesive on top of a thick volume
Probiotics	8.2	Hard, adhesive on top of unchanged volume
control sample	7.9	Thick top of the liquid at the bottom

Conclusions

1. Replacing the sludge structure is relevant at the intermediate stage of the technological process when it is necessary to change the properties of sludge supplied to the drying unit, to improve its structure and application on the surface of the strip. The idea of switching properties with the addition of a stabilizer will be attractive if the properties change practically immediately or within an hour. The experiment showed that even after mixing with 25 % of the stabilizer, the sludge was still more sticky, tight, had no tendency to crumble, to split from a solid mass.

2. It can be argued that storing sludge with a 15–20 % stabilizer for a longer period with a relatively thin layer of 5–8 cm in the warm, ventilated room in addition to the activated water adsorption and evaporation due to the porosity of the heated bedrocks, could improve the structure of the sludge. The use of such technology in the wastewater treatment plant is not feasible as several thousand square meters of sludge should be spread and dried, as well as appropriate equipment.
3. In the sludge structure change experiment, the smell's variation was difficult to assess, since the specimens were kept outdoors and the measurements of odors both by statistical olfactometry and measurements of gas emissions by devices were highly dependent on environmental conditions. Obviously, the olfactometric measurements of the first day after mixing showed a strong ammonia odor.
4. Olfactometric assessment within 50 days, was attended by about 20 of the respondents and their subjective view of measuring the average change in odor samples.
5. After reagent insertion, unpleasant odors occurs 2–5 day when the effects of intensive sludge. Odor intensity and duration of the change is related to the ambient temperature, which is directly correlated with the chemical and biological processes sludge activity.
6. Sludge structure changes even with the addition of 30 % reagent, in the absence of intense dryness unsuitable for the technological process.
7. Fourth offered by manufacturers in a ratio of 20 % when mixed with the sludge reagent Polonite Stabilizer, move the t 15–18 C average temperature subjective assessment:
   - after a few hours have a strong odor of ammonia;
   - after two days emits a particularly strong odor of ammonia;
   - after 50 days, emits a weak, difficult to describe, but an unpleasant odor.

References

[1] Restoring Waters in the Baltic Sea Region, 2015. Access via internet: Raceforthebaltic.com/news-events/new-report [Accessed: 2017-10-17].

[2] Brogowski Z., Renman G. Characterization of opoka as a basis for its use in wastewater treatment. 2004, Polish J. Environ. Stud. 13, pp. 15–20.

[3] Kwon H.–B., Lee C.-W., Jun B.–S., Yun J.–D., Weon S.–Y., Koopman B. Recycling waste oyster shells for eutrophication control. 2004, Resour. Conserv. Recycl. 41, pp. 75–82.

[4] Li Y., Liu C., Luan Z., Peng X., Zhu C., Chen Z., Zhang Z., Fan J., Jia Z. Phosphate removal from aqueous solutions using raw and activated red mud and fly ash. 2006, J. Hazard. Mater. B137, pp. 374–383.

[5] Renman G., Kietlinska A. Treatment of phosphorus and bacteria by filter media in onsite wastewater disposal systems, Deutsche Gesellschaft für Technische Zusammenarbeit, 2004, pp. 573–576.

[6] Hylander LD., Kietlińska A., Renman G., Simán G. Phosphorus retention in filter materials for wastewater treatment and its subsequent suitability for plant production, Bioresource technology, Elsevier, 2006.

[7] Westholm LJ. Substrates for phosphorus removal – potential benefits for on-site wastewater treatment? Water research, Elsevier, 2006.

[8] Pratt C., Shilton A. Active slag filters – simple and sustainable phosphorus removal from wastewater using steel industry byproduct – Water Science and Technology, 2010.

[9] Renman A., Renman G. Long-term phosphate removal by the calcium-silicate material Polonite in wastewater filtration systems, Chemosphere, Elsevier, 2010.

[10] Vohla C., Kõiv M., Bavor HJ., Chazarenc F. Filter materials for phosphorus removal from wastewater in treatment wetlands – A review. Ecological Engineering, Volume 37, Issue 1, January 2011, pp. 70-89.

[11] Driver J., Lijmbach D. and I. Steen. Why Recover Phosphorus for Recycling, and How? Environmental Technology, Volume 20, 1999, Issue 7, pp. 651–662.

[12] Elisabeth M., Morel Ch. A. L., Krogstad T. Plant-availability of phosphorus in filter substrates derived from small-scale wastewater treatment systems, Ecological Engineering,Volume 22, Issue 1, 1 February 2004, pp. 1–15.

[13] Cucarella V., Renman G. Phosphorus Sorption Capacity of Filter Materials Used for On-site Wastewater Treatment, Journal of Environmental Quality, Vol. 38, No. 2, pp. 381–392.

The 12th International Scientific Conference Intelligent Technologies in Logistics and Mechatronics Systems (ITELMS'2018), 26–27 April 2018, Panevėžys, Lithuania

Crack Velocity of 904L Steel and Welded Joints

Barbara Nasiłowska[a]*, Zdzisław Bogdanowicz[b]

[a]*Military University of Technology, Institute of Optoelectronics, st. Gen. Witolda Urbanowicza 2, 00-908, Warsaw 49, Poland,*
[b]*Military University of Technology, Faculty of Mechanical Engineering, st. Gen. Witolda Urbanowicza 2, 00-908, Warsaw 49, Poland*

Abstract

The paper presents the results of experimental tests of crack velocity and cracking development of joints made by GTAW method and laser beam of 904L austenitic steel. The crack development period was the longest in the parent material, whereas the shortest in laser welded joints. It was proved that shot peening of the surface layer of the welded joints caused retardation of the initiation of fatigue cracks and an increase of a number of cycles to damage.

© 2018 B. Nasiłowska, Z. Bogdanowicz
Peer-review under responsibility of the Kaunas University of Technology, Panevėžys Faculty of Technologies and Business

Keywords: austenitic steel, crack velocity, welded joints, shot peening, GTAW, laser beam

1. Introduction

Development of a destructive phenomenon, namely fatigue of material, in the constructive material is a characteristic feature of structure exploitation at time-varying stress. This multi-step fatigue phenomenon influences a change in physical and mechanical properties of the material, change of the structure (separation of intermetallic phases, etc.), susceptibility of the element to cracking and crack growth, which consequently results in increased fatigue life of the structure [1, 2].

Potential areas of material fatigue initiation are local stress concentration in the material (structural heterogeneity), geometrical notches (in the point of cross-section changes) or change of element curvature. In these places/points, the yield point of the material is locally exceeded and permanent plastic deformations are formed – the basic condition for the development of the fatigue phenomenon of the material and structure.

In chemical installations, in addition to material defects, the main causes of damage due to long-term exploitation are, among others, notches and welding defects leading to the inintion of fatigue fractures [3, 4, 5]. Sample

* Corresponding author. Tel.
E-mail address: barbara.nasilowska@wat.edu.pl

© 2018 B. Nasiłowska, Z. Bogdanowicz
Peer-review under responsibility of the Kaunas University of Technology, Panevėžys Faculty of Technologies and Business

B427H0027 . Ed*Learning*.it

photographs of pipeline sections after a failure due to fatigue fracture, initiated in the welded joint under the influence of long-term exploitation [4].

The cracking rate of 1.4435 and 1.4404 austenitic steel welded with GTAW (TIG) method was tested by Kusko, Dupont and Marder [6]. Through microfractographic analysis of the surface of the fatigue fracture, they proved that that the fracture growth depends on the stress amplitude.

Maddox [7], Parry et al. [8] pointed out that the increase in weld fatigue fractures can be described with Paris's formula which links the fatigue fracture development rate da/dN with the range of stress intensity factor ΔK [9] (1).

$$\frac{dl}{dN} = C(\Delta K)^n. \qquad (1)$$

In this formula, C and n are constants, whereas:

$$\Delta K = \Delta\sigma_a \sqrt{\pi l}; \qquad (2)$$

where $\Delta\sigma_a$ – a range of cycle stress amplitude.

According to the equation (1) the rate of fatigue cracks in the elements is determined by the value of the range of changes in the stress intensity coefficient and, additionally, on the R asymmetry coefficient of the load cycle [10].

Microscopic observations of the fracture development allowed the authors of the paper [11] to refer to the mechanism of crack propagation and fatigue life. The tests were carried out on round samples with a diameter of 44.4 mm and 25.4 with a cyclic loading frequency of 0.1 Hz to 5.0 Hz at room temperature. In the case of R=-1, a transient suppression of a fracture rate growth was observed.

Predicting the fatigue life of the laser welded stainless steel joints is presented in [12, 13, 14]. The observations of the fatigue fractures of 1.4301 austenitic steel proved that fatigue fractures initiate in the vicinity of the phase σ. The weld structure is fully dendritic, which results from fast cooling.

The analysis of the problem status included in work [15] demonstrated a considerable improvement of stress distribution, utility properties, i.e. hardness, fatigue strength, yield point, impact reduction and austenitic steel extension after the shot peening process.

The available literature lacks a full definition of an influence of joints made of 904L "super"-austenitic steel and welded with GTAW method and with a laser beam on the cracking rate. The information presented in the literature [16] inconsistently indicates a drop in the cracking rate, which is confirmed by the fact of necessity to conduct tests concerning an influence of shot peening of welded joints on the crack development and a cracking rate.

2. Experimental research

Specimens of 904L steel with dimensions 500x60x5 mm cut out with Waterjet technique from 250x500x5 mm steel sheets welded with GTAW method and with a laser beam were subjected to own tests (Fig. 2 and 3). GTAW welding was carried out with a non-consumable tungsten electrode ϕ 2.5mm MTC MT-904L and with weld G/W 20 25 5 CuL (20 % Cr, 25 % Ni, 4.5 % Mo, 1.5 % Cu) in Zakłady Budowy Aparatury Chemicznej – Grupa Azoty in Tarnów according to a production technology applied in practice to produce, among others, chemical apparatus.

Laser welding was carried out in the Centre for Laser Technologies of Metals of Kielce University of Technology with the use of CO_2 TRIUMF 1005 laser at the distance of 260 mm of the focal length of lens with a spot on the surface of the specimen ϕ 0.4 mm with power P = 4.5 kW and welding speed v = 1.4 m/min in the helium shield.

A batch of specimens after welding was subjected to the mechanical shot peening process. Two-sided shot peening of the surface layer of the joints of 904L steel welded with a laser beam and with GTAW method was conducted in the Institute of Precision Mechanics with the use of a jet of a spring steel shot (hardness ca 640 HV0,1), with a diameter of ϕ 0.8 mm under the pressure of 5bar. The exposition time was 6 mins, specimen coverage was 100%. Shot peening intensity defined with the use of Almen Strips plate was f_A = 0.246 mm. Such parameters were applied to a large number of welded structures [16].

Tests of fatigue fractures development were carried out on the specimens (Fig. 2 and 3) with an edge notch on hydraulic pulsator INSTRON 8802 under the conditions of unilaterally alternating tension at a cycle asymmetry coefficient R = 0.1 with frequency equal to 3 Hz (Fig. 1).

The record of the fracture development was conducted with the use of an optical microscope with zoom 4x (Projektiv k 4:1, Carl Zeiss Jena) and with a stuck millimeter straight edge with record accuracy 0.1 mm.

A geometric shape of the notch was carried out according to standard [17] and literature analysis [18, 19, 20, 21, 22]. The angle between the edges of the notch cut out to the depth of 5.5 mm was equal to 30° [17, 20].

Tests of fractures of specimens made of the parent material, joints welded with GTAW method and with a laser beam were conducted on three levels of nominal stress amplitude σ_{an} = 250, 300, 350 MPa.

The Crack Opening Displacement was recorded with the use of Instron S 2620-601 extensometer with a measuring base 5 mm. It was used to determine the beginning of the fracture initiation.

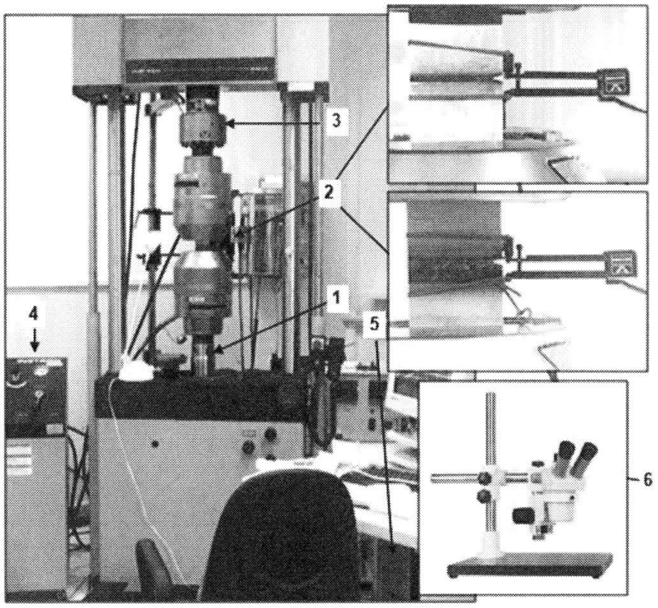

Fig. 1. Hydraulic pulsator INSTRON 8802 (hydraulic actuator – 1, specimens – 2, dynamometer – 3, power supply – 4, control unit – 5, optical microscope – 6, extensometer – 7)

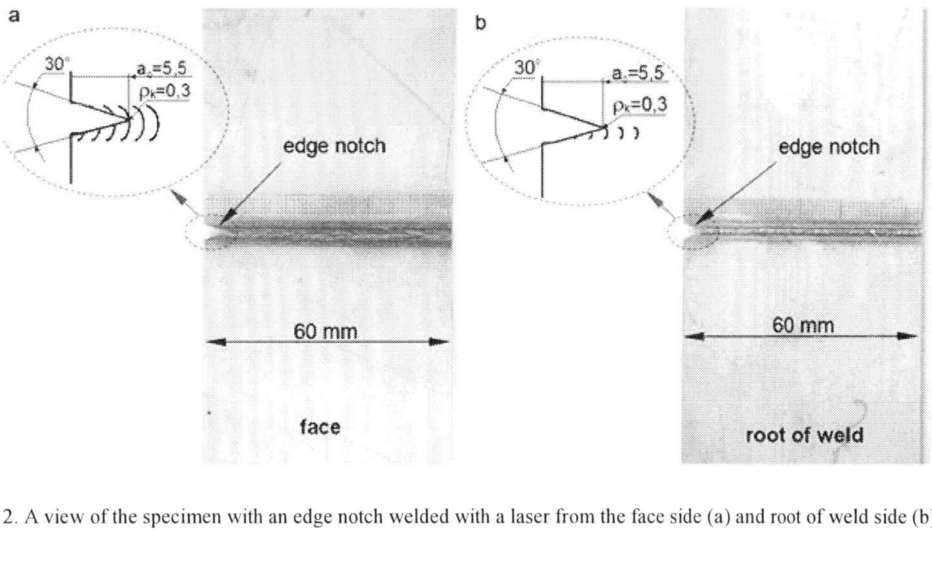

Fig. 2. A view of the specimen with an edge notch welded with a laser from the face side (a) and root of weld side (b)

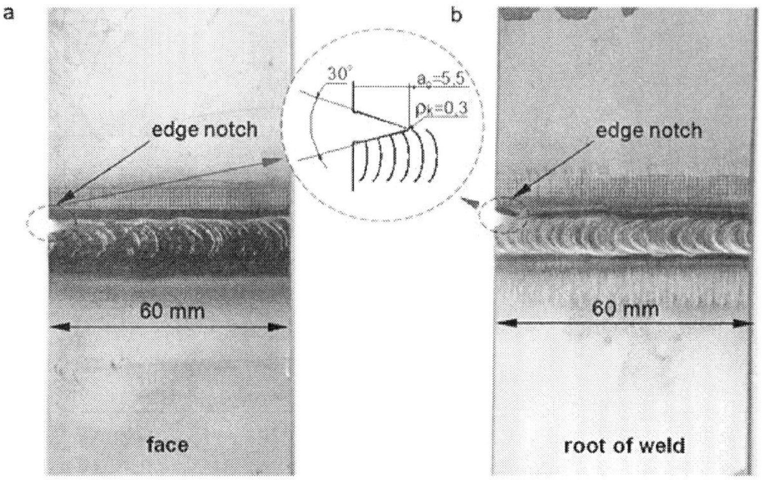

Fig. 3. A view of the specimen with an edge notch welded with GTAW method from the face side (a) and root of weld side (b)

Analysis of specimens transverse metallographic microsetions after fatigue tests was conducted with a light microscope Nicon MA 200, whereas microfractography of the fractures with scanning electron microscope Quanta 3D FEG.

3. Process of base material fatigue fracture

During the tests, both a shape and size of the plastic zone around the peak of the fracture (visible even to a naked eye) was changing with a growth of cycles and a size of its opening (Fig. 4 and 5).

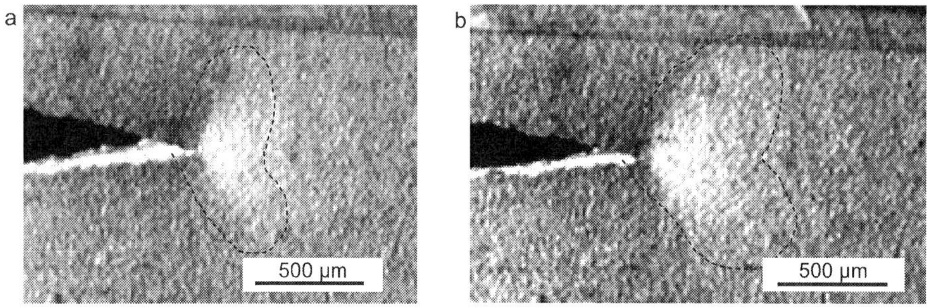

Fig. 4. Shape of the plastic zone in a front of the gap

In the final stage of fracturing, there were generated a visible slide lines arranging along the front of the parent material flow (Fig. 5).

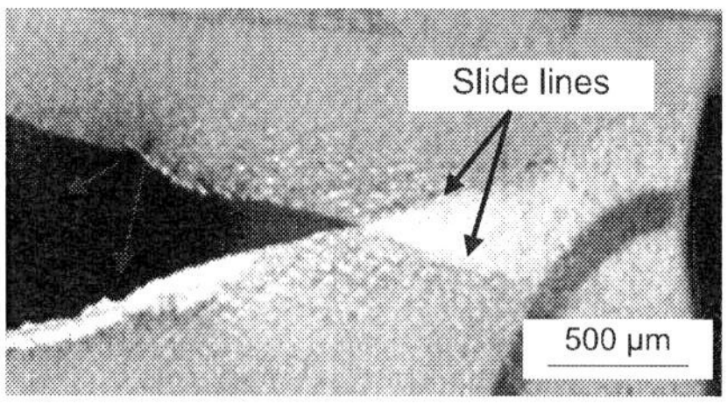

Fig. 5. The shape of the plastic zone in the final stage of specimen fracturing

A growth of the fracture occurred in the slide planes, which is rather typical for austenitic steels with a yield point $\geq$ 230 MPa (according to [23]).

The metalographic analysis (Fig. 6) showed a typical development and process of fatigue fracture for specimens with a one-side notch cut out. On the surface of the fatigue fracture, there were observed numerous secondary fractures (Fig. 6, *a*). Along the specimen axis, perpendicularly to the line of the front of the cut notch, there occurred a visible focal slide (Fig. 6, *b*).

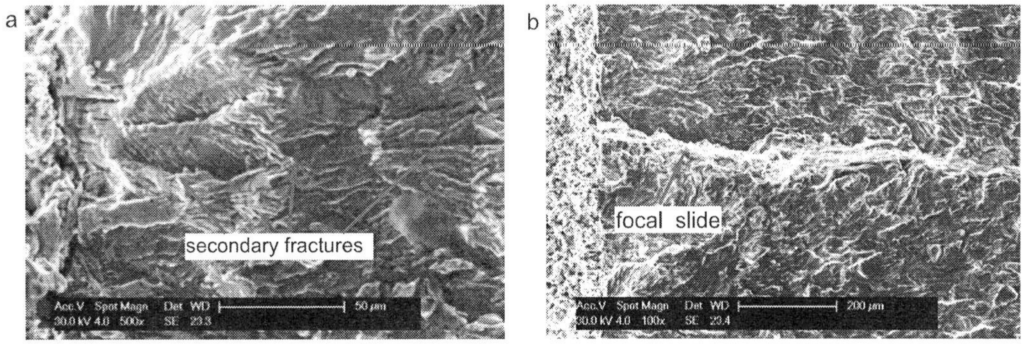

Fig. 6. Fatigue fracture of the parent material near the centre of the fracture zone

The fracture was spreaded successively from the zone near the centre of the fracture, while a number of cycles was increasing, creating a fatigue zone. The size of fatigue stripes arranged perpendicularly to the front of the peak was increasing successively while fracturing was developing (Fig. 7, *a*, *b*, *c*, *d*).

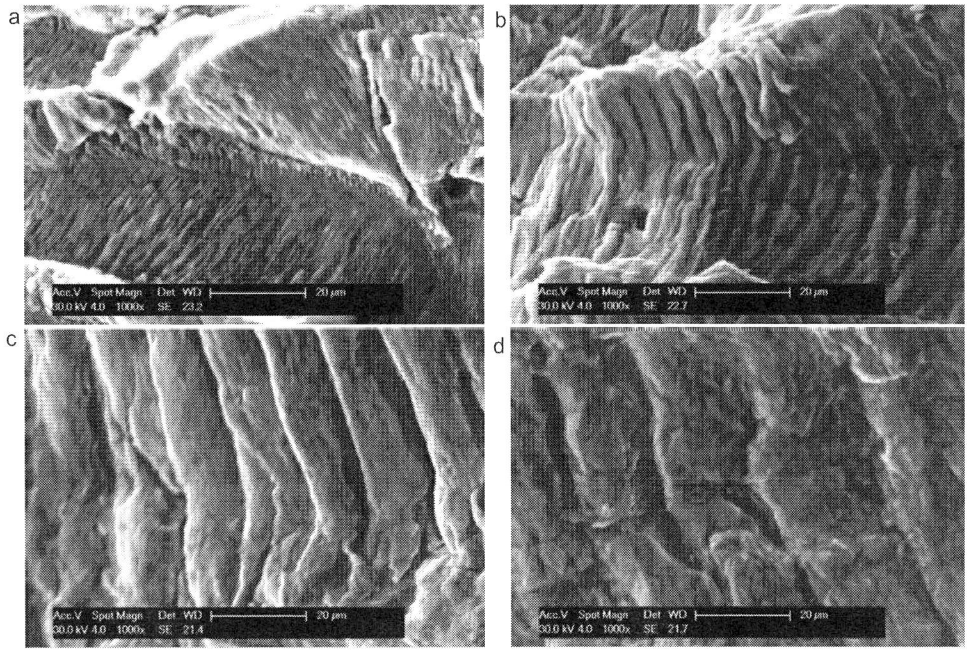

Fig. 7. Changes of fatigue stripes while the crack rate was increasing

A transitional zone (Fig. 7, *a*) occurred on the border of the fatigue zone which was transferring into a residual zone (Fig. 7, *b*) in which a typical sliced structure with numerous holes was observed.

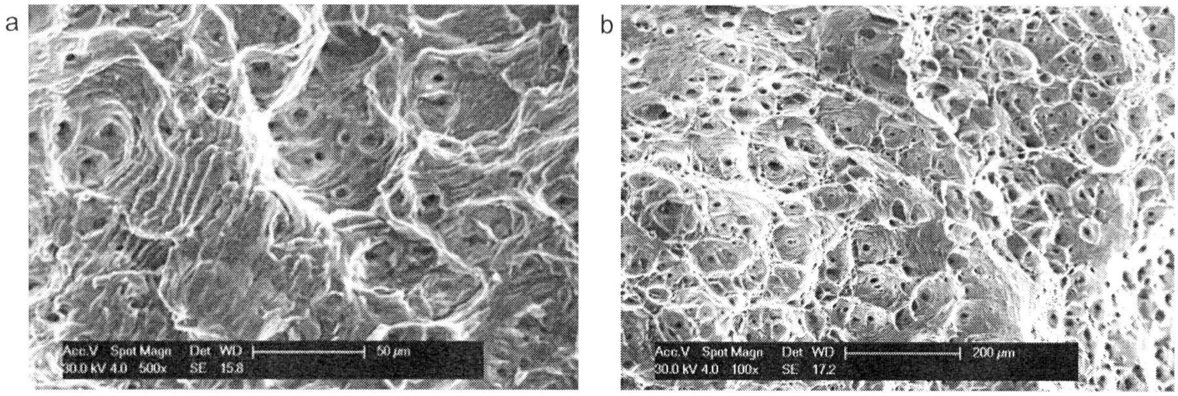

Fig. 8. The transitional (a) and residual (b) zone of the parent material tested at constant stress amplitude σ_{an} = 300 MPa

4. Development of fatigue fractures of joints welded with a laser beam and GTAW method

The development of fatigue fractures of joints welded with a laser beam and GTAW method in each of the three considered levels of nominal stress amplitude (σ_{an} = 250, 300, 350 MPa) took place in a similar way (Fig. 9).

The longest period of the fracture development occurred in the parent material, whereas the shortest in the joints welded with a laser beam, which was indicated by a number of cycles in the conditions of fatigue fractures development N_R, that is, a difference between a number of cycles to damage the specimens N_f, and a number of cycles to initiate the fatigue fractures N_i. It is caused by, among others, structural parameters generated during the welding process. In the weld made with a laser beam there occurred a one-stitch weld with a flat front of solidification of crystallites arranged in a herringbone form towards heat dissipation. Whereas during generating the weld made with GTAW method, there were carried out three passages of the weld, two of them from the face side, then the root of weld was undercut and pre-welded. It probably contributed to a slower development of the front of the fatigue fracture than in the one-stitch welds made with a laser beam. Development of fracturing of the specimens with a front fracture weld is not as regular as in the parent material. It is caused by passing/moving of the fracturing through different zones of the welded joint (as arrows indicate).

The period of fatigue fractures development was dependent on the working cycle stress amplitude (Fig. 9).

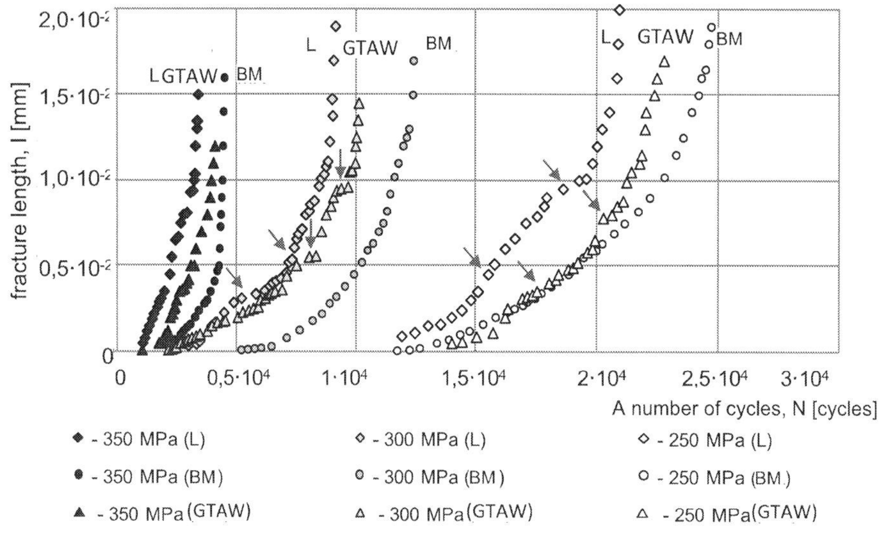

Fig. 9 An increase in fracture length in specimens made with GTAW method and with a laser beam, as well as made of the parent material, tested at a different level of stress amplitude 250, 300 and 350 MPa

Shot peening of the surface layer of the analyzed welded joints caused retardation in fatigue fractures initiation by, for example, 50% at the level of nominal stress amplitude σ_{an} = 300 MPa (Fig. 10).

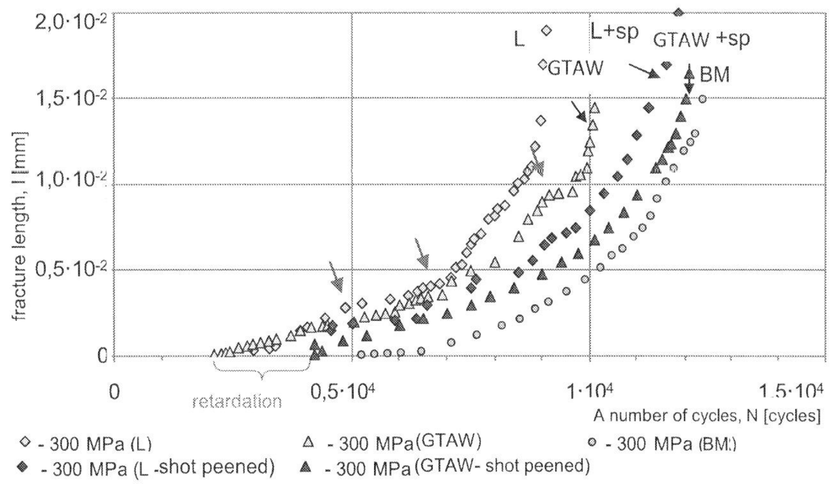

Fig. 10. An increase in fracture length of specimens made of the base material and the specimens with a weld made with GTAW method and with a laser beam as well as not shot peened specimens tested at nominal stress amplitude 300 MPa

A strengthened surface layer favourably influenced also development of fracturing. A number of cycles to damage the specimens with a weld made with GTAW method and with a laser beam during development of fracturing was greater by approximately 20 % and 25 %, respectively, than in the case of not shot peened specimens. It is mainly a result of the generated compressive residual stress [24] generated after the shot peening process.

In the fatigue zone of the joints welded with a laser beam, there were observed visible fatigue stripes which revealed fragments of dendrites generated during the process of weld crystallization (Fig. 11).

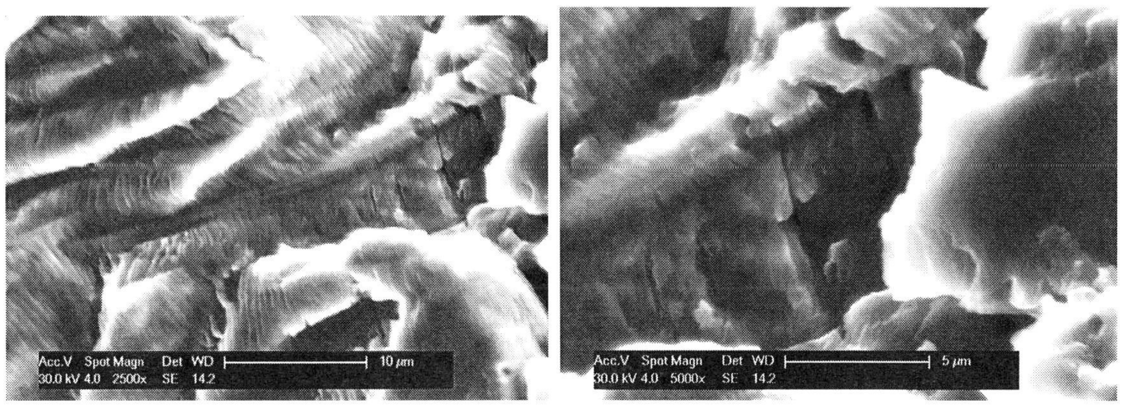

Fig. 11. Fragments of dendrites, generated during the process of weld crystallization, revealed on the fatigue stripes

5. Crack velocityof the parent material and welded joints

The crack velocityof joints welded with GTAW method and with a laser beam was described by Paris formula (1.1). According to publication [23] "the crack rate in the welded steel elements, i.e., welds and heat affected zones is generally slightly higher than in the parent material", which was confirmed in our own tests and is presented in Fig. 12.

The author of [23] noticed also that the crack rate is defined by separation of secondary phases generated as a result of the welding process, their distribution, distance between separations and coherence with a matrix.

The crack rate of the welded joints and 904L steel parent material was slightly diverse independently on the method of manufacturing. The lowest crack rate was characteristic for the welded joints subjected to shot peening (Fig. 12). Reduction in crack rate in shot peened elements resulted from a favourable system of compressive residual stress described in work [3].

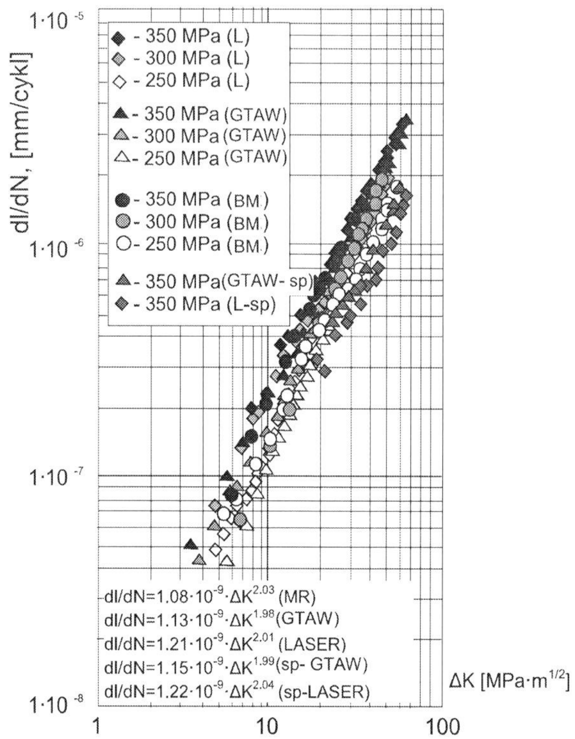

Fig. 12. Fatigue crack rate of the parent material, elements with a weld made with a laser beam and GTAW method and shot peened joints

Tests [24] proved a high metallurgic purity of 904L austenitic steel and the considered welded joints, which was likely to influence slight diverse in crack rate in not shot peened elements.

Figure 12 presents a collective graph of crack rate for the parent material, joints made with GTAW method and with a laser beam after shot peening of the surface layer as well as for not shot peened joints. C and n coefficients are presented in Table 1.

Table 1. C and n coefficient values according to Paris formula

	C $\left[\frac{m}{\left(MPa\sqrt{m}\right)^{n}\cdot cykl}\right]$	n
Parent material	$1.08\cdot10^{-9}$	2.03
Specimen with a laser beam – not shot peened	$1.21\cdot10^{-9}$	2.01
Specimen with with GTAW method – notshot peened	$1.13\cdot10^{-9}$	1.98

Conclusion

The process of fatigue fracture of the parent material, joints welded with a laser beam and with GTAW method in each of the three considered levels of nominal stress amplitudes (σ_{an} = 250, 300, 350 MPa) was run in a similar way.

Development of fatigue fractures was dependent on the working cycle stress amplitude. The longest period of fatigue fractures development occurred in the parent material, whereas the shortest in the joints welded with a laser beam. The shot peening of the surface layer of the analyzed welded joint retarded fatigue fractures initiation by, for example, 50% at the level of nominal stress amplitude σ_{an} = 300 MPa. These joints were also characterized by the lowest fracture rate.

A strengthened surface layer favourably influenced also development of fracturing . A number of cycles to damage of the specimens with a weld made with GTAW method and with a laser beam during development of fracturing was greater by approximately 20% and 25%, respectively, than in the case of not shot peened specimens.

The crack rate of the joints welded with a laser beam and with GTAW method was similar, which can be conditioned by metallurgic purity of steel, residual stress and the structure of the welded joint. The lowest crack rate was demonstrated by the welded joints after shot peening.

References

[1] Kocańda S. Zmęczeniowe pękanie metali. WNT, Warszawa: 1985.

[2] Kocańda D. Analiza rozwoju krótkich pęknięć zmęczeniowych, rozprawa habilitacyjna, WAT, Warszawa, 1996.

[3] Bogdanowicz Z., Jóźwik P., Nasiłowska B. Microstructure and mechanical behavior of a CO2 laser and TIG welded 904L steel, *Metallurgy and Foundry Engineering* 2014, 40 (2): 69–81.

[4] Palmer AC., King RA. Subsea Pipeline Engineering, *PennWell Books* 2008, 624 p.

[5] Fazzini PG., Belmonte JC., Chapetti MD., Otegui JL. Fatigue assessment of a double submerged arc welded gas pipeline International, *Journal of Fatigue* 2007, 29: 1115–1124.

[6] Kusko CS., Dupont JN., Marder AR. Influence of Stress Ratio on Fatigue Fracture Propagation Behavior of Stainless Steel Welds. Fracture initiation and growth rates in relation to residual stresses were studied in gas metal arc welds of 316L, *Welding Research* 2004.

[7] Maddox SJ. Fatigue design rules for welded structures. Fracture and Fatigue of Welded Joints and Structures 2011, pp. 168–207.

[8] Parry M., Nordberg H., Hertzberg RW. Fatigue fracture-propagation in A514 base plate and welded-joints, *Welding Journal* 1972, 51(10): 485–490.

[9] Paris PC. The Fracture Mechanics Approach to Fatigue, Fatigue, An Interdisciplinary Approach, Proceedings Tenth Sagamore Army Materials Research Conference, *Syracuse University Press* 1964, pp. 107–132.

[10] James LA. Fracture Propagation Behavior in Type 304 Stainless Steel Weldments at Elevated Temperature, Nuclear Technology 14: pp. 163–170, 1972.

[11] Dattoma V. Prediction of the fatigue life of laser welded stainless steel joints, *Fatigue and Fracture of Engineering Materials and Structures* 1996, 17 (11): 1335–1342.

[12] Goss Cz., Nasiłowska B., Śnieżek L. Niskocyklowa trwałość zmęczeniowa połączeń spawanych ze stali X5CrNi18-8, *Biuletyn WAT* 2012, 4: 249–263.

[3] Bogdanowicz Z., Nasiłowska B. Wpływ kulowania na trwałość zmęczeniową stali austenitycznej 904 L spawanej metodą TIG, *Sympozjon Podstaw Konstrukcji Maszyn* 2015.

[14] Slbabe S., Fuentes M. Characterization of low-cycle fatigue and creep-fatigue induced damage in a 316L stainless steel (Characterisierung von durch niedrigperiodische Ermudang lind durch schleichende Ermudung verursachte Schaden im Falle van Nichtrustendem Stahl A 316) *Z. Metallkd* 1994, 85 (4): 273–281.

[15] Shaikh H. Corrosion of Austenitic Stainless Steels Mechanism, Mitigation and Monitoring A volume in Woodhead Publishing Series in Metals and Surface Engineering, 7 – Stress Corrosion Fracturing of Austenitic Stainless Steel Weldments 2002; pp. 166–189.

[16] Nakonieczny A. Dynamiczna powierzchniowa obróbka plastyczna – kulowanie (shot peeling), *IMP* 2002.

[17] Designation: E647 – 13a Standard Test Method for Measurement of Fatigue Fracture Growth Rates1.

[18] Bochenek A. Elementy mechaniki pękania- podręcznik dla materiałoznawców, *Częstochowa: Wydaw. Politechniki Częstochowskiej*; 1998.

[19] Neimitz A. Mechanika pękania, *PWN* 1998.

[20] Lepionka M. Badania rozwoju pęknięć zmęczeniowych przy obciążeniach z przeciążeniami, *ITWL* 1998.

[21] Kłysz S. Rozwój pęknięć zmęczeniowych w materiałach lotniczych i stali konstrukcyjnej z uwzględnieniem przeciążeń, *ITWL* 2001.

[22] Kocańda S., Kocańda A. Niskocyklowa wytrzymałość zmęczeniowa metali, *PWN* 198, Warszawa.

[23] PN-EN 10088-1:2014-12 – wersja angielska. Stale odporne na korozję – Część 1: Wykaz stali odpornych na korozję.

[24] Nasiłowska B., Bogdanowicz Z., Mońka G., Szymański W. Wpływ kulowania na właściwości użytkowe połączeń spawanych stali austenitycznej 1.4539 wykonanych metodą TIG i wiązką lasera. *Przegląd Spawalnictwa* 2015; 87(3): 36–39.

The 12th International Scientific Conference Intelligent Technologies in Logistics and Mechatronics Systems (ITELMS'2018), 26–27 April 2018, Panevėžys, Lithuania

Corrosion Protection for S235 JR Steel with Graphene Oxide Covering

Barbara Nasiłowska[a*], Zdzisław Bogdanowicz[b], Michał Wojucki[c], Bartosz Bartosewicz[a], Małgorzata Djas[d]

[a] *Military University of Technology, Institute of Optoelectronics, st. Gen. Witolda Urbanowicza 2, 00-908, Warsaw 49, Poland*
[b] *Military University of Technology, Faculty of Mechanical Engineering, st. Gen. Witolda Urbanowicza 2, 00-908, Warsaw 49, Poland*
[c] *Institute of Precusion Mechanics, Department of Protective Coatings, st. Duchnicka 3, 01-796 Warsaw, Poland*
[d] *Institute of Electronic Materials Technology, Department of Chemical Technologies, Wólczyńska 133, Warsaw 01-919, Poland*

Abstract

Work presents positive impact of usage of Graphene Oxide on steel S235 JR surface protection against development of general corrosion. The surface of samples was analysed after 1000h exposition to atmospheric environment and after 5h combined exposition to aqueous C_3H_6O solution with ultrasound generated by converter. The volume ratio of H_2O /C_3H_6O was 50/50 Studies showed that despite the action of ultrasound on the surface, the Graphene Oxide cover was good prevention against corrosive factors.

© 2018 B. Nasiłowska, Z. Bogdanowicz, M. Wojucki, B. Bartosewicz, M. Djas
Peer-review under responsibility of the Kaunas University of Technology, Panevėžys Faculty of Technologies and Business

Keywords: graphene oxide, corrosion, S235 JR steel

1. Introduction

Despite many numbers of solutions of anticorrosion protection, economic losses related to failures and downtime are still large [1, 2]. This forces continuous research for new solutions in the field of fundamental research, development works and implementations to cause decreasement of direct and indirect losses resulted from corrosion of metals [3]. At present intensive studies in application of graphene oxide (GO) underway in many fields, among others as corrosion proofing. Most vulnerable to corrosion are steels which chemical composition does not provide

* Corresponding author. Tel.
E-mail address: barbara.nasilowska@wat.edu.pl

© 2018 B. Nasiłowska, Z. Bogdanowicz, M. Wojucki, B. Bartosewicz, M. Djas B427H0028 . EdLearning.it
Peer-review under responsibility of the Kaunas University of Technology, Panevėžys Faculty of Technologies and Business

cathodic protection and without additional surface treatment i.e. shot peening [4].

In the article [5] exhibited enhancement of the corrosion protection performance and cathodic delamination resistance of epoxy coating through treatment of steel substrate by a novel nanometric sol-gel based silane composite film filled with functionalized graphene oxide nanosheets. Results reveal that incorporation of fGO nanosheets into the silane film significantly enhances the corrosion protection performance of the epoxy coating and reduced cathodic delamination. Graphene oxide significantly enhanced corrosion resistance of silane film and GO/silane film significantly improved corrosion protection performance of epoxy coating [5].

An amino functionalized graphene oxide (FGO) was synthesized and characterized by Fourier transform infrared spectroscopy (FTIR) and X-Ray diffraction analysis (XRD). Then, FGO/epoxy composite was prepared through dispersing 0.1 wt.% of FGO in an epoxy coating through wet transfer method (WTM). The GO/epoxy and FGO/epoxy composites were applied on the mild steel substrates and their barrier and corrosion protection performance were characterized by salt spray test and electrochemical impedance spectroscopy (EIS). Incorporation of 0.1 wt.% of FGO nanosheets into the epoxy coating significantly enhanced the corrosion resistance of the coating through improving its ionic resistance as well as barrier properties [6]. FGO significantly enhanced the corrosion resistance of the epoxy composite. The graphene coating on stainless steel can enhance its anti-corrosion property and exhibits a low interfacial contact resistance [7].

In scientific work [8] the present work investigates the barrier protection mechanism of functionalized graphene when coated on mild steel substrate. Electrochemical parameters of the coating such as pore resistance, coating capacitance and water uptake characteristics were evaluated using electrochemical impedance spectroscopy. Additionally, breakpoint frequencies were deduced to evaluate the electrochemical activity at the surface over time. These electrochemical results in combination with the structural analysis of the coating help in developing a systematic understanding of the barrier properties rendered by graphene [8].

2. Methodology

Items submitted to GO deposition were cut from construction steel sweet for general applications S235JR, this material is applied to welded construction, supporting structures, bearing and dynamically loaded i.e. pylons, towers, parts of machines and equipment, housings, wheels and platforms.

Deposition of graphene oxide was performed according to procedure developed in the Institute of Optoelectronics of the Military University of Technology. Graphene oxide was produced in Institute of Electronic Materials Technology.

Studies of the structure were perform with high-resolution scanning electron microscope Quanta 250 by FEI (Fig. 1, *a*), confocal microscope Zeiss LSM 800 and optical microscope Smartcoom 5.

Presence of GO was confirmed by typical spectroscopy spectrum FT-IR (System Spectrum GX) in the extent of the 0–2000 cm^{-1} (laser wave length 532 nm and 633 nm).

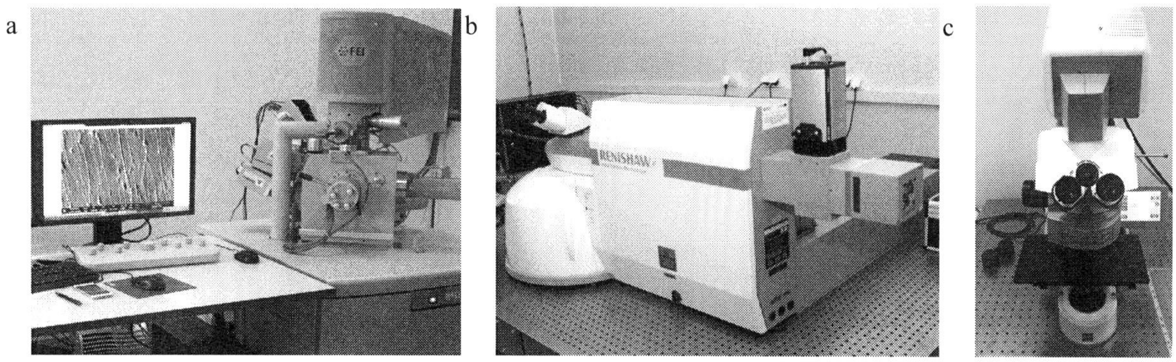

Fig. 1. (a) Scanning microscope Quanta 250 FEG by FEI (b) Raman spectrometer (c) confocal microscope

After deposition of GO, samples were placed in aqueous C_3H_6O solution and subjected to ultrasound generated by converter. Mixing process of the aqueous C_3H_6O solution induced alternately formulation of high and low pressure spheres. Micro-bubbles occurred in aqueous C_3H_6O solution, were imploding in high pressure phase. Subsequently studies of surface topography on confocal microscope and analysis of spectrum in Raman Spectroscopy were performed.

3. Results of studies

During the experiment S235 JR steel was immersed in C_3H_6O solution and subjected to ultrasound generated by converter. Test samples were covered with GO according to the technology developed in Institute of Optoelectronics of Military University of Technology. Pictures of GO deposited on sample with spectroscopic spectrum before the studies are show on Figure 2. On the picture there are two typical peaks for graphene oxide D, G 2D and D+G are marked [9].

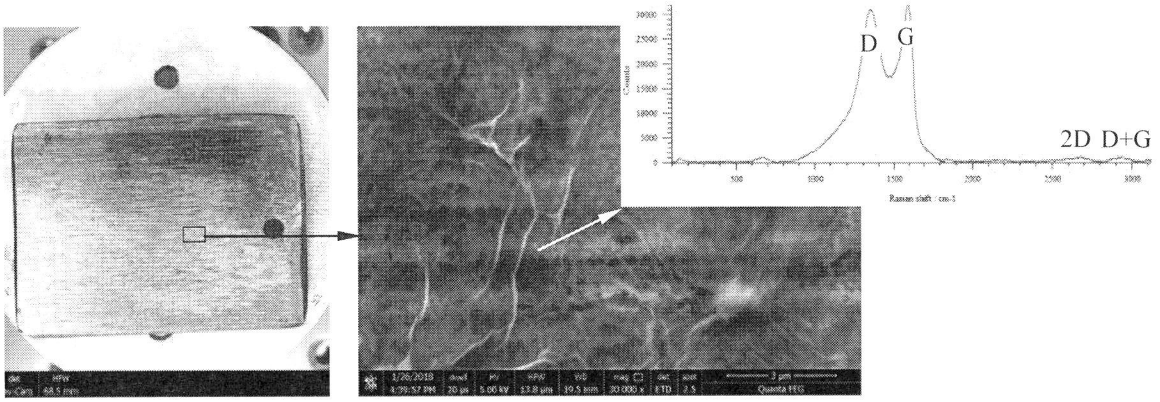

Fig. 2. Surface of steel S235 JR covered with graphene oxide (GO)

Studies have shown that surface covered with GO is prevented against corrosive factors such a humidity.

In the figures below there are pictures of surface topography of steel S235 JR as delivered (Fig. 3, *a*), after 1000h exposition to atmospheric environment (Fig. 3, *b*) and after exposition to aqueous C_3H_6O solution with ultrasound treatment (Fig. 3, *c*).

The largest corrosion increment was observed on the surface of the samples not covered with GO and subjected to aqueous C_3H_6O solution with ultrasound (Fig. 3, *c*, *f*).

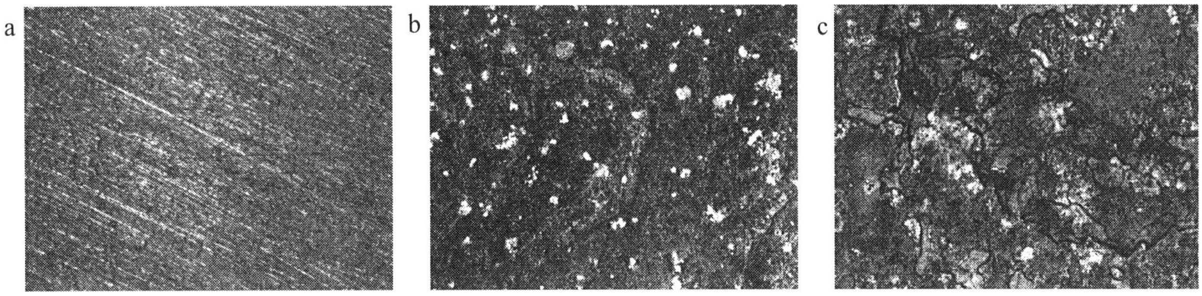

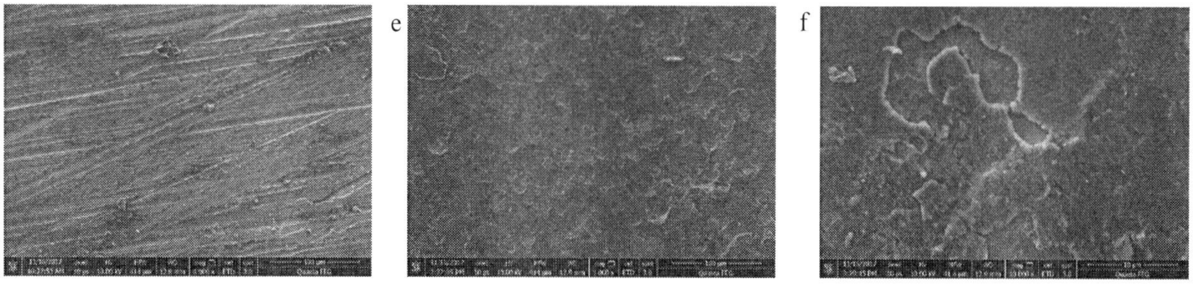

Fig. 3. Surface morphology of S235 JR steel as delivered (a) after 1000h of exposition of atmospheric environment (b), after 5h of exposition of aqueous C3H6O solution with ultrasound (c) performed by optical microscope (a, b c) and scanning microscope (d, e, f) 35 x

With usage of confocal microscope study of surface topography was done (Fig. 4). In case of samples without GO cover emerging corrosion pits caused larger surface development and decease the grade of roughness (Table 1). Surface roughness in comparison to delivered samples (Table 1).

Table 1. Surface roughness of studied samples

	Ra	Rz	Roughness grade
Surface of sample as delivered	0.63	3.2	8
Surface of sample covered with GO after 5h in aqueous C_3H_6O solution with ultrasound	0.672	2.313	8
Surface of sample non covered with GO after 5h in aqueous C_3H_6O solution with ultrasound	1.439	8.391	6

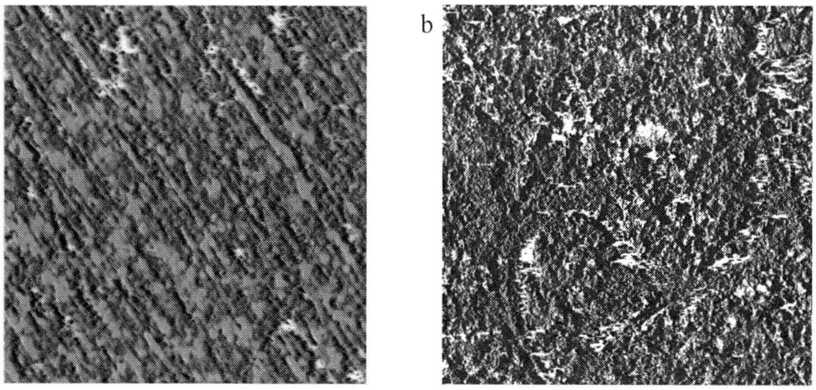

Fig. 4. Surface topography of S235 JR steel after 5h of exposition to aqueous C3H6O solution with ultrasound (a) covered with GO (b) non covered with GO

The Raman spectroscopy of the surface of samples was done after 1000h of exposition to atmospheric environment. Studies confirmed presence of C-C bonds in the extent of the 150–450 in aliphatic chain. The confirmed presence of spectra originating from corrosion products on the samples non covered with (Fig. 5) General corrosion occurred on entire surface non covered with GO reaching 100 %, whereas studies performed on samples covered with GO showed insignificant development of corrosion, in approximation 2 %.

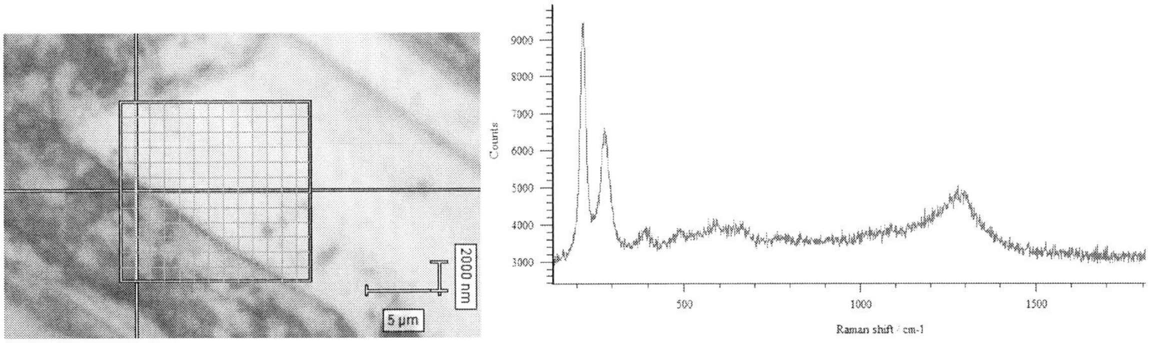

Fig. 5. Raman spectra of surface of steel non covered with GO after 1000h of exposition of atmospheric environment

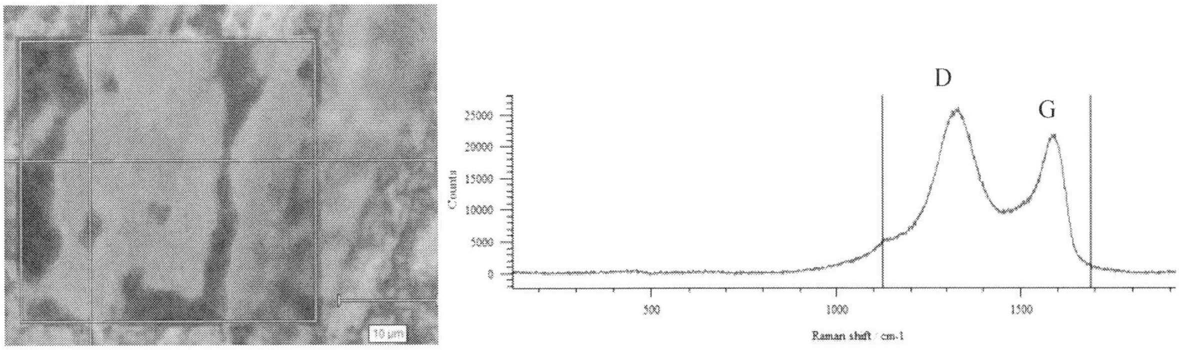

Fig. 6. Raman spectra of surface of steel covered with GO after 1000h of exposition of atmospheric environment

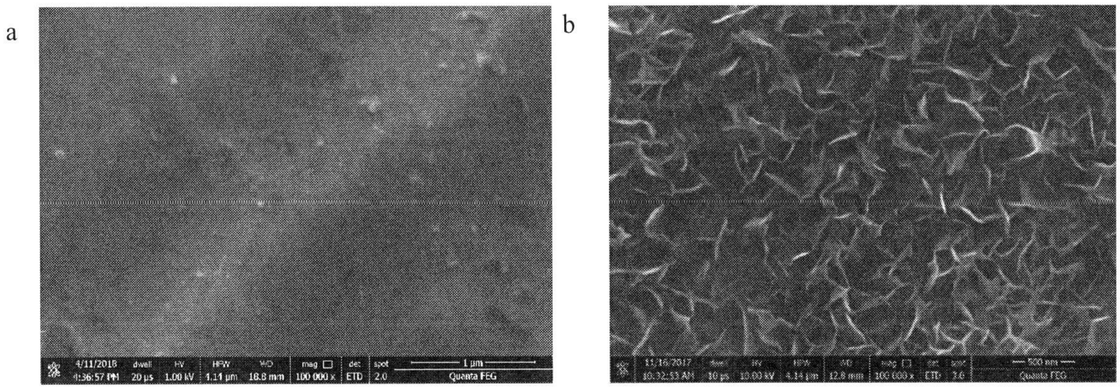

Fig. 7. Structure of surface GO after exposition to C3H6O solution and ultrasound

Under the affection of aqueous C_3H_6O solution with ultrasound on the surface of the sample covered with GO no corrosion changes were observed, whereas under magnification 100 000x the deformation of topography of graphene oxide lamellas is visible. Initially before the studies graphene lamellas were homogenic structure covering surface of steel (Fig. 7, *a*) whereas after exposition to aqueous C_3H_6O solution with ultrasound in places where lamellas are connected they apparently changed their locations (Fig. 7, *b*). It is related to action of acoustic ultrasound waves on GO planes.

Conclusion

Basing on performed studies it was found adequate deposition of GO on surface of steel S235 blocks development of surface corrosion induced by influence of atmospheric environment.

Effect of C_3H_6O solution with ultrasound on the sample non covered with GO caused acceleration of general corrosion. Whereas in case of sample covered with GO the same conditions did not induce the corrosion process. The only observed effect was change in topography of GO lamellas. Those lamellas were barrier for corrosive factors.

References

[1] PHMSA (Pipeline and Hazardous Materials Safety Administration). Access via internet: http://www.phmsa.dot.gov [Accessed: 2018-01-24].

[2] Krishnan KN. Effect of microstructure on stress corrosion cracking behaviour of austenitic stainless steel weld metals, *Materials Science and Engineering* 1991, 142 (1): 79–85.

[3] Zahrani EM., Alfantazi AM. Molten salt induced corrosion of Inconel 625 superalloy in $PbSO_4$–Pb_3O_4–$PbCl_2$–Fe_2O_3–ZnO environment, *Corrosion Science* 2012, 65: 340–359.

[4] Nasiłowska B., Bogdanowicz Z., Wojucki M. Shot peening effect on 904 L welds corrosion resistance, *Journal of Constructional Steel Research* 2015, 115: 276–282.

[5] Ramezanzadeh B., Ahmadi A., Mahdavian M. Enhancement of the corrosion protection performance and cathodic delamination resistance of epoxy coating through treatment of steel substrate by a novel nanometric sol-gel based silane composite film filled with functionalized graphene oxide nanosheets, *Corrosion Science* 2016, 109: 182–205.

[6] Ramezanzadeh B., Niroumandrad S., Ahmadi A., Mahdavian M., Moghadam M. H. Enhancement of barrier and corrosion protection performance of an epoxy coating through wet transfer of amino functionalized graphene oxide, *Corrosion Science* 2016, 103: 283–304.

[7] Pu NW., Shi GN., Liu YM., Sun X., Chang JK., Sun CL., Ger MD., Chen CY., Wang PC., Peng YY., Wu CH., Lawes S. Graphene grown on stainless steel as a high-performance and ecofriendly anti-corrosion coating for polymer electrolyte membrane fuel cell bipolar plates *Journal of Power Sources* 2015, 282: 248–256.

[8] Karanveer S. Aneja. H.L. Mallika Böhm, A.S. Khanna, Siva Böhm Functionalised graphene as a barrier against corrosion, FlatChem 2017, 1: 11–19.

[9] Grodecki K. Spektroskopia ramanowska grafenu, Materiały Elektroniczne (Electronic Materials) 2013, 41: 47–53.

ITMS'2018

The 12th International Scientific Conference Intelligent Technologies in Logistics and Mechatronics Systems (ITELMS'2018), 26–27 April 2018, Panevėžys, Lithuania

Laboratory Investigation of Kinematic Discrepancy Compensation Ability in Multi-Axial All-Wheel Drive Teleoperated Unmanned Ground Vehicles with Hydrostatic Drivetrain

Marian J. Łopatka[a], Mirosław Przybysz[a*], Arkadiusz Rubiec[a]

[a]*Military University of Technology, Kaliskiego 2, 00 -908 Warsaw, Poland*

Abstract

Drivetrains of Unmanned Ground Vehicles (UGVs) are one of the key systems which are responsible for ensuring them high mobility. Such a capability is very important especially when UGVs are dedicated to perform activities in rough terrain. In this applications negotiating with obstacles is quite challenging. Specific requirements during moving over obstacles caused that UGVs drivetrain are exposed on occurring high kinematic discrepancy and in extreme cases circulating power phenomena. It is desired that UGVs drivetrains have ability to kinematic discrepancy compensation. This paper presents laboratory research of hydrostatic drivetrain main parameters influence on kinematic discrepancy compensation ability. This issue was analyzed in context of it use in multi – axial all drive wheel UGVs.

© 2018 M. J. Łopatka, M. Przybysz, A. Rubiec
Peer-review under responsibility of the Kaunas University of Technology, Panevėžys Faculty of Technologies and Business

Keywords: kinematic discrepancy, Unmanned Ground Vehicles, hydrostatic drivtreins

1. Introduction

The tasks requested from the Unmanned Ground Vehicles (UGV) during rescue activities and military missions put forward high requirements in terms of their mobility and manoeuvrability. They include, among others, the capacity of UGV to move on rough terrain (with obstacles high 20–30cm), hills with the inclination angle of 40% (22°), low carrying capacity areas = 150kPa, overcoming typical field obstacles, i.e.: debris, logs, ditches [1, 2, 3, 4]. The above requirements are possible by applying a hydrostatic drive system and suspension ensure well distributed of wheels – ground pressure [5].

* Corresponding author. Tel.: +48 261–837–107
E-mail address: miroslaw.przybysz@wat.edu.pl

© 2018 M. J. Łopatka, M. Przybysz, A. Rubiec
Peer-review under responsibility of the Kaunas University of Technology, Panevėžys Faculty of Technologies and Business

B427H0029 . EdLearning.it

Driving with UGVs on very rough terrain or with different values of dynamic radiuses of road wheels, which may be caused by uneven load or different value of pressure in tires, results in wheels rotating with different rotary speeds (Fig. 1). The phenomenon is known as kinematic discrepancy and it is defined by a degree of kinematic discrepancy. The degree of kinematic discrepancy Δ of a system is the ratio of the result of substracting the numbers of wheels rotation ω', ω" performed on a given road section (while rotating without slipping), to the number of rotations of the wheel rotating with greater speed [6, 7, 8, 9]

$$\Delta = \frac{\omega' - \omega''}{\max(\omega', \omega'')}. \tag{1}$$

Kinematic discrepancy of the drive system may be examined between the wheels located on the same axis or between different axes (Fig. 1). According to the results published in [10, 11, 12] during vehicle negotiating with significant obstacle, kinematic discrepancy peak values could reach even 300 %.

a) b) c)

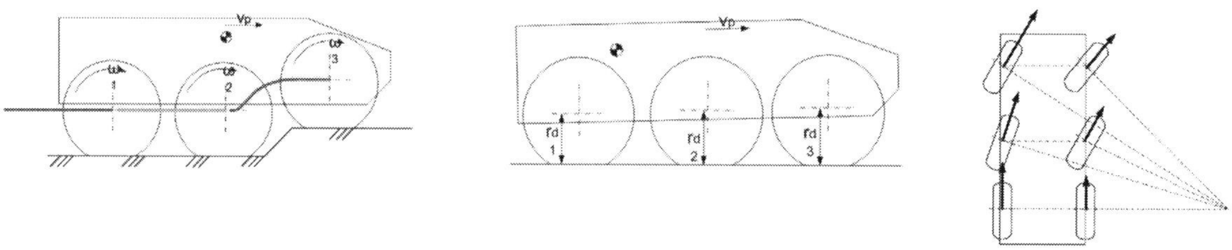

Fig. 1. Kinematic discrepancy in six – wheels vehicles: a) obstacle negotiating; b) different wheels radius; c) different turning radius

Mobility requirements for UGVs caused increase ground – drive system interaction requirements. High value of drive system stiffness during obstacles overcoming, caused highest values of wheels – ground slips. These slips have positive values when wheel rotate with smaller angular velocity relative to ideal movement (without kinematic discrepancy) and slips have negative values when wheel rotate with higher angular velocity relative to ideal movement. Wheel with positive slip is breaking. Breaking wheels caused extra power input to drive system – circulating power. As an effect of circulating power phenomena are: significant limitation of obstacle overcoming ability and decrease of total efficiency of drivetrain (pressure increase in hydraulic system). When circulating power is long-term, UGV drive train could be even damaged.

Hydrostatic drive systems (HDS) are characterized by great kinematic rigidity, which is associated with small compressibility of the hydraulic oil, i. e. a medium in HDS that ensures power transfer from the pressure generator to the receivers, which are the hydraulic motors responsible for the road wheels drive. This feature results in the fact that HDS provides a great kinematic accuracy of performed movements, but it is also sensitive to the occurring kinematic discrepancies. The accuracy depends mainly on internal leaks resulting from the construction features of the used elements, pressure drops on particular elements and dividing accuracy of flow divider [13, 14, 15, 16]. The aspects enable, to some extent, the differentiation of rotational speed of wheels necessary for compensation of kinematic discrepancy – which is defined as factor of kinematic discrepancy compensation FKDC.

Representative example of all wheel drive HDS of multiaxial UGV was shown on Figure 2. In this circuit each wheel is driven by separate motor. In HDS with parallel structure – hydraulic oil always flow to wheel which have minimal external load. In such a system – during obstacle overcoming it is possible that one of the wheel will lose contact with the ground. This will cause that all flow from pump will go to wheel which have no contact with the ground and UGV will not negotiate with obstacle. From this reason in such a systems flow dividers are used. Kinematic flexibility of such a systems depends on: dividing accuracy of flow dividers and hydraulic motors volumetric efficiency.

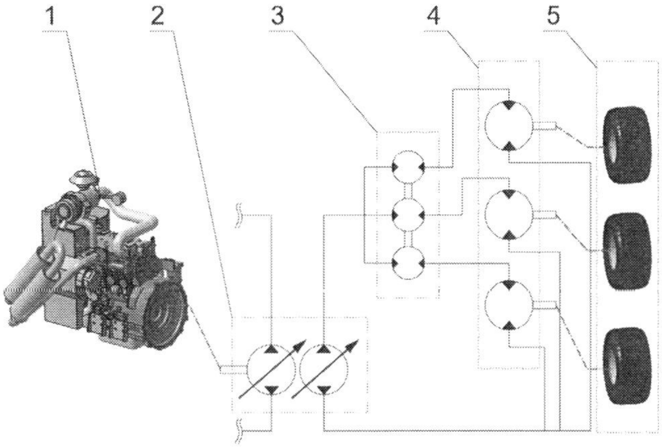

Fig. 2. Hydrostatic drive system of multiaxial all – wheel drive UGV: 1 – engine, 2 – variable displacement pump, 3 – flow divider, 4 – hydraulic motors, 5 – UGV wheels

Literature reviews shows that there is a limited number of papers describing research results of kinematic flexibility of all wheel drive UGVs with hydrostatic drive systems. That's why in Military University of Technology was taken research on influence of hydraulic drive system components properties on factor of kinematic discrepancy compensation FKDC.

2. Method

2.1. Scope of the research

Research was focusing on identification of volumetric efficiency of different types of hydraulic motors and dividing accuracy of different types flow dividers in hydrostatic drivetrain. Basic on this results were calculated FKDC of UGV drivetrain.

During research were tested three types of fixed displacement hydraulic motors with similar shaft angular velocities and generated torque:

- axial piston motor with displacement $q_s = 34.2\ cm^3/rev$ connected with reduction gear with transmission ratio $i = 5.77$;
- radial piston motor with displacement $q_s = 211\ cm^3/rev$;
- orbital motor with displacement $q_s = 237\ cm^3/rev$;

Additionally were tested two types of flow dividers;

- spool type with nominal flow $Q = 80\ dm^3/min$;
- gear type with nominal flow $Q = 80\ dm^3/min$.

Research had beenconducted with motors shaft angular velocities which corresponds to UGV velocity during negotiate with significant obstacles 1 to 3 m/s. When wheels dynamic rolling radius is equal to 0,3 m, flow within hydraulic motors is from 10 to 30 dm^3/min. However because of expected flow dividers high sensitivity on flow value (influence on dividing accuracy) – for flow divider testing flow range has been increased (10 to 40 dm^3/min). Research had been conducted with pressure from 2 to 16 MPa.

2.2. Test stand and research methodology

Research of factor of kinematic discrepancy compensation have been divided in two groups:

- influence of hydraulic motor load on its volumetric efficiency;
- influence of pressure increase in one output lines of flow divider on dividing accuracy.

Both research groups had been conducted with the same hydraulic oil temperature which was equal to $50 \pm 1^{\circ}C$.

Factor of kinematic discrepancy compensation FKDC depends on working conditions. For hydraulic motors FKDC has been calculated according to relation:

$$FKDC = 1 - \eta_v \quad (2)$$

where η_v – hydraulic motor volumetric efficiency.

For flow dividers FKDC has been calculated according to relation:

$$FKDC = \frac{Q_{max} - Q_{min}}{Q_{max}} \quad (3)$$

where Q_{max} – maximum flow value on one of the flow dividers output;
Q_{min} – minimum flow value on one of the flow dividers output (in the same time as Q_{max}).

Laboratory test stand (Fig. 3) for research on influence of hydraulic motor load on its volumetric efficiency consist of two separate subsystems: investigated system and loading system. Investigated system was consist of variable displacement pump (1), and hydraulic motor (2). Output shaft of investigated hydraulic motor has been connected with shafts of loading hydraulic motor (3). External load value depend on set of dumping value (5). To avoid cavitation in loading system has been located additional charge pump (4). Charge pump maximum pressure was limited to 2MPa.

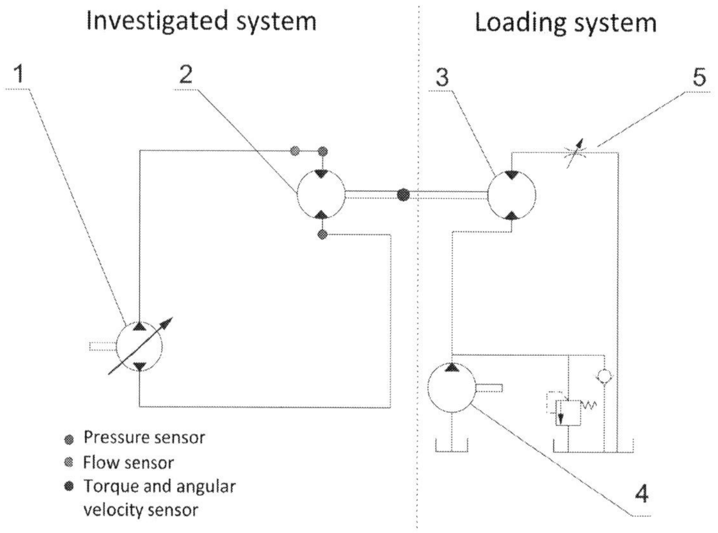

Fig. 3. Laboratory test stand diagram for research on influence of hydraulic motor load on its volumetric efficiency

On laboratory stand were used: a – flow sensor, b – pressure sensors, c – torque and angular velocity sensor. Detailed technical data of sensors which were used during research was shown in table 1 and the structure of measurement and control system has been shown on Figure 4.

Laboratory test stand (Fig. 5) for research on influence of pressure increase in one of the output lines of flow divider on dividing accuracy was also consist of two separate subsystems: investigated system and loading system. Investigated system was consist of: variable displacement pump (1), flow divider (2), investigated hydraulic motors (3). Output shaft of investigated hydraulic motors (3) was connected with shafts of loading hydraulic motors (5). External load value depend on set of dumping value (6). To avoid cavitation in loading system has been also located additional charge pump (4). Charge pump maximum pressure was also limited to 2 MPa.

On laboratory stand were used: a – flow sensors on the output lines of flow divider output, b – pressure sensors, c – torque and angular velocity sensors. Detailed technical data of sensors which were used during research was shown in table 1 and the structure of measurement and control system was shown on Figure 6.

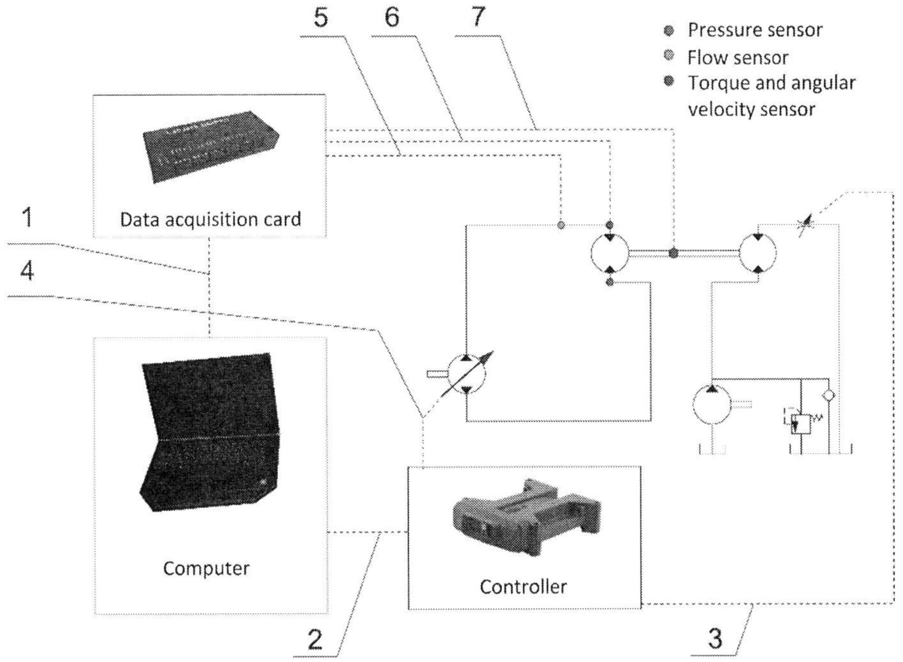

Fig. 4. Measurement and control system structure for research on influence of hydraulic motor load on it volumetric efficiency: 1, 2 – universal serial bus, 3 – control signal for adjusting dumping valve, 4 – control signal for adjusting pump displacement, 5 – output signal from flow sensor, 6 – output signal from pressure sensors, 7 – output signal from torque and angular velocity sensor

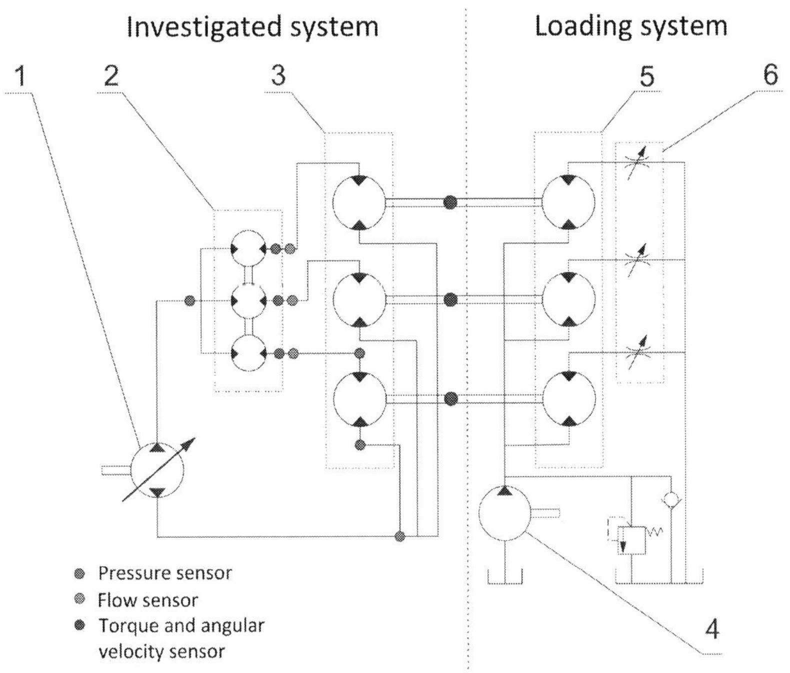

Fig. 5. Laboratory test stand diagram for research on influence of pressure increase in one of the output lines of flow divider on dividing accuracy

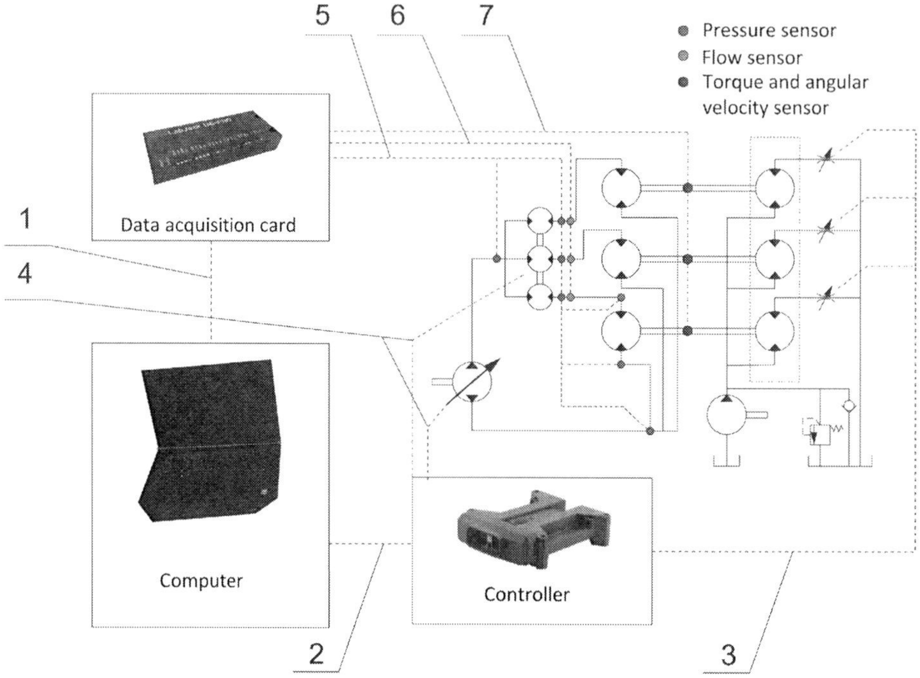

Fig. 6. Measurement and control system structure for research on influence of pressure increase in one of the output lines of flow divider on dividing accuracy: 1, 2 – universal serial bus, 3 – control signal for adjusting dumping valve, 4 – control signal for adjusting pump displacement, 5 – output signal from pressure sensors, 6 – output signal from flow sensors, 7 – output signal from torque and angular velocity sensor

Table 1. Sensors used during research

Type of sensor	Sensor range	Sensor accuracy
Torque sensor	1	2
Angular velocity sensor	3	4
Flow sensor	5	6
Pressure sensor		

3. Results and discussion

Laboratory research on influence of hydraulic motor load on volumetric efficiency allow to plot characteristics of volumetric efficiency in a function of load torque:

- for axial piston motor with reduction gear (Fig. 7);
- radial piston motor with displacement (Fig. 8);
- orbital motor with displacement (Fig. 9).

The biggest factor of kinematic discrepancy compensation have orbital motor (from three types of motors that were investigated). Its volumetric efficiency at about value of loading torque equal to 500 Nm (which correspond to motor pressure drop $\Delta p = 15$ MPa) were $\eta_v = 0{,}78–0{,}93$.Volumetric efficiency depend on oil flow through motor. At flow Q = 10 dm^3/min (which correspond to UGV velocity about 1 m/s), FKDC is equal 21 %. FKDC of axial piston motor (Fig. 7) connected with reduction gear is about 65 % less than FKDC of orbital motor. Factor of kinematic discrepancy compensation for piston radial motor was about 80 % when we compare it with orbital motor.

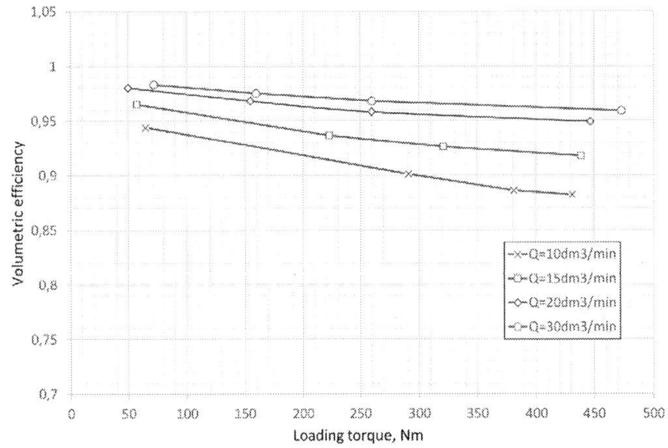

Fig. 7. Characteristic of volumetric efficiency of axial piston motor with displacement $q_s = 34{,}2$ cm^3/rev connected with reduction gear with transmission ratio i = 5,77

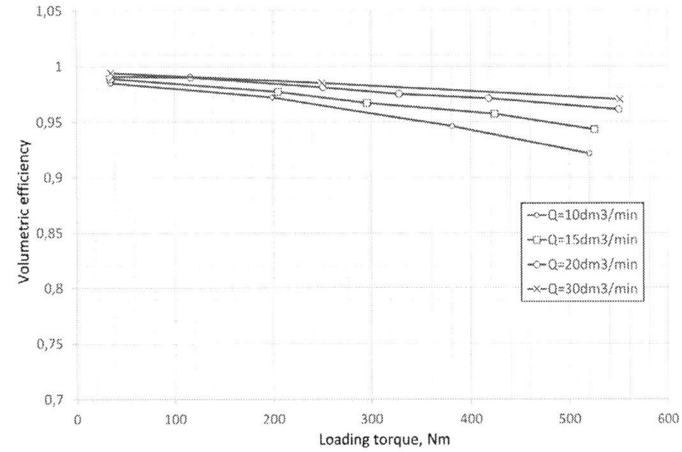

Fig. 8. Characteristic of volumetric efficiency of radial piston motor with displacement $q_s = 211$ cm^3/rev

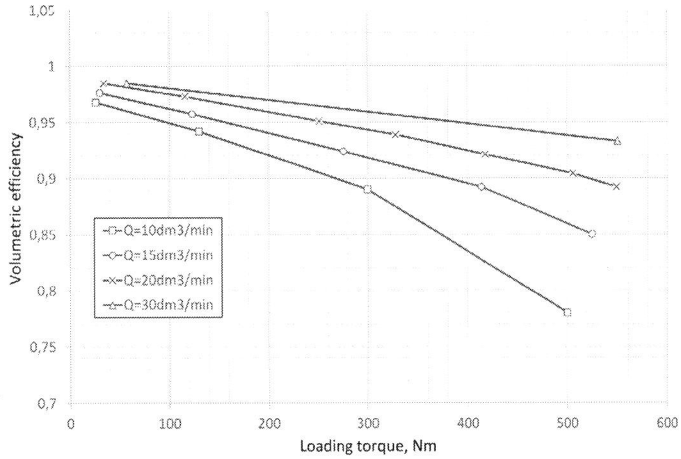

Fig. 9. Characteristic of volumetric efficiency orbital motor with displacement $q_s = 237$ cm^3/rev

On Figures 10 (spool type flow divider) and 11 (gear type flow divider) was shown results of influence of pressure increase in one output lines of flow divider on dividing accuracy. According to (3) in this case dividing accuracy should also be considered as FKDC.

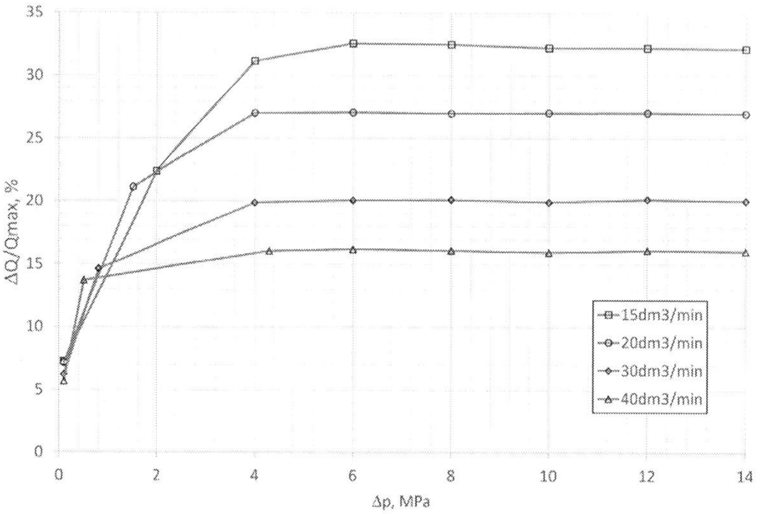

Fig. 10. Characteristic of dividing accuracy (FKDC) of spool type flow divider for different values of input flow rate

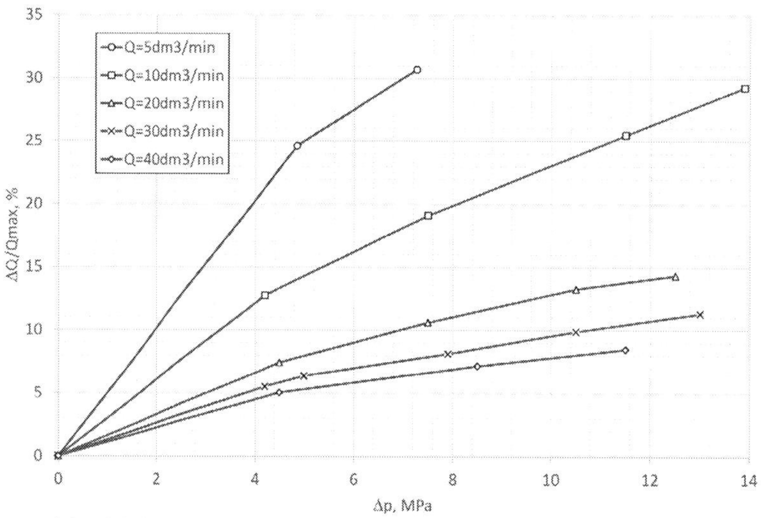

Fig. 11. Characteristic of dividing accuracy (FKDC) of gear type flow divider for different values of input flow rate

FKDC (dividing accuracy) of spool type flow divider (Fig. 10) for output pressure difference bigger than 4 MPa is almost constant. For each value of pressure difference between flow divider outputs dividing accuracy increase with input flow increase. This mean that UGV drivetrain will be more rigidly and factor of kinematic compensation ability will decrease. FKDC for small input flows (15 dm^3/min) is about 33 % (for pressure difference bigger than 4 MPa). For bigger flow values (40 dm^3/min) FKDC is equal to 15 % (for the same pressure difference between flow divider outputs).

Dividing accuracy of gear type flow divider (Fig. 11) is changing with pressure difference between outputs changing (unlike the spool type flow divider). Independently from flow divider input flow rate dividing accuracy decrease (FKDC increase) with pressure difference between outputs increase. For gear type flow divider - FKDC for small input flows (5 dm^3/min) is about 25 % (for pressure difference equal to 4 MPa). For bigger flow values

(40 dm^3/min) FKDC is equal to 5 % (for the same pressure difference between flow divider outputs). This mean that UGV drivetrain with gear type flow divider will be more rigidly than drive train with spool type flow divider.

Conclusions

Kinematic flexibility of UGVs hydrostatic drivetrain depends on hydraulic motors volumetric efficiency and flow divider, dividing accuracy as well. Based on hydraulic motors volumetric efficiency research results it has been found that its construction type have significant influence on UGV drivetrain to compensation of kinematic discrepancy ability. Proper selection of hydraulic motors for drivetrains allow to have influence on its flexibility (in some range). This is very important aspect especially for rough terrain vehicles. However when we compare kinematic discrepancy compensation ability between motors and flow divider it turns out that flow dividers (independently form type) have bigger FKDC than motors.

Comparing each other research results of motors and flow dividers, it has been found out, that (for investigated flow values) the biggest rigidly will have drivetrain with radial piston motor and gear type flow divider. Drive train with this components configuration have also the smallest kinematic discrepancy compensation ability. It allowed to achieve about 10% angular velocity difference between UGV wheels. Smallest rigidly will have drivetrain with orbital motors and spool type flow divider. Drive train with this components configuration have also the biggest kinematic discrepancy compensation ability It allows to achieve about 52 % angular velocity difference between UGV wheels.

It also should be noticed that drivetrain flexibility is only one of the aspects which should be considered during UGVs drivetrain designing. This process should be complex and include aspects such as: drivetrain price, drivetrain weight, components reliability, total efficiency and above negotiate obstacles ability by UGV.

Acknowledgements

We would like to thank of Polish Ministry of Science and Higher Education for research founding (PBS 936/2016–2018) within have been conducted investigation described in this paper.

References

[1] Sprawka P. The methods of evaluation the mobility of off - road vehicles. Solid State Phenomena Vol. 180, Trans Tech Publications, Switzerland, 2011. pp. 355–365.
[2] Dabrowska A., Konopka S., Przybysz M. Ability to negotiate terrain obstacles by lightweight six-wheeled unmanned ground vehicles, Intelligent Technologies In Logistics And Mechatronics Systems – ITELMS'2015, Panevezys, Lithuania, May 21–22, 2015, pp. 102–109.
[3] Bartnicki A., Łopatka M., J., Muszyński T., Jaskółowski M. The Researches of Stability Evaluation Articulated Loader while Working on Slopes, ITELMS'2013, Panevėžys.
[4] Łopatka M. J. Możliwości robocze bezzałogowych platform lądowych. WAT, Warszawa 2010.
[5] Dabrowska A., Przybysz M., Rubiec A. Hydropneumatic suspension efficiency in terms of the teleoperated unmanned ground vehicle tests, Intelligent Technologies In Logistics And Mechatronics Systems – ITELMS'2015, Panevezys, Lithuania, May 21–22, 2015. pp. 110–116.
[6] Studziński K.: Teoria I sterowanie. Samochód. WKŁ, Warszawa 1980.
[7] Konopka S., Łopatka M. J., Przybysz M. Kinematic Discrepancy of Hydrostatic Drive of Unmanned Ground Vehicle, ITELMS'2013, May 23–24, 2013, Panevėžys, Lithuania.
[8] Wong J. Y. Theory of ground vehicle, ISBN: 978-0-470-17038-0, August 2008.
[9] Konopka S., Sprawka P., Muszyński T., Spadło K. Investigating resistance of turn six-wheel skid-steer vehicle, ITELMS'2013, Panevėžys, Lithuania, May 23–24, 2013.
[10] Studziński K. Teoria I sterowanie. Samochód. WKŁ, Warszawa 1980.
[11] Wong J. Y. Theory of ground vehicle. ISBN: 978-0-470-17038-0, August 2008.
[12] S. Konopka, M. J. Lopatka, M. Przybysz. Kinematic Discrepancy of Hydrostatic Drive of Unmanned Ground Vehicle, Intelligent Technologies In Logistics And Mechatronics Systems, ITELMS 2013, pp. 116–125, Panevezys, Lithuania, May 23–24 2013.
[13] Budny E. Napęd i sterowanie układów hydraulicznych w maszynach roboczych. ITE, Radom 2001.
[14] Stryczek S. Napęd hydrostatyczny. Tom I. Wydawnictwa Naukowo – Techniczne, Warszawa 2005.
[15] Comellas M., Pijuan J., Potau X., Nogue´s M., Roca J. Analysis of a hydrostatic transmission driveline for its use in off-road multiple axle vehicles. Journal of Terramechanics, No 49, 2012.
[16] Łopatka, M. J. Praca zbiorowa, Inżynieryjny Robot Wsparcia IOD/EOD – usuwania ładunków niebezpiecznych, Sprawozdanie z projektu rozwojowego Nr OR00001205/PBR, Warszawa 2011.

ITMS'2018

panevėžys faculty of technologies and business

1922

The 12th International Scientific Conference Intelligent Technologies in Logistics and Mechatronics Systems (ITELMS'2018), 26–27 April 2018, Panevėžys, Lithuania

Utilization of S355J2 Mild Steel in Highly Loaded Welded Structures

Tomasz Ślęzak[a*], Janusz Torzewski[a]

[a] *Military University of Technology, Faculty of Mechanical Engineering, ul. gen. W. Urbanowicza 2, 00-908 Warsaw, Poland*

Abstract

Numerous modern grades of steel are now inducted in design and produce of transportation means or other structures. Nevertheless, conventional mild steels are commonly utilized in production what is caused by perfect weldability and hitherto practice. Therefore, the research of S355 steel are widely conducted in many scientific centers.
The aim of this research was to determine the fatigue properties of welded joints made of S355J2 steel tested at extreme loadings. There was found that V-joints have slightly better fatigue properties in comparison with I-joints and Morrow formula describes predicted fatigue life of welds in sufficient accuracy. On the basis of conducted fractographic observation was stated that the nature of crack propagation is ductile in all stages of crack growth.

© 2018 T. Ślęzak, J. Torzewski
Peer-review under responsibility of the Kaunas University of Technology, Panevėžys Faculty of Technologies and Business

Keywords: mild steel; welded joints; fatigue life; strain mode

---

1. Introduction

A design process of advanced and highly complex structures needs extensive knowledge concerning the properties of structural materials. For this reason, new structural materials are developing or specific properties of existing materials are determined [1–5]. Nonetheless, different grades of steel are as before one of the most widely utilized materials in transportation means [6–7]. In an automotive industry numerous modern, high strength grades of steels are used like the dual phase steels DP, complex phase steels CP or TRIP steels with an effect of transformation induced plasticity. Mentioned groups of steels are produced in the form of thin sheets and then are

* Corresponding author. Tel.: +48 261-837-685
E-mail address: tomasz.slezak@wat.edu.pl

© 2018 T. Ślęzak, J. Torzewski
Peer-review under responsibility of the Kaunas University of Technology, Panevėžys Faculty of Technologies and Business

B427H0030 . EdLearning.it

exploited mainly on the elements of cars [8–10]. Therefore, in design of highly loaded structures, like the jibs of mobile crane, truck frames,

Nomenclature

A	percentage elongation at fracture
E	Young's modulus
N	number of cycles
N_f	number of cycles to failure (fatigue life)
Z	percentage reduction of area
σ_U	ultimate tensile strength
σ_Y	yield strength
b	fatigue strength exponent
c	fatigue ductility exponent
ε	strain
σ	stress
ε_f'	fatigue ductility coefficient
σ_f'	fatigue strength coefficient
Subscripts:	
a	amplitude
ac	total amplitude
ae	amplitude of elastic component
ap	amplitude of plastic component

trailers or railway wagons are used other high strength structural steels, for example fine-grained martensite-bainitic or martensite. They are utilized in other structures: in military bridges, offshore platforms or pipeline systems [11–14].

In order to use advanced steels successfully particular requirements should be fulfilled. This is especially connected with the manufacturing which demands joining technique by welding. Modern steels need to employ the low-hydrogen fusion technologies and consumables, a high quality welds realization, a proper value of heat input and an adequate ductility of heat affected zone [15–16]. If above conditions are not implemented the danger of a hydrogen embrittlement, cold or hot cracking can occur. For this reason, the common grades of steels i.e. carbon steels are still used in welded structures. One the most popular of them is S355J2 mild steel in tempered state. It is characterized by very good weldability and what is more important widely examined. Nevertheless, presently new challenges appear connected with development of modern structures aimed to fulfill specious requirements. Then, even popular and commonly used materials developed many years ago are the purposes of researchers' interest [17–19].

The aim of this paper is to present the results of study performed on the S355J2 structural mild steel. Almost all real constructions are exploited in the conditions of variable loadings caused fatigue damage. This steel and their welded joints were widely examined taking into consideration the alternating loadings. However, published results concern mainly the strength of studied materials in the range of fatigue life above 10^4 numbers of cycles. Moreover, all investigations of welded joints were performed under the stress mode, both the low- and the high-cycle-fatigue [20–21]. For this reason, there is lack of investigation results in the area of mild steels welded connections examination performed under strain mode with extreme values of strains induced plastic strains and very low fatigue life.

Two types of welded joints made of S355J2 steel were investigated under strain mode. Fatigue properties of the welds were determined. Additionally, microfractographic investigations of the failure ruptures were conducted. This part of study enables to indicate the origins of fatigue cracks and the mechanisms of fatigue crack propagation. Finally, the crack propagation description was achieved.

2. The results of investigation

2.1. The subject of investigation

The study was carried out on the structural mild steel S355J2 in tempered state. A chemical composition and mechanical properties of this steel are shown in Table 1 and 2. Chemical composition was verified during own measurements using scanning electron microscope SEM equipped with energy dispersive spectroscopy detector (EDS) X-Max by Oxford, whereas the properties were conducted by tensile test conducted on the Instron hydraulic pulsator 8802 equipped with an extensometer.

Table 1. Chemical composition (wt %) of S355J2 steel

Alloying elements	C	Mn	Al	Si	Cr	Ni	Cu
The certificate	0.17	1.45	0.06	0.02	0.04	0.03	0.06
Own measurements	0.17	1.54	0.07	0.04	0.02	0.04	0.08

Table 2. Mechanical properties of S355J2 steel

Property	E [MPa]	σ_Y [MPa]	σ_U [MPa]	A [%]	Z [%]
The certificate	---	456	562	24.8	---
Own measurements	192 150	385	531	30.7	64.4

The results of structural investigation are shown in the Figure 1.

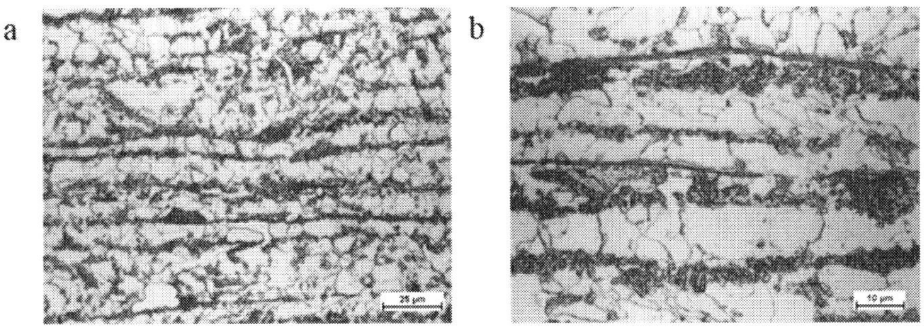

Fig. 1. The structure of S355J2 steel at two magnifications with visible texture (a) and sulfide bands

Investigated steel has ferritic-pearlite microstructure with noticeable texture (Fig. 1, *a*) as a result of rolling. This grade is contaminated by some amount of sulfur or phosphor causing the presence of undesired non-metallic compounds in the form of inclusions or bands (Fig. 1, *b*).

The study was performed on two types of butt welded joints, namely "I" and "V". They were accomplished using a sheet with a thickness of 6 mm by GMAW technique. Parameters of welding are tabulated in Table 3 and the outline of prepared joined edges with a sequence of made welding stitches are presented in Figure 2.

Table 3. Welding parameters

Joint	Stitch	Welding current [A]	Arc voltage [V]	Welding speed [mm/min]	Wire feed rate [m/min]	Flow rate [dm^3/min]	Heat input [kJ/mm]
"I"	1	260	27.0	450	7.5	14	0.75
	2	260	27.0	450	7.5	14	0.75

Table 3 (Continued)

Joint	Stitch	Welding current	Arc voltage	Welding speed	Wire feed rate	Flow rate	Heat input
		[A]	[V]	[mm/min]	[m/min]	[dm^3/min]	[kJ/mm]
"V"	1	150	19.5	210	5.5	14	0.67
	2	260	27.0	450	7.5	14	0.75

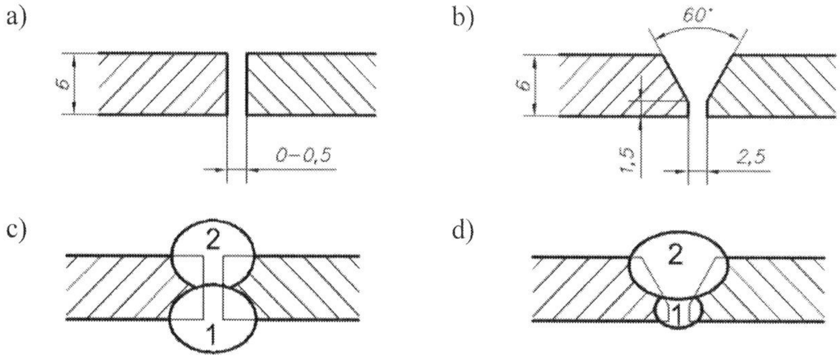

Fig. 2. The edges before welding (a, b) and a sequence of welding (c, d) for "I" joint (a, c) and "V" joint (b ,d), respectively

It is worth noting that the heat input during welding was below 1 kJ/mm. The welding was done GMAW technique using shielding active gas EN ISO 14175-M21-ArC – 18 containing 18% of Ar and 82% of CO_2. The OK Autrod 12.51 (EN ISO 14341-A-G3Si1) wire with 1.2 mm diameter was utilized. The first stitch signed in Fig. 2d was made manually as opposed to the others were made in a robotized way.

2.2. Fatigue research methodology

Fatigue tests of S355J2 welded joints were conducted in the low cycle range. The tests were carried out in the conditions of load controlled by the value of total strain amplitude ε_{ac} (strain mode) when the strain rate of 10^{-2} s^{-1} was stable (the frequency differed) and strain asymmetry rate R = 0.1. These assumptions were made in order to avoid heat release under high strains. Samples were tested at fife values of strain amplitude, namely: 0.15 %, 0.20 %, 0.25 %, 0.30 % and 0.40 %. The tests were made using Instron hydraulic pulsator 8802 equipped with dynamic extensometer Instron 2620-603 with measuring base of 50 mm to control strain. Geometric dimensions of tested samples are shown in Figure 3.

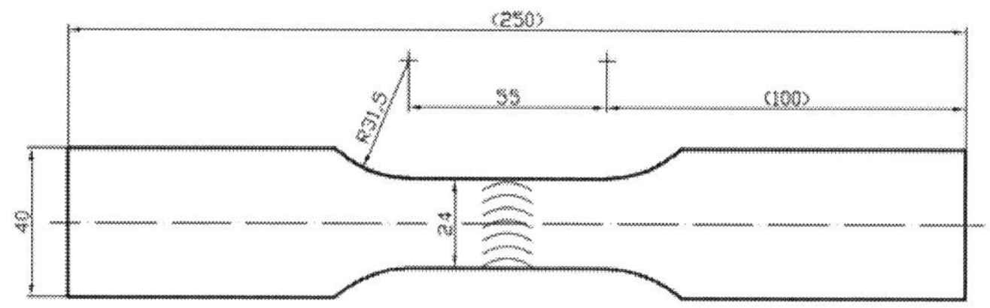

Fig. 3. Dimensions of sample for fatigue test

The drop of maximum stress by 25 % in comparison with first half-cycle was defined as the failure criterion. It is one of the failure criteria descripted in ASTM E606 [22]. The surfaces of samples were not machined due to reflect the test conditions to real as close as possible. Moreover, it allowed to identify and indicate the fatigue origins.

2.3. Fatigue tests results

Selected hysteresis loops obtained as result of low cycle fatigue tests of S355J2 steel welded joints are presented in Figure 4 and 5. Each plot includes fife representative loops of first cycle (1), a tenth cycle, the cycle of 10-percentage drop of maximum stress (N90 %), half-life cycle (Nf/2) and cycle at failure, respectively.

Both types of butt joints have similar courses of cracking parameters. On the basis of shown hysteresis loops (Fig. 4–5) and Fig. 7 it can be stated that the welded joints utilized from S355J2 steel succumb to cyclic weakness. The results for joints of "V" type tested under ε_{apl} of 0.20 % and 0.25 % indicate that an initial strengthen has place and the weakness after them (Fig. 7, *b*). It is an evidence that this steel grade has tendency to cyclic weakness. The stabilization of fatigue features was observed after 5–10 cycles by fixation of characteristic trend. Additionally, shown hysteresis loops have significant participation of plastic component in total strain, what is also proofed in. Figure 8.

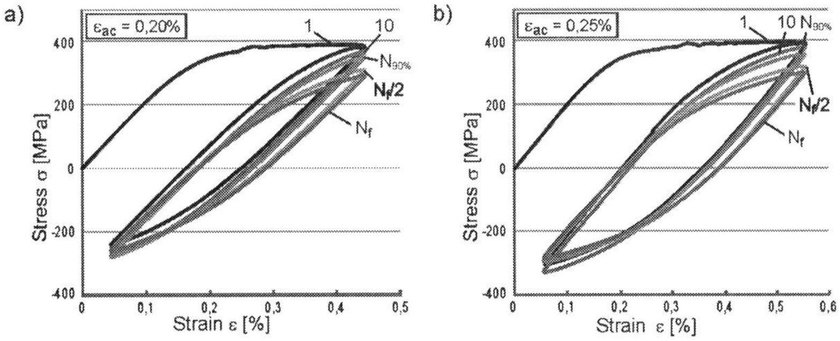

Fig. 4. Hysteresis loops obtained for "I" butt joints of S355J2 steel tested under two levels of total strain amplitude 0.20 % (a) and 0.25 % (b)

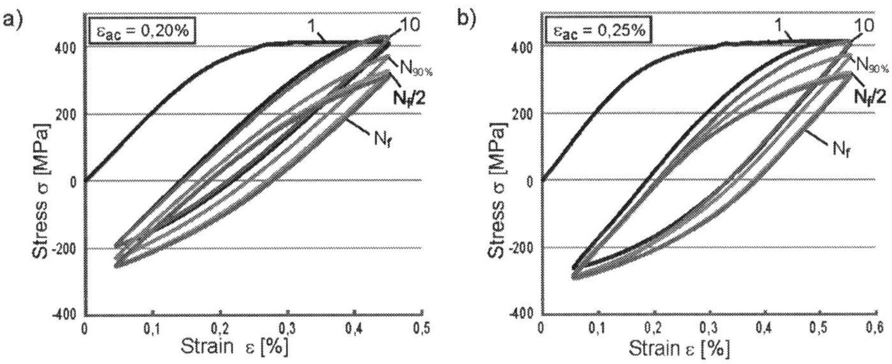

Fig. 5. Hysteresis loops obtained for "V" butt joints of S355J2 steel tested under two levels of total strain amplitude ε_{ac} 0.20 % (a) and 0.25 % (b)

The plots of changes for stress amplitude σ_a, maximum stress σ_{max} and plastic strain amplitude ε_{apl} obtained during examination of S355J2 steel welded joints are presented in Figures 6–8.

On the basis of data from individual fatigue tests and reading the values of parameters from stabilized hysteresis loops, mean values of parameters were determined. These data were used for further analysis to conduct a fatigue analysis using the Morrow formula (1), whose graphical interpretation is presented in Figure 9.

The values of fatigue strength coefficients σ'_f and fatigue ductility coefficient ε'_f and the corresponding exponents *b* and *c* are also shown onto the diagram and in Table 4 (index *(w)* means that result relates to the weld joint). During determination of the coefficients the results were omitted if buckling or bending of the samples occurred.

$$\varepsilon_{ac} = \varepsilon_{ae} + \varepsilon_{ap} = \frac{\sigma'_f}{E} \cdot \left(2N_f\right)^b + \varepsilon'_f \cdot \left(2N_f\right)^c. \quad (1)$$

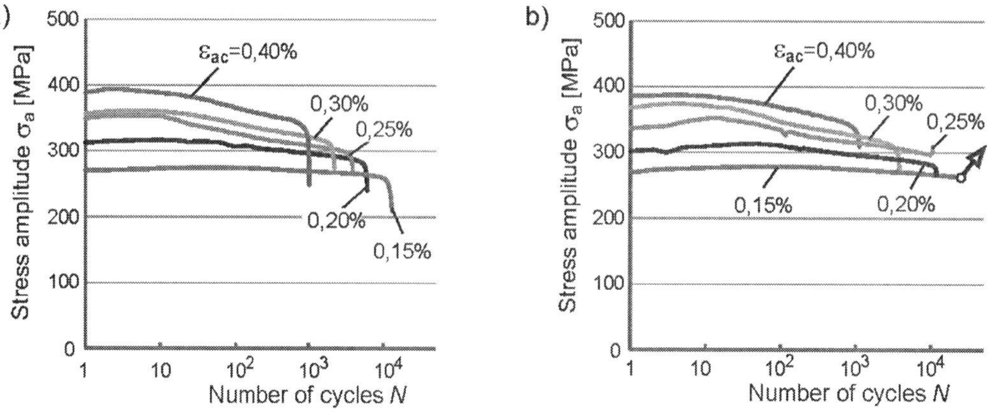

Fig. 6. Changes of stress amplitude σ_a obtained for "I" (a) and "V" (b) butt joints of S355J2 steel tested under different levels of ε_{ac}

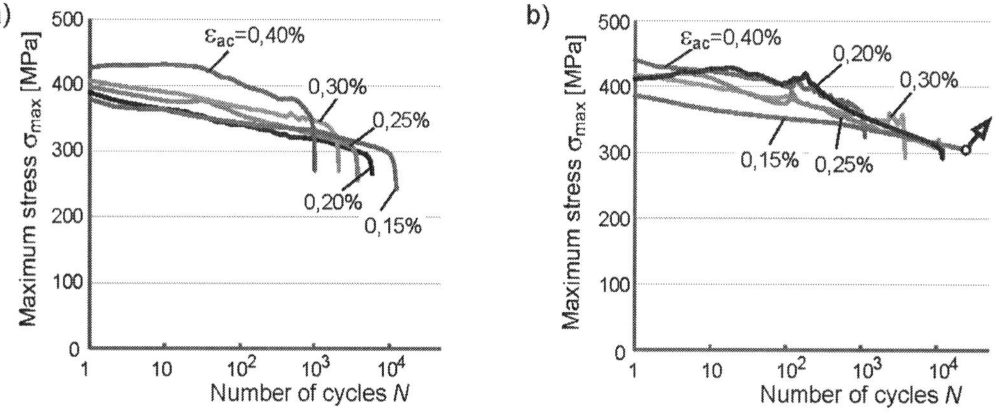

Fig. 7. Changes of maximum stress σ_{max} obtained for "I" (a) and "V" (b) butt joints of S355J2 steel tested under different levels of ε_{ac}

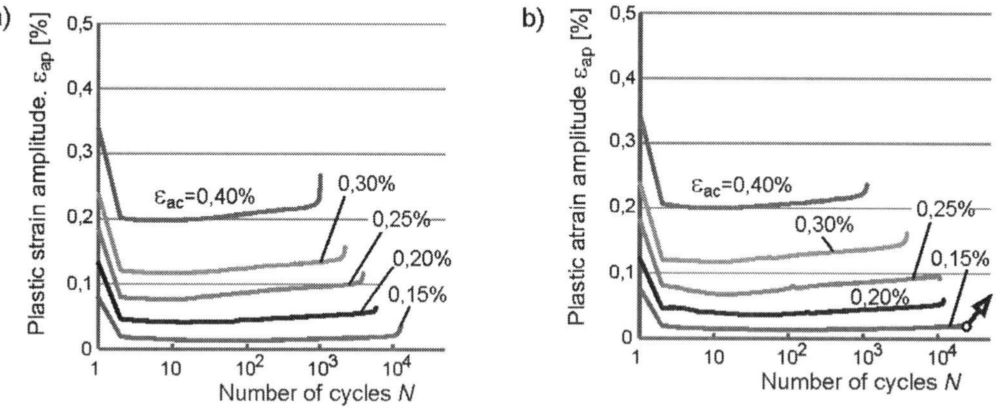

Fig. 8. Changes of plastic strain amplitude ε_{apl} obtained for "I" (a) and "V" (b) butt joints of S355J2 steel tested under different levels of ε_{ac}

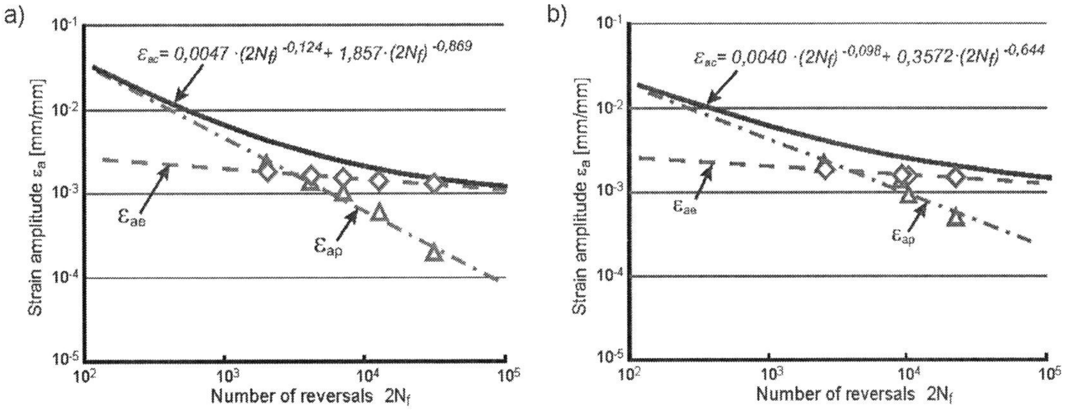

Fig. 9. Strain-life graphs of welded joints of S355J2 steel with an I-joint (a) and V-joint (b)

Table 4. Fatigue properties of S355J2 steel welded joints

Weld type	$\sigma'_{f(w)}$	$b_{(w)}$	$\varepsilon'_{f(w)}$	$c_{(w)}$
	[MPa]	[-]	[mm/mm]	[-]
"I" joint	923.6	-0.124	1.857	-0.869
"V" joint	786.0	-0.098	0.357	-0.644

Complete strain-life graph comparing experimental results of the fatigue life for both types of welded joints made of S355J2 steel and the curves describing fatigue properties if tested joints using Morrow formula are shown in Figure 10.

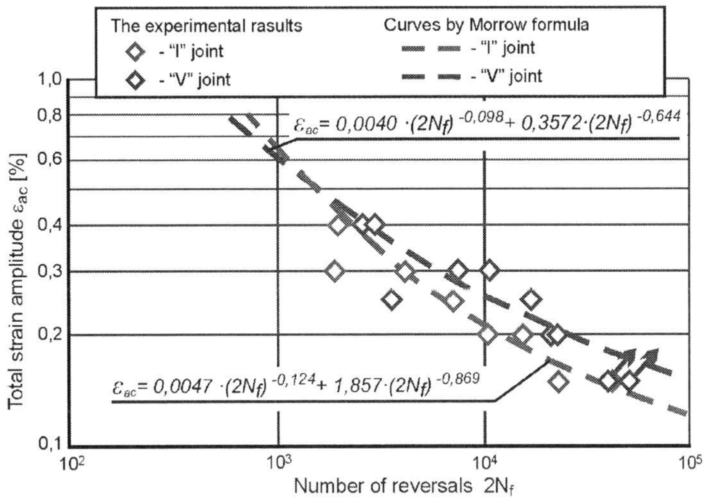

Fig. 10. Strain-life graphs of welded joints of S355J2 steel with an I-joint (a) and V-joint (b)

The obtained results indicate slightly better fatigue properties of welded joints made of S355J2 steel using V-joints. It causes greater fatigue life and on the lowest load (ε_{ac} = 0.15 %) the samples did not failure up to the moment when the limit of $2.5 \cdot 10^5$ cycles was reached. Moreover, the dispersion of results obtained for "V"-joint is rather large thus the Morrow curve is worse fitted in comparison with data for "I"-joint. This is significant factor for predicting the fatigue life of structures.

2.4. SEM investigation

Fractographic investigation of the fatigue fractures was performed using scanning electron microscope Jeol JSM-6610 equipped with the detectors of secondary electrons (SE) and back scattered electrons (BSE). Fractographic analysis was based on fatigue fractures in welded samples tested at total strain amplitude of 0.25 %. The exemplary macro-photos of fractures are shown in Figure 11.

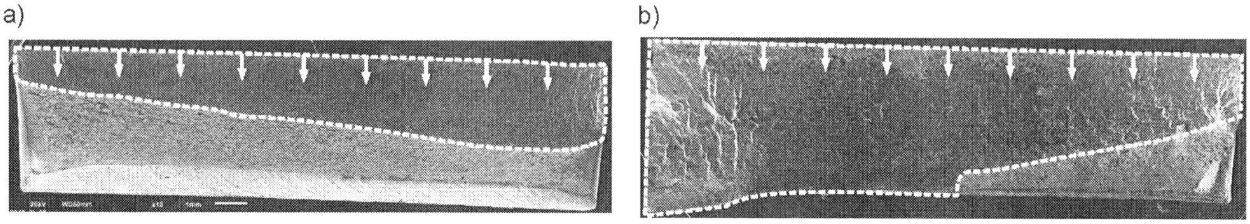

Fig. 11. Fatigue fractures of welded joints made of S355J2 steel with I-joint (a) and V-joint (b) tested at $\varepsilon_{ac} = 0.25$ %

Dashed lines indicate the area of fatigue cracking and the arrows visualize the direction of crack propagation. Lighter regions are the areas of final rupture – each sample was disrupted after reaching the defined failure criterion to enable observation of the fatigue fractures. Different areas of the rupture surface showing the mechanism of failure in tested weldments are presented in Figure 12. There were selected representative regions of cracks surfaces from zones of a crack initiation, an initial stage of crack propagation and a stage of high crack rate and presented in Figure 11. They ensure a sufficient data to state the nature of cracking.

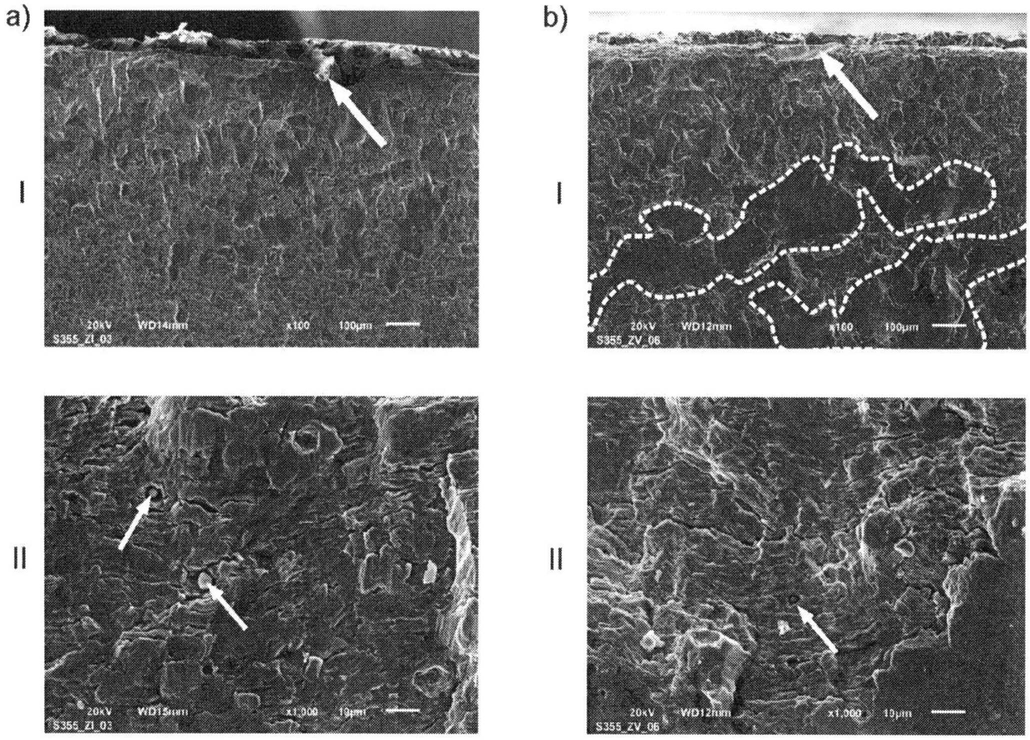

III 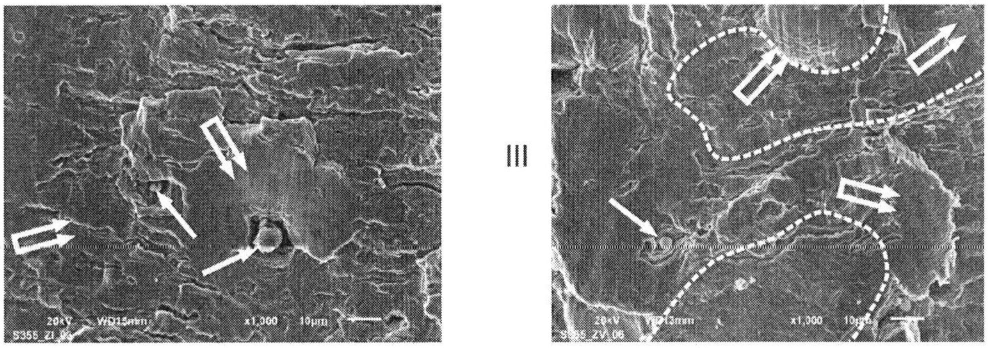 III

Fig. 12. Fatigue fractures of welded joints made of S355J2 steel with I-joint (a) and V-joint (b) tested at $\varepsilon_{ac} = 0.25$ % where the initiation (I), stable growth (II) and high crack rate (III) stages are shown

The origins of cracking cannot be unambiguously indicated in both cases of tested butt joints. Cracks always start in the immediate vicinity of the fusion lines and took place on some length but in result on whole length of tested joint (Fig. 11). In the case of V-joint the cracking starts only on the root site. Hard and brittle layer of material is present directly the surface near the fusion line (photos No. I). The objects which could accelerate decohesion of material were observed rare (arrows at photos No. I). Some elements pressed in the surface were for the I-joint and welding imperfection for V-joint, respectively. The nature of fracture is ductile at each propagation stages. The cracking of V-joint is more disturbed what is provided through the large places of shear in photos b-I and b-III (highlighted areas). The process of fracture is locally enhanced by the presence of inclusions (pointed in the photos II and III). The most important factor which evidence of the ductility character of cracking is observation of fatigue striations at the surface of rupture (double arrows). Many times, they are squashed what is caused by the high value of induced plastic strains.

Conclusions

The study in the form of fatigue tests was performed on the structural mild steel S355J2 with welded joints realized by GMAW technique have indicated that V-joints have slightly better fatigue properties in comparison with I-joints. It was evidenced by the results of fatigue durability of these weldments and exceeding the fatigue limit in some cases. The prediction of fatigue life of welded joints made of mild steel can be described using Morrow formula with sufficient accuracy. Nevertheless, the dispersion of results obtained for "V"-joint is rather large thus the curve is worse fitted in comparison with data for "I"-joints which were made by fully automated welding process.

Fractographic investigation of obtained fatigue fractures has shown the range and nature of crack propagation. The cracks initiated always from a surface in the vicinity of fusion line but in V-joints on the root site each time. Any unambiguous origins were found. The crack growth in I-joints was uniform and regular contrary to the second joint type cracking more disturbed. The most important fact is that under extreme loadings the crack propagation in welded joints made of S355 structural steel has ductile character what provide high safety of welded structures.

References

[1] Bloniarz R., Majta J., Trujillo C. The mechanisms for strengthening under dynamic loading for low carbon and microalloyed steel. *Int J Impact Eng* 2018, 114: 53–62.

[2] Zasimchuk E., Turchak T., Baskova A, et. el. Structural Transformations in Metallic Materials During Plastic Deformation. *J Mater Eng Perform* 2017, 26: 1293–1299.

[3] Zasimchuk E., Gontareva RG., Baskova A., et. el. On Possibility to Use Single Crystal Sensors of Deformation Damage under Dynamical Loading. *Metallofizika i Noveishie Tekhnologii* 2012, 34: 509–520.

[4] Sniezek L., Szachogluchowicz I., Hutsaylyuk V. Testing of AA2519/TI6AL4V laminate for resistance on dynamic impact of the targets. *Proceedings of the 10th International Conference on Intelligent Technologies in Logistics and Mechatronics Systems ITELMS 2015*; Panevezys, Lithuania, May 21–22, 2015, pp. 238–244.

[5] Mazancova E., Ostroushko D., Wachowski M., et. al. Chosen metallographic and microfractographic parameters of explosively wealded bimetal 304 SS and Ti. *Proceedings of 20th Anniversary International Conference on Metallurgy and Materials*; Brno, Czech Republic, May 18-20, 2011, pp. 543–548.

[6] Ha DW., Jeong CY. Effect of Pre-Straining on the High-Cycle Fatigue Properties of Hot-Rolled Steel Sheets for Automotive Structural Uses. *Korean J Met Mater* 2018, 56: 177–186.

[7] Daehn GS. Sustainable design and manufacture of lightweight vehicle structures. In: Folkson R, editor. *Alternative Fuels and Advanced Vehicle Technologies for Improved Environmental Performance.* Woodhead Publishing, 2014, pp. 433–461.

[8] Wiewiórowska S., Muskalski Z. The Application of Low and Medium Carbon Steel with Multiphase TRIP Structure in Drawing Industry. *Procedia Manufacturing* 2015, 2: 181–185.

[9] Sugimoto K., Mukherjee M. TRIP aided and complex phase steels. In: Radhakanta Rana, Shiv Brat Singh, editors. *Automotive Steels.* Woodhead Publishing, 2017, pp. 217–257.

[10] Kuziak R., Kawalla R., Waengler S. Advanced high strength steels for automotive industry. *Arch Civ Mech Eng* 2008, 8: 103–117.

[11] Mazanek K., Śnieżek L., Ślęzak T. Fatigue research of welded joints of high strength S960QL steel. *Biuletyn Wojskowej Akademii Technicznej* 2013, 62: 253–269.

[12] Mandal S., Tewary NK., Ghosh SK., et. al. Thermo-mechanically controlled processed ultrahigh strength steel: Microstructure, texture and mechanical properties. *Mat Sci Eng A-Struct* 2016, 663: 126–140.

[13] Collin P., Johansson B. Bridges in high strength steel. Access via internet: http://stalbyggnadsinstitutet.se/uploads/source/files/Artiklar/Bridges_in_High_ Strenght_Steel_article.pdf [Accessed: 2017-10-17].

[14] Zou T., Li D., Wu G., et. al. Yield strength development from high strength steel plate to UOE pipe. *Mater Design* 2016, 89: 1107–1122.

[15] Gao W., Wang D., Cheng F., et. al. Microstructural and mechanical performance of underwater wet welded S355 steel. *J Mater Process Tech* 2016, 238: 333–340.

[16] Rozumek D., Marciniak Z., Lesiuk G., et. al. Mixed mode I/II/III fatigue crack growth in S355 steel. *Procedia Structural Integrity* 2017, 5: 896–903.

[17] Soltysiak R. Effect of Laser Welding Parameters of DUPLEX 2205 Steel Welds on Fatigue Life. *Solid State Phenom 2015*, 223: 11–18.

[18] Chausov M., Pylypenko A., Berezin V., et. al. Influence of dynamic non-equilibrium processes on strength and plasticity of materials of transportation systems. *Transport-Vilnius* 2018, 33: 231–241.

[19] Sniezek L., Szachogluchowicz I., Wachowski M., et. al. High cycle fatigue properties of explosively welded laminate AA2519/AA1050/ Ti6A14V. *Procedia Structural Integrity* 2017, 5: 422–429.

[20] de Jesus AMP., Matos R., Fontoura BFC., et. al. A comparison of the fatigue behavior between S355 and S690 steel grades. *J Constr Steel Res* 2012, 79: 140–150.

[21] Ulewicz R, Szataniak P, Novy F, et. al. Fatigue characteristics of structural steels in the gigacycle region of loading, *Mater Today-Proc* 2017, 4: 5979–5984.

[22] ASTM E 606-04. *Standard Practice for Strain-Controlled Fatigue Testing*, 2005.

ITMS'2018

The 12th International Scientific Conference Intelligent Technologies in Logistics and Mechatronics Systems (ITELMS'2018), 26–27 April 2018, Panevėžys, Lithuania

Investigation of Microstructure and Mechanical Properties Joints cp-Ti Obtained by FSW

Urszula Sobczak[a*], Marcin Wachowski[a], Lucjan Śnieżek[a], Janusz Mierzyński[a]

[a]*Military University of Technology, Faculty of Mechanical Engineering, Warsaw, Poland*

Abstract

Aim of this research was investigation of microstructure and mechanical properties joints cp-Ti obtained by Friction Stir Welding. Friction stir welding (FSW), a solid – state welding process developed by The Welding Institute in Cambridge. In this method alloys are joined in a solid state at temperature lower than melting points of joined materials. Conducted work includes analysis of microstructure and examining mechanical properties of joints. Fracture surfaces of specimens used in impact test were subject to microfractographic research.

© 2018 U. Sobczak, M. Wachowski, L. Śnieżek, J. Mierzyński
Peer-review under responsibility of the Kaunas University of Technology, Panevėžys Faculty of Technologies and Business

Keywords: titanium, welding, mechanical properties, friction stir welding

1. Introduction

Friction Stir Welding is a solid state welding process patented in 1991 by The Welding Institute in UK. Welding of clad materials by traditional fusion welding techniques presents several difficulties due to high material reactivity with oxygen, hydrogen and nitrogen and brittle intermetallic inclusions such as FeTi, $FeTi_2$. In this way FSW represents high quality solutions.

To make the connection rotary tool, finishing shoulder and the pin are necessary. This equipment allows to perform penetration and mixing of the materials along the line of contact elements. Rotating tool set material into rotational motion. Movement of the tool, after a moment of contact leads to the heat and plasticize of the materials. The pin has a conical threaded shape, which allows mixing of the material and the creation of consistent and compact joint. The tool both heats of the workpiece and moves the materials to produce of the joint [1]. As the tool traverses, the probe stirs the locally plasticized area and forms the solid – phase joint [2]. Shoulder tool, scrape the surface

* Corresponding author.
E-mail address: urszula.sobczak@wat.edu.pl

© 2018 U. Sobczak, M. Wachowski, L. Śnieżek, J. Mierzyński
Peer-review under responsibility of the Kaunas University of Technology, Panevėžys Faculty of Technologies and Business

protects the welding area from oxidation [3]. Figure 1 presents scheme of the friction stir welding process. Rotary tool with a threaded pin and shoulder which is inserted into the abutting edges and plates move along the line of the joint. Welding parameters, tool geometry and joint design have a significant effect on the material flow pattern and temperature distribution, thereby influencing on microstructure of the material. The uniformity of microstructure and properties as well as process loads are governed by the tool design [4]. Tool rotation rate (ω, rpm) and tool traverse speed (v, mm/min) are very important parameters for FSW. The rotation of tool influence on stirring and mixing of the material around the rotated pin while motion of the tool transports the stirred material from the front to the back of the pin and finishing welding process [5–6].

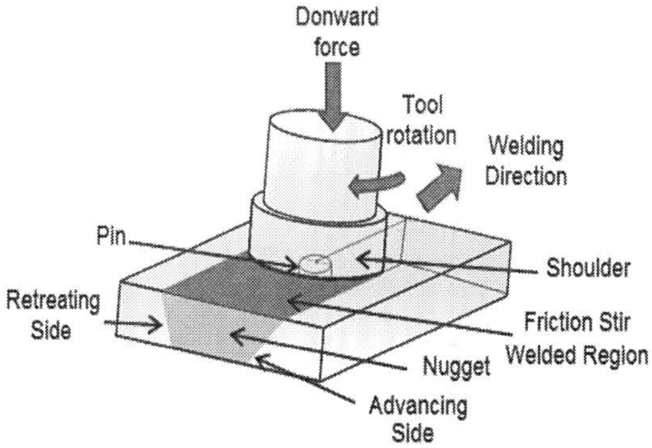

Fig. 1. Scheme of the friction stir welding process

Another important process parameter is an angle of spindle or tool tilt relative to the workpiece surface. A suitable tilt of the spindle towards trailing direction ensures holding of the stirred material by threaded pin and its efficiently movement from the front to the back of the pin. For titanium, which had a high melting point and high conductivity the heat produced by friction and stirring may be not sufficient to soften and plasticize the material around the rotating tools. In this case, the titanium must be preheated, because preheating from external heating source, allows to flow of the material. The process of Friction Stir Welding results in intense plastic deformation and temperature increase within and around the stirred zone. This results in significant microstructural evolution, including grain size, grain boundary character and dissolution of precipitates [7–8].

2. Experimental studies

This paper presents possibilities of using Friction Stir Welding (FSW) method to combine 3.2 mm thick cp-Ti (Table 1) titanium alloy plates. Joints were made from appropriately prepared 250 x 1000 mm plate samples, which were bound frontally, along rolling direction. Before pressure welding was commenced, surface layers of aforementioned samples were cleaned mechanically in order to negate inadvertent oxide influence.

Table 1. Composition of cp-Ti

Element	N	C	H	Fe	O	Al	V	Ti
Content [%]	0,03	0,10	0,015	0,02	0,18	-	-	rest

Pressure welding process was performed on ESAB Legio 4 UT stand designed to join metals via FSW method. In order to perform this process, the stand was equipped with gas supply devices. Plates were welded using heavy sinter tool with tungsten matrix. This tool features microstructure (Fig. 2) typical for sintered materials, and as such this microstructure is composed of highly fusible phase particles of regular shape, which are surrounded by bonding phase. Thanks to this microstructure, this tool exhibits a combination of high hardness, high ductility and high impact strength [9–10]. Additionally, it features high thermal conductivity and high resistance against corrosive agents.

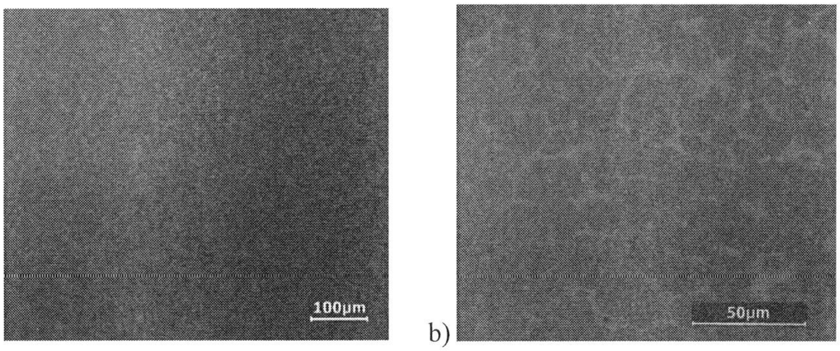

a) b)

Fig. 2. Microstructure of the tool

There is a 62.3% grain content on sample's surface in our studied material. The rest is made up of matrix. Geometrical parameters of our tool were adjusted to the thickness of welded sheets: tool shoulder diameter – 15.5 mm, pin diameter – 5–6 mm, tool length – 150 mm. This tool is shown on Figure 3.

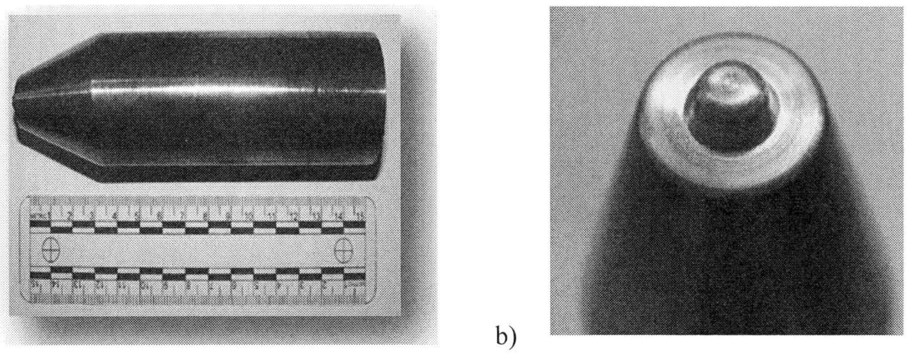

a) b)

Fig. 3. Tunsgten matrix heavy sinter tool used in FSW process to weld cp-Ti sheets: a) entire tool; b) pin

In order to check the tool's durability, first friction stir welding tests were performed on AA2519 aluminium sheets. Their composition is presented in Table 2, whereas microstructure of used material is presented on Figure 4.

Table 2. Composition of aluminium AA2519

Element	Si	Fe	Cu	Mg	Zn	Ti	V	Zr
Content [%]	0,06	0,08	5,77	0,18	0,01	0,04	0,12	0,2

Using the aforementioned tool we achieved overly satisfying results (Fig. 5). Despite local microstructure heterogeneities, we obtained a fully crystalized, equiaxed, fine-grained microstructure created in weld nugget due to plastic deformation in high temperature.

After weld test had been performed, tool's surface examination via confocal microscope was conducted, in order to identify flaws, however no visible signs of fatigue were noticed, as well as tool's geometry was not ompromised. After performing aforementioned process, cp-Ti titanium alloy binding was attempted. Parameters were selected based on consecutive trials of friction binding using various radial pin velocities, various tool movement speed and adding additional protection in the form of argon. Figure 6 presents the pictures of welds obtained during first pilot trials.

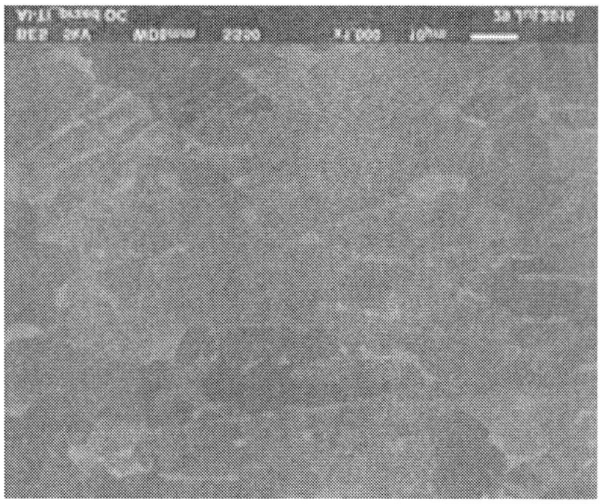

Fig. 4. AA2519 Aluminium alloy microstructure

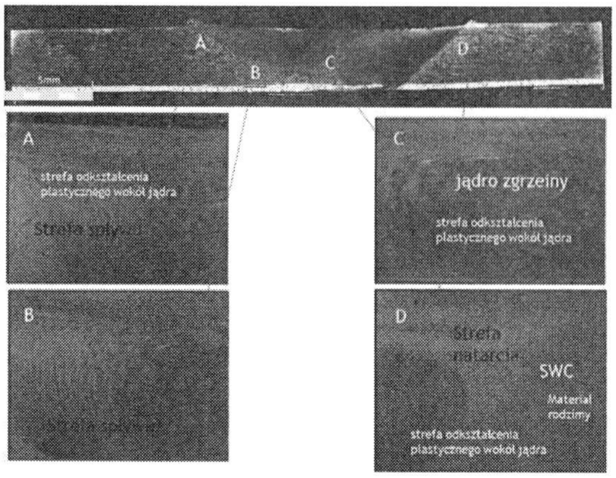

Fig. 5. Weld microstructure confronted with advancing side, weld nugget and retreating side

Fig. 6. Pictures of obtained welds with different selected parameters: a) 300 rpm/min, 50mm/min; b) 350 rpm/min, 25mm/min, argon; c) 350 rpm/min, 50 mm/min, argon

As one can notice from attached pictures, surface quality was not satisfactory, and due to that fact, welding parameters were changed, as well as shielding gas was switched to a mixture of argon and hydrogen. Applying these changes allowed to obtain joints of better quality (Fig. 7).

Fig. 7. Pictures of welds obtained during selection of welding parameters: a) 250 rpm/min, 100 mm/min; b) 250 rpm/min, 200 mm/min; c) 250 rpm/min; 75 mm/min; argon

For every foregoing joint, shielding atmosphere of argon and hydrogen was supplied. Studies, which consisted of microstructural studies, mechanical properties tests, such as: microstifness and crack resistance tests, were conducted on welds obtained during welding.

Cross section conducted through joint section in cp-Ti titanium alloy (Fig. 8) has shown empty spaces in weld nugget and discontinuity (0.57 mm height) in joint. This discontinuity is a result of the fact that thickness of joined elements was greater than pin height. Three microstifness measurements series were performed on studied microsections. Hardness of native material is lower than that of obtained weld. Existing variations of hardness in some points are a result of microstructure changes and temperature influence.

Whereas cross section through joint section in cp-Ti for parameters made with parameters: 250 rpm/min; 100 mm/min (Fig. 9), showed that there are much smaller discontinuity in joint. Also, in this case we have technological crack (high: 0.65 mm). Figure 10 presented cross section through joint section made with parameters: 250 rpm/min; 75 mm/min; distribution of microstiffness of studied cross section.

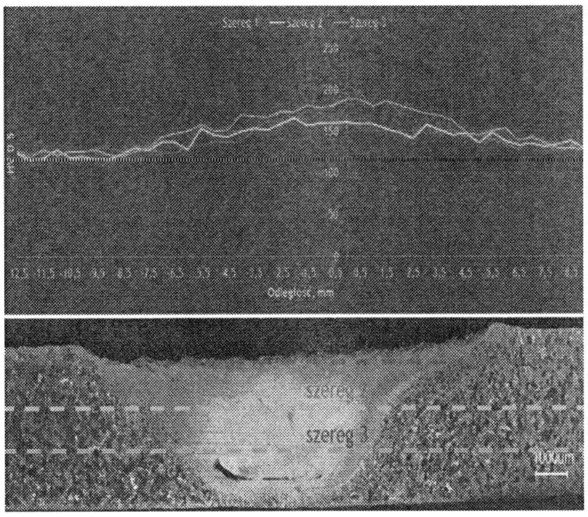

Fig. 8. Cross section through joint section made with parameters: 250 rpm/min; 100 mm/min; distribution of microstiffness of studied cross section

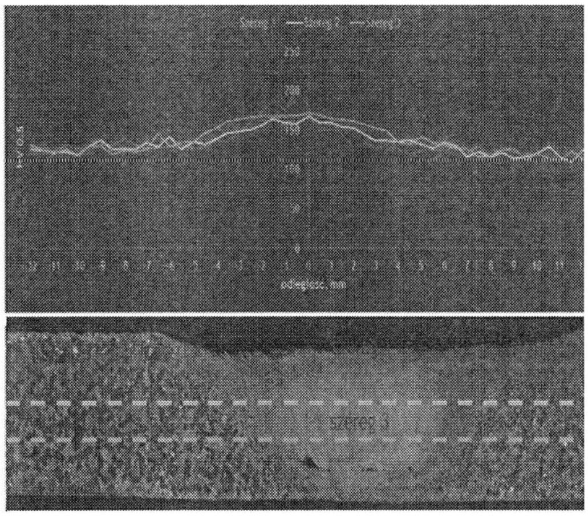

Fig. 9. Cross section through joint section made with parameters: 250 rpm/min; 200 mm/min; distribution of microstiffness of studied cross section

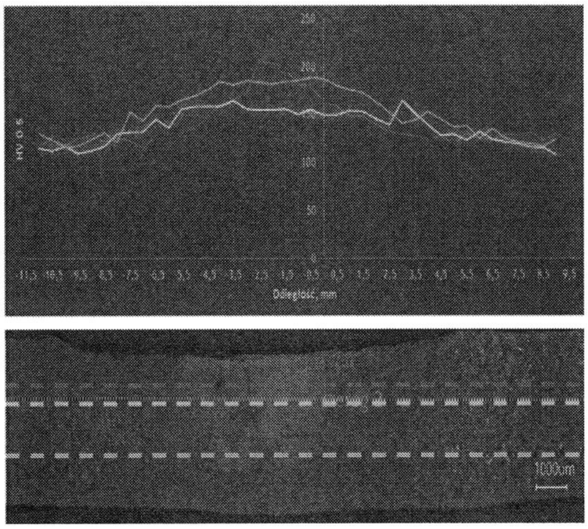

Fig. 10. Cross section through joint section made with parameters: 250 rpm/min; 75 mm/min; distribution of microstiffness of studied cross section

In the next stage of our studies, Charpy's method of impact strength test was used, in accordance to the obligatory standards regarding destructive tests of welded metal joints. Samples dimensions are presented in Table 3.

Table 3. Dimensions of samples for impact strength testing

	Dimension [mm]					
Probe	l	l_1	h_1	h	b	S_0
N04	57.0	27.6	7.8	10.0	3.5	27.30
N05	57.0	27.8	7.9	10.0	3.5	27.65
N06	57.0	27.6	8	10.0	3.5	28.00

The study was conducted by breaking samples with a pendulum hammer and measuring breakdown energy. The samples were braced without restraints on both ends. These samples also had a notch, which was placed on the opposite side of hammer strike. In the Figure 11 represent the amount of work needed to break the sample.

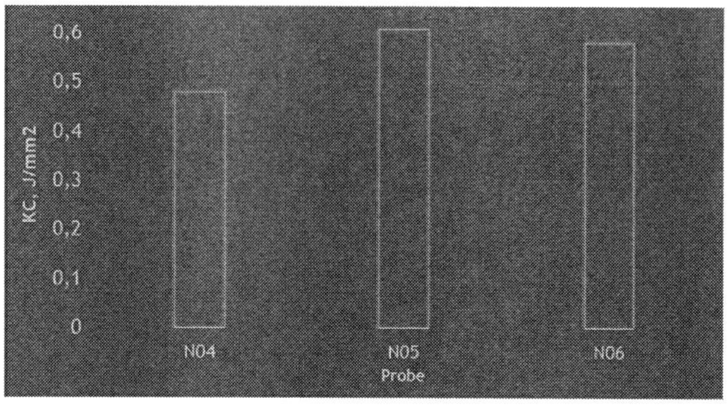

Fig. 11. Impact strength test results: N04 – 250 rpm/min; 100 mm/min, N05 – 250 rpm/min; 200 mm/min, N06 – 250 rpm/min; 75 mm/min

As it is shown on the Figure 11, impact strength for samples procured with bonding via 250 rpm/min radial velocity and 200 mm/min movement speed (designated as N05 sample in the Figure 11) is almost 25 % higher than that achieved for sample N04, which was procured at the same radial velocity, but at 100 mm/min movement speed. It is also 15 % higher than impact strength results for N06 sample (same 250 rpm/min radial velocity and 75 mm/min movement speed). After impact tests were carried out, the crack were analysed (Fig. 12).

a.

b.

c.

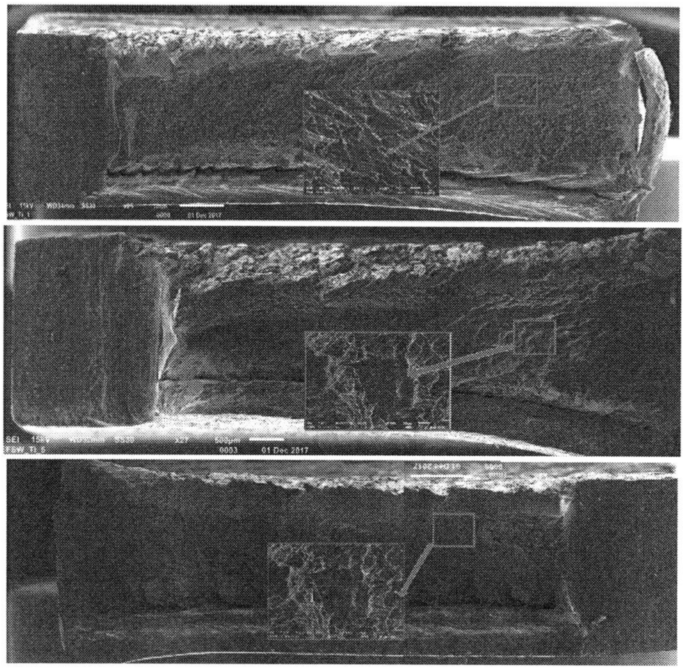

Fig. 12. Examples of fracture surfaces joints: 250 rpm/min; 100 mm/min – a; 250 rpm/min; 200 mm/min – b; 250 rpm/min; 75 mm/min – c;

In each case we analyzed, breach had plastic nature with distinctive expanded surface. Full breach was not obtained in any of analyzed cases.

Conclusions

1. Titanium cp-Ti alloy can be friction stir welded successfully using a specially designed welding system composed of tungsten tool and shielding gas shroud.
2. Excellent weld formation can be obtained under proper welding conditions. The cp-Ti plates can be friction stir welded at a welding speed from 75-200 mm/min and the overall mechanical properties was obtained at welding speed of 200 mm/min.
3. Cross section conducted through joints section in cp-Ti titanium alloy has shown empty spaces in weld nugget in joints.
4. In the all joints obtained full breach was not obtained in any of analyzed cases.

References

[1] Thomas W. M., Nicholas E. D., Needham J. C., Murch M. G. Improvements Relating to Friction Stir Welding. European Patent Specification 0615 480B1, 1995.

[2] Polmear I. J. Control of precipitation processes and properties in aged aluminium alloys by trace element additions. The Japan Inst. of Light Metals, Tokyo, Japan. 1998. pp. 75–86.

[3] Bellini L., Giumelli R. Welding of titanium clad components Welding International 2. 1986, pp. 155–160.

[4] Dawes C. J., Thomas W. M. Friction Stir Welding. Proc. 11th Annual North American Welding Research Conference, Columbus, Oh, USA. 1995.

[5] Mishra R. S., Ma Z. Y. Friction Stir Welding and Processing. Materials Science and Engeenering. R50. 2005, pp. 1–78.

[6] Rai R., De A., Bhadeshia H. K. D. H., DebRoy T. Review: friction stir welding tools. Science and Technology of Welding and Joining. 4. 2011, pp. 325–342.

[7] Kim J. D., Jin E. G., Yeong-Do Park S. P. M. Recent Advances in Friction-stir Welding Process and Microstructural Investigation of Friction Stir Welded Pure Titanium. Journal of Welding and Joining 2017. 35(4): 6–15.

[8] Russell M. J. Friction Stir Welding of titanium alloys - a progress update. Paper presented at 10th World Conference on Titanium. Hamburg. 13–18 July 2003. [1] Van der Geer J, Hanraads JAJ, Lupton RA. The art of writing a scientific article. *JSci Commun* 2000; 163: 51–9.

[9] Fratini L., Buffa G., Shivpuri R. On the material flow in FSW of Ti-joints: Influence of geometrical and technological parameters, *Int.J. of Adv. Manuf. Technl.* 44: 570, 2009.

[10] Sato Y. S., Miyake M., Kokawa H., Omori T., Ishida K., Imano S., Park S. H. C., Hirano S. Development of a cobalt-based alloy FSW tool for high-softening-temperature materials. *TMS Annual Meeting*. 2011, pp. 3–9.

ITMS'2018

ktu 1922 panevėžys faculty of technologies and business

The 12th International Scientific Conference Intelligent Technologies in Logistics and Mechatronics Systems (ITELMS'2018), 26–27 April 2018, Panevėžys, Lithuania

Anti-Slip System Parameters Influence on Robot Mobility in Rough Terrain

Kacper Spadło[a], Karol Kończalski[a], Marian Janusz Łopatka[a]

[a]*Military University of Technology, Warsaw, Gen. Witolda Urbanowicza 2 str., 00-908, Poland*

Abstract

The article includes analysis of influence of anti-slip system parameters to ability to over-come selected obstacles by an unmanned ground vehicle with hydrostatic drive system. There is described construction of model and simplifying assumptions. Is shows models of anti-slip system and ideas of their actions. It was determined parameters that have influence to ability to overcome selected obstacles.

Conducted research indicates that the use of anti-slip system increases the ability to move in terrain and overcoming obstacles. The use of external components that control the wheel speed and as a consequence, the slip of individual wheels does not reduce the efficiency of the drive system. During the study it was examined several configurations of anti-slip model. It was changed limiting value of slip and braking torque. Braking torque was dependent and independent for drive system loads. Conducted analysis showed that the anti-slip system operates in some extent their values. It is recommended to carry out tests on realistic conditions in order to verify the simulation model

© 2018 K. Spadło, K. Kończalski, M. J. Łopatka
Peer-review under responsibility of the Kaunas University of Technology, Panevėžys Faculty of Technologies and Business

Keywords: unmanned ground vehicles, anti slip system, overcome obstacles

1. Introduction

Unmanned ground vehicles are type of vehicle from which requires high mobility and ability to move in very diverse terrain [1]. In contrast to the other vehicles they are characterized by the absence of the operator who reacts online at what is happening with the vehicle. In the case of unmanned ground vehicles control is realized by teleoperation control system, which have limits the number of stimulus which is able to pick up operator [3, 7]. As a consequence, this limits ability to respond to the situation, e.g. sinks platform. In the case of unmanned ground vehicle is recommended to develop the systems to replace the operator that will provide a high ability to move in rough terrain. An important parameter for the mobility is to reduce excessive wheel slippage. In vehicles it is realized by the operator in the form of differential locks or by electronic systems that control slippage [1, 4, 9. 10]. These systems can not be used in unmanned ground vehicle due to other permitted slippage and necessity to overcome obstacles. Systems used in vehicles allow 30 % of the individual wheel speed differences, while unmanned ground vehicles require different limits on the level of 350 %. It is necessary to develop a anti slip system for unmanned ground vehicles.

The most common drive system used in the unmanned ground vehicles is the hydrostatic drive system. In order to maximize the efficiency of the desired configuration is a parallel configuration [2, 6]. It allows to use the full potential of hydraulics actuators. In the case of hydrostatic systems, it is possible to control the speed of individual hydraulic receivers by throttling the flow of the working fluid. Unfortunately, this reduces the efficiency of the overall hydraulic system, which is highly disadvantageous.

The solution is to use an external system regulates the speed of individual receivers. In the case of hydraulic drive system, they may be brakes in all wheels. These systems are successfully used in classical vehicles [8, 11].

Conducted a preliminary analysis shows that it is possible to use braking system in unmanned ground vehicles. The most important for the operation of the system is to determine point of system activation, and the value of braking torque for individual wheels. For this purpose there was built a simulation model with a hydrostatic drive system on all wheels and active slip control system.

2. Simulation model

It was simulation model built based on the analysis of light unmanned ground platforms. Based on platforms weighing up to 300 kg it have been adopted mass of main components. On that basis it was built a 3D model which allows to specify the position of the center of the gravity of the front and rear section (Fig. 1).

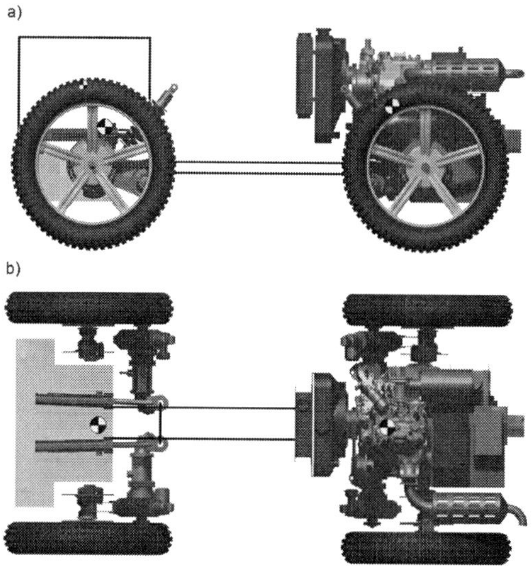

Fig. 1. Components center of gravity accepted in the deployment platform: a) side view; b) top view

Based on the position of the model's centers of gravity a dynamic model was built in MSC Adams software (Fig. 2). In the model, the weight of individual parts was replaced by the masses associated in form of spheres with a certain mass.

In this model is was used a flexible model of tire which takes into account wheel slip [5]. This model allows the reproduction of profiling cooperation wheels with the ground. Features of the slip obtained from tests wheel model is similar to the characteristics obtained from literature [1]. On this basis, the model correctly reflects the nature of wheel slip (Fig. 3).

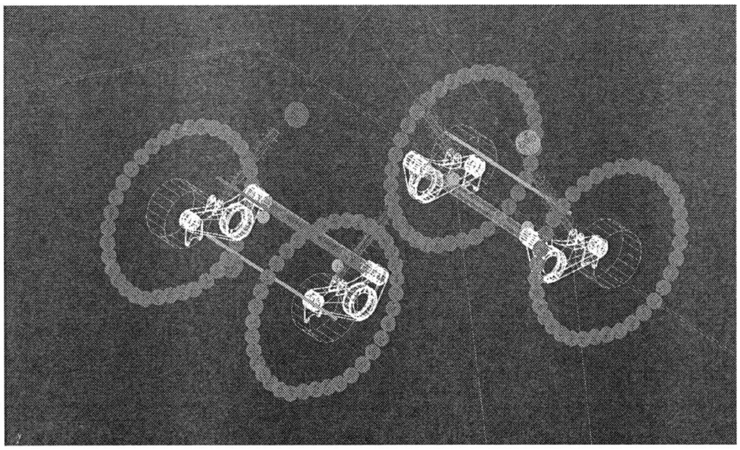

Fig. 2. Dynamic model of platform built in MSC Adams

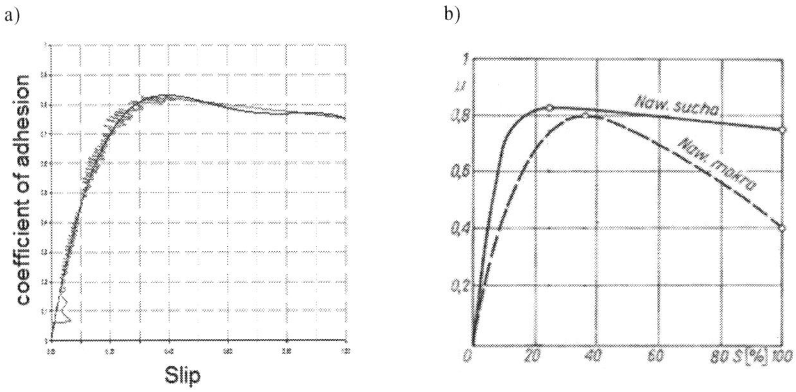

Fig. 3. Coefficient of adhesion as a function of slip: a) simulation results; b) characteristics book [1]

The wheel model was built with the specified quantity of spheres connected to the ring by elastic-damping elements. Elements image stiffness and damping in radial, circumferential and transverse direction. This model allows to change the dynamic radius depending on the load and enables profiling cooperation with the ground (Fig. 4).

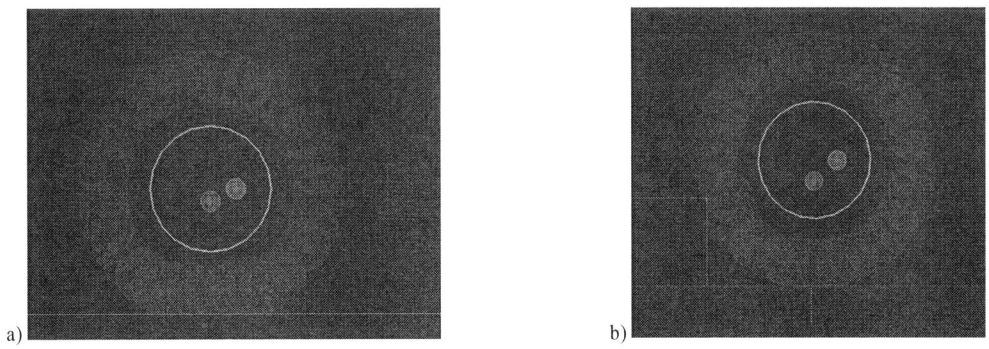

Fig. 4. Dynamic model of wheel: a) dynamic radius; b) overcome obstacle

In the model of the platform as a driving gear uses three differential gears connected together in a mapper the hydrostatic function in parallel configuration. As part of forcing the movement has been attributed to the speed on the input shaft of the main differential gear. The speeds were chosen as to allow the movement speeds 1, 2 and 5 km/h. As selected for the research it were adopted to overcome obstacles such as:

- log – diameter 20 cm (Fig. 5, *a*) and 30 cm (Fig. 5, *b*);
- curb – high 10 cm (Fig. 5, *c*) and 20 cm (Fig. 5, *d*);
- wall – high 30 cm (Fig. 5, *e*).

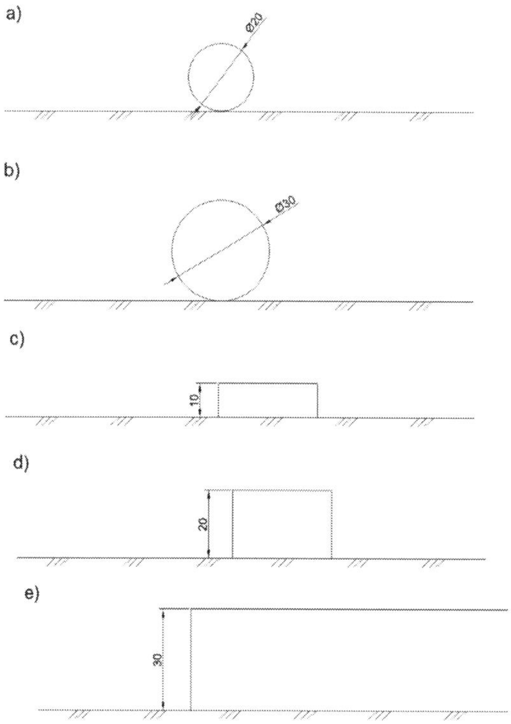

Fig. 5. Geometry of obstacle type: a) log 20cm, b) log 30cm, c) curb 10cm, d) curb 20cm, e) wall 30cm

It was envisaged that the obstacles are going to overcome at 90°, 66° and 45°. During the study coefficient of friction it was assumed on the highest value at which the model was not able to overcome obstacles without anti-slip system. These values ranged 0.3–0.45. During the tests it was assumed that the ground with the obstacle is undeformable.

3. Anti-slip system

In the model it was applied model of anti-slip system which increasing the ability to overcome obstacles.

The model is based on the measurement of linear velocity V of the wheels, the wheels angular velocity ω and the measurement of the dynamic wheel radius R_d (Fig. 6). Based on these parameters, the slip S is computed for each wheel using the formula [1]:

$$S = \frac{V - \omega \cdot R_d}{V}. \qquad (1)$$

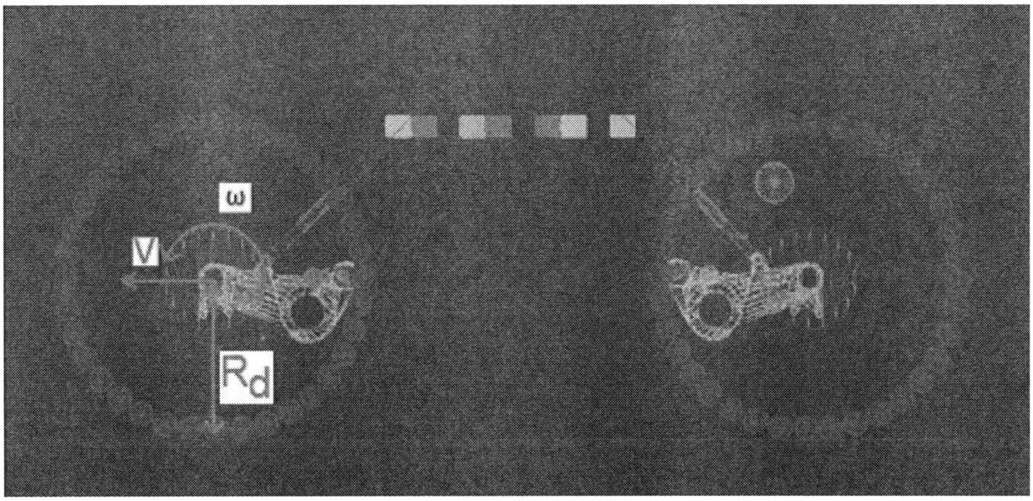

Fig. 6. Measured parameters in anti-slip model

The speed *V* is the velocity of which should have a theoretical circle while maintaining the specified slip. Slip is calculated by comparing the forward and circumferential velocity. When it exceeds limit set by program it follows the activation of the braking system on wheel.

During the study, it was examined two models of anti-slip system:

- system with constant braking torque and variable moment of system activation;
- system with variable braking torque and variable moment of system activation.

In both cases, subject to the application of the anti-slip was not an obstacle for the model without anti-slip system.

3.1. System with constant braking torque and variable moment of system activation

Considered model of anti-slip system have constant braking torque adapted to the conditions under which the move to model. Variable parameter was the limit value at which the slip followed by activation of the wheel brake (Fig. 7).

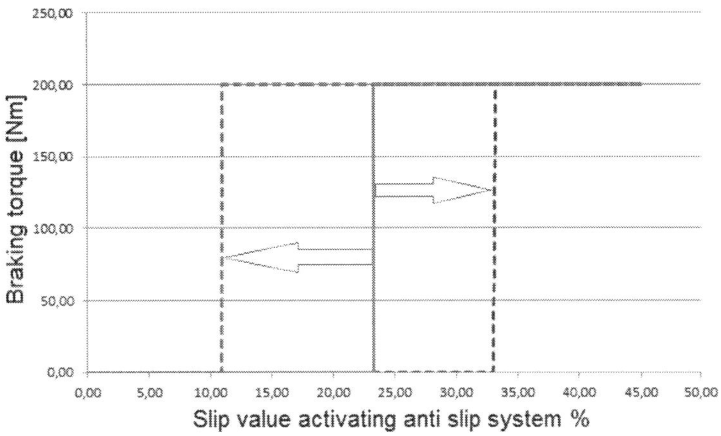

Fig. 7. Variable threshold value of the slip activating braking system

Figure 8 shows the course of the different speed of the selected wheel. Analyses were performed for the case of overcoming obstacles such curb with a height of 20 cm, speed of 1 km/h at 90°. The limit allowable slip is 30 %.

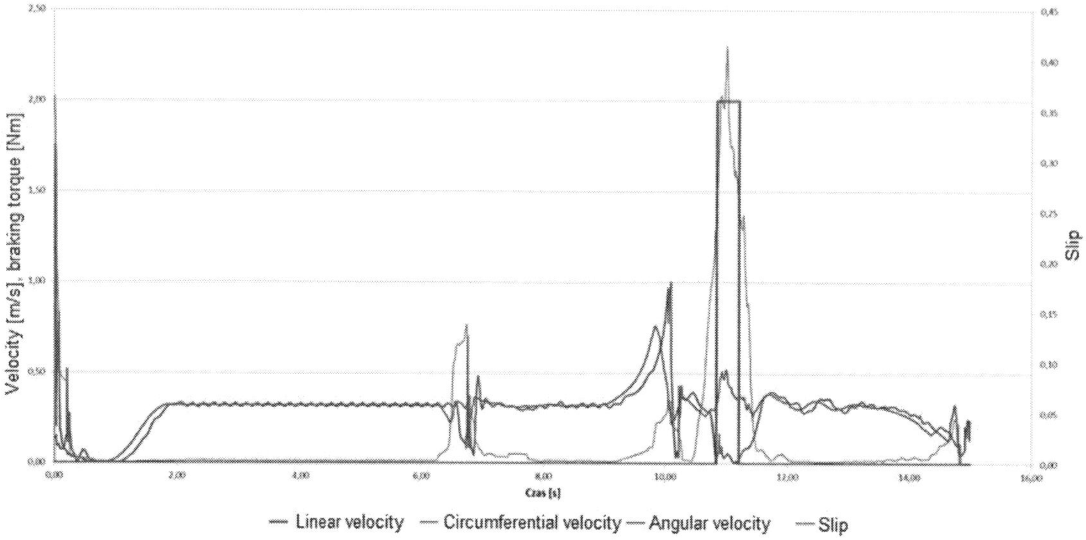

Fig. 8. The course of velocity, brake torque, and slip while overcoming obstacles such curbs

In about 11 second slip exceeds a predetermined value and activated braking system on wheel. Is decreased the wheel speed values providing less than acceptable slippage. At the same time driving force possible to achieve increased to the other wheels. When the slip falls below the limit system switches off braking.

3.2. *System with variable braking torque and variable moment of system activation*

The second consideration is anti-slip model of the system where braking torque is proportional to the input torque M0 in central differential. This value during the research was varied in the range of 0–100 % (Fig. 9).

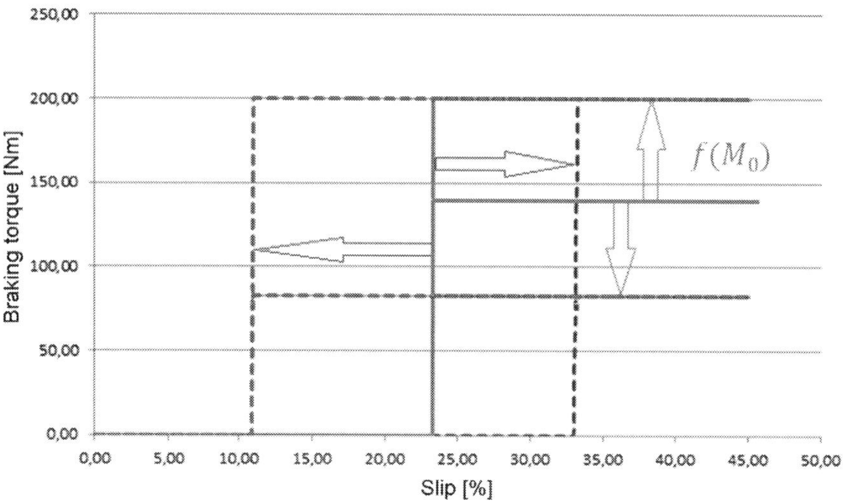

Fig. 9. Variable threshold value of the slip and activating braking system and variable value of braking torque

Figure 10 shows the progress of individual velocity for the selected wheel. Analyses were performed for the case of overcoming obstacles such curb with a height of 20 cm, speed of 1 km/h at 90°. The limit allowable slip is 30 % The value of the braking torque has been determined as 20 % of the input torque.

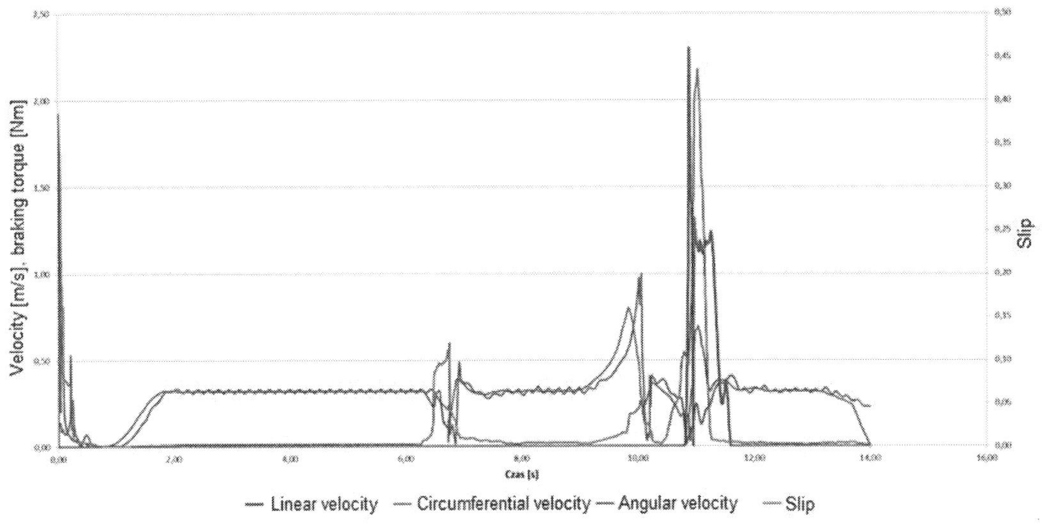

Fig. 8. The course of velocity, brake torque, and slip while overcoming obstacles such curbs

Model of the platform passing an obstacle in the same conditions as before. In about 11 second slippage with a value greater than the limit. Application occurs braking system. In this case, this value is not fixed and changes dynamically depending on the load model. When the slip decreases, braking system is switched off.

Conclusions

Conducted research indicates that the use of anti-slip system increases the ability to move in terrain and overcoming obstacles. The use of external components that control the wheel speed and as a consequence, the slip of individual wheels does not reduce the efficiency of the drive system.

During the study it was examined several configurations of anti-slip model. It was changed limiting value of slip and braking torque. Braking torque was dependent and independent for drive system load.

In the case of the model with constant braking torque:

- the system activated by slip value below 15 % resulted creation kinematic discrepancy;
- use the larger braking torque caused an increase wheel slip, which is not in excessive slip – less than the limit.

In the case of model with variable value of braking torque:

- at braking torque below 10 % of input torque, system is not improved ability to overcome obstacles;
- the braking torque over 50 % of the input torque cause creation excessive wheel slip which is not in in excessive slip.

Conducted analysis showed that the anti-slip system operates in some extent their values. It is recommended to carry out tests on realistic conditions in order to verify the simulation model.

Acknowledgment

The work presented in this article has been supported by the Polis National Center for Research and Development – Grant No. DOBR-BIO4/083/13431/2013.

References

[1] Andreev A., Kabanau V., Vantsevich V. Driveline Systems of Ground Vehicles: Theory and Design, CRC Press, 2010.

[2] Cieślik K., Konopka S. Concept and preliminary research of anthropomorphic manipulator with hydrostatic drive system for mobile robot, 21st International Conference on Methods and Models in Automation and Robotics, Miedzyzdroje, Poland 2016, pp. 1154–1159.

[3] Cieślik K., Przybysz M., Rubiec A. Concept for a vision system for an extreme mobility unmanned ground vehicle, 10th Intelligent Technologies in Logistics and Mechatronics Systems, Panevezys, Lithuania 2015, pp. 75–79.

[4] Cieślikowski B., Długosz A. Koncepcja systemu elektronicznej kontroli poślizgu granicznego kół ciągnika rolniczego, Inzynieria Rolnicza 2008.

[5] Konopka S., Łopatka M. J., Przybysz M. Kinematic discrepancy of hydrostatic drive of Unmanned Ground Vehicle. Proceedings of 8th International Conference ITELMS'2013.

[6] Konopka S., Łopatka M. J. and P. Krogul. Simulation Research of Hydrostatic Power System Control of Engineer Robot Manipulator, Proceedings of 8th International Conference on Intelligent Technologies in Logistics and Mechatronics Systems ITELMS 2013, Kaunas University Technology Press, Panevezys 2013, pp. 126–135.

[7] Krogul P., Typiak R. Issues with controling hydrostaticly driven manipulator using an intuitive human – machine interface, Proceedings of 8th International Conference on Intelligent Technologies in Logistics and Mechatronics Systems ITELMS 2016, Kaunas University Technology Press, Panevezys 2016, pp. 101–109.

[8] Li Z., Xiong L., Zhuoping Y. A research on Anti-slip regulation for 4WD electric vehicle with in-wheel Motors. 2nd International Conference on Computer Science and Electronics Engineering, ICCSEE 2013.

[9] Salat A. Układy mikroprocesorowe i sensory w maszynach rolniczych – aktualne badania, innowacyjne rozwiązania, SGGW Katedra Podstaw Inzynierii, Warszawa 2011.

[10] Volvo Articulated Haulers – *Autoamtic Traction Control*, Sweden 2009.

[11] Qian-nan W., Ziai L., Wang J. *Study on acceleration slip regulation torque distribution control strategy for four in-wheel-motors electric vehicle steering*. 25th Chinese Control and Decision Conference 2013.

ITMS'2018

panevėžys faculty of technologies and business

The 12th International Scientific Conference Intelligent Technologies in Logistics and Mechatronics Systems (ITELMS'2018), 26–27 April 2018, Panevėžys, Lithuania

Analysis of the Strain Distribution Made With the Digital Image Correlation Method in a Static Tensile Test of an Explosive Welding Composite Layer AA2519-AA1050-Ti6Al4V

Ireneusz Szachogluchowicz[a*], Lucjan Sniezek[a], Janusz Torzewski[a], Janusz Mierzynski[a]

[a]*Military University of Technology, gen. S. Kaliskiego 2 str., 00-908 Warsaw, Poland*

Abstract

Monotonic testing, uniaxial tensile base materials allowed to define their basic strength parameters. The resulting diagrams tensile tests of base are needed to assess the effects of explosive bonding methods for tensile strength. The tests were carried out for titanium Ti6Al4V alloy, aluminum AA2519 alloy and aluminum AA1050 after the rolling process and after heat treatment. An attempt of monotonic stretching of layered composites, whose base materials were compared with the theoretical composite strength calculated on the basis of the law of compound. Verification of the process of deformation of the laminate surface in the monotonic uniaxial stretching test was performed using the digital image correlation method DIC. Observations were conducted on three sides of laminate: aluminum alloy, titanium alloy and connectors. The results of the tests carried out for the laminate in the initial condition and for the laminate after heat treatment.

© 2018 I. Szachogluchowicz, L. Sniezek, J. Torzewski, J. Mierzynski
Peer-review under responsibility of the Kaunas University of Technology, Panevėžys Faculty of Technologies and Business

Keywords: explosive welding, cladding, layered composites, alluminum, titanum, alloy

1. Introduction

Materials used for mechanical constructions are most often subjected to cyclically variable loads. Under these conditions, various damage may occur, both in the base materials of the laminates and in the zone of their connection. Therefore, the shape and quality of the bonding zone made by the explosive bonding method has a significant impact on the strength properties of the material produced. One of the basic parameters for assessing the correct performance of a joint is its structure, which, among others, determines the susceptibility of a given

* Corresponding author. Tel.: +48-261-837-285; fax: +48-261-837-211
E-mail address: ireneusz.szachogluchowicz@wat.edu.pl

© 2018 I. Szachogluchowicz, L. Sniezek, J. Torzewski, J. Mierzynski
Peer-review under responsibility of the Kaunas University of Technology, Panevėžys Faculty of Technologies and Business

composite for further use. For the purpose of correct verification of connection parameters and evaluation of the obtained connection, the "weldability window" is most often used. Considering the nature of the connection surface and an extremely important issue due to the application possibilities, it is necessary to determine the material behavior under variable load conditions in the range of a small number of cycles. Works in this area are particularly important due to the possibility of investigating the mechanisms of cyclic deformation, which mechanisms lead to the location of deformation, material damage and, consequently, destruction of the construction element. Supplementing the research of the above studies should be microfractographic analyzes of breakthroughs and descriptions of mechanisms of cracking explosively joined lamellas. A common example of damage in laminates is the occurrence of delamination in the joint zone called delamination in the nomenclature of laminar materials. Damages of this type locally reduce the fatigue strength of the composite. Delamination created in the middle of the wall thickness, that is dividing the laminate into two sublaminates, can even double the fatigue strength for bending the test piece [1–3]. Particularly exposed to this type of cracks are laminates with a flat bonding boundary, or those with a transition zone formed as a result of an explosive bonding. In the case of a flat joining zone, the adhesive nature of the anastomosis is characterized by a limited range of allowable stresses as compared to the joining materials. However, connections in which the transition zone is visible despite the range of permissible stresses exceeding the parameters of one of the joined materials also pose a risk of delamination. This is due to the variety and discontinuity of the structure of the interlayer formed. The concentration of stresses in native materials and in the transition zone can also act as initiators of cracking and delamination of the laminate [4–6].

Another important issue is the impact of the joining zone on ballistic resistance of the material. In this case, the occurrence of a transitional zone causing local strengthening of the material affects the improvement of ballistic resistance. Testing of strength properties of AA2519-AA1050-Ti6Al4V laminate and its base materials under axial tensile conditions was carried out in accordance with PN EN ISO10002-5: 2004 on the Instron 8802 pulsator. The deformation measurement was performed using an Instron 2630-112 with a 50mm measuring base. The dimensions of the samples are shown in Figure 1.

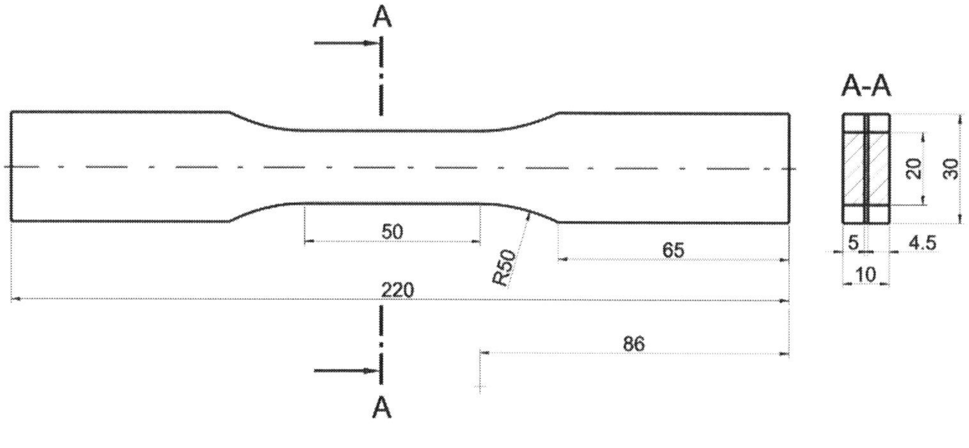

Fig. 1. The dimensions of the test specimens uniaxial tension

Monotonic stretching tests were carried out for base materials and the produced laminate in the state before and after heat treatment. At least 6 samples were made of each material (Fig. 2). The direction of cutting samples relative to the base material was random.

Adopted a more practical work included examination of selected mechanical properties of layered composite material composed of alloys: Ti6Al4V / AA1050 / AA2519 with the resulting diffusion layer Al3Ti type [7]. It was assumed that such material will be characterized by unique properties that combine the beneficial properties of titanium and aluminum alloys (high strength, high plasticity and low specific gravity) and alloys, intermetallic matrix Ti-Al (high hardness and stiffness) [8].

The composite layer made of base materials in the form of aluminum alloy AA2519 (AlCuMgMn + ZrSc) and the titanium alloy Ti6Al4V. Chemical composition and mechanical properties of base materials are provided in Table 1 and Table 2.

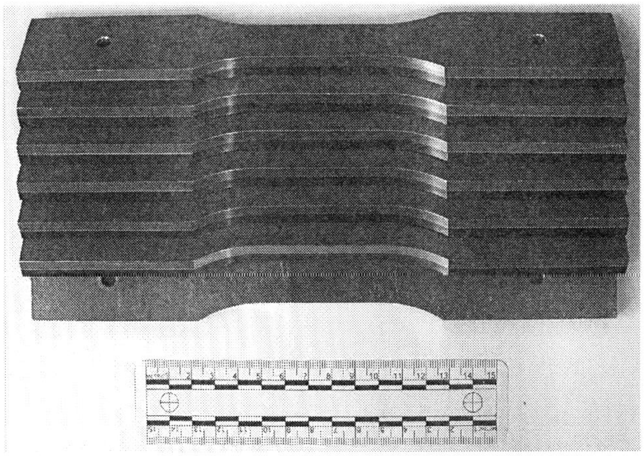

Fig. 2. Samples of AA2519-AA1050-Ti6Al4V laminate for monotonic stretching tests

Table 1. Mechanical properties and chemical composition of the Ti6Al4V alloy

Mechanical properties			Chemical composition (wt %)							
Rp0,2 (MPa)	Rm (MPa)	A (%)	O	V	Al	Fe	H	C	N	Ti
950	1020	14	<0,20	3,5	5,5	<0,30	<0,0015	<0,08	<0,05	rest

Table 2. Mechanical properties and chemical composition of the AA2519 alloy

Mechanical properties			Chemical composition (wt %)								
Rp0,2 (MPa))	Rm (MPa)	A (%)	Si	Fe	Cu	Mg	Zn	Ti	V	Zr	Sc
312	335	6,5	0,06	0,08	5,77	0,18	0,01	0,04	0,12	0,2	0,36

Combining explosive was realized at the Department of Technology Energetic "EXPLOMET". Cooperation with the plant allowed to undertake research aimed at obtaining and determining the performance of a construction material in the form of composite Al-Ti layer with specific properties, which does not have a solid material [9, 10].

Nomenclature

A	elongation sample
Rm	maximum rupture resistance
$Rp_{0,2}$	ductility limit

2. Procedures and research

Conducting tests of monotone, uniaxial stretching of base materials allowed to determine their basic strength parameters. Obtained graphs of the course of tensile tests of base materials are necessary to assess the impact of the explosive bonding method on tensile strength. The tests were carried out for Ti6Al4V titanium alloy, AA2519 aluminum alloy and AA1050 aluminum after the rolling process (Fig. 3).

Based on the results obtained (Fig. 4) it was found that the material performing the role of reinforcement in the laminate layer AA2519-AA1050-Ti6Al4V should be made of Ti6Al4V titanium alloy. Titanium alloy, achieving maximum material tensile strength Rm = 910 Mpa exceeds the maximum tensile strength of alloy AA2519 more than 2.5 times. The relative elongation A for the Ti6Al4V titanium alloy is about 12 %, while for the aluminum alloy AA2519 – about 8%. Axial stretching tests were also carried out for base materials subjected to the same heat treatment as the AA2519-AA1050-Ti6Al4V laminate (Fig. 4). The blue curve indicates the b-laminate in the initial state, while the red curve is marked with the laminate after heat treatment.

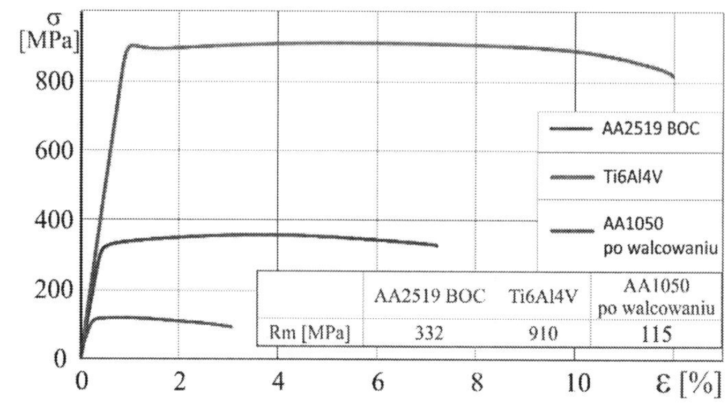

Fig. 3. Monotonic stretching of flat samples from the AA2519-AA1050-Ti6Al4V layered composite in the initial state

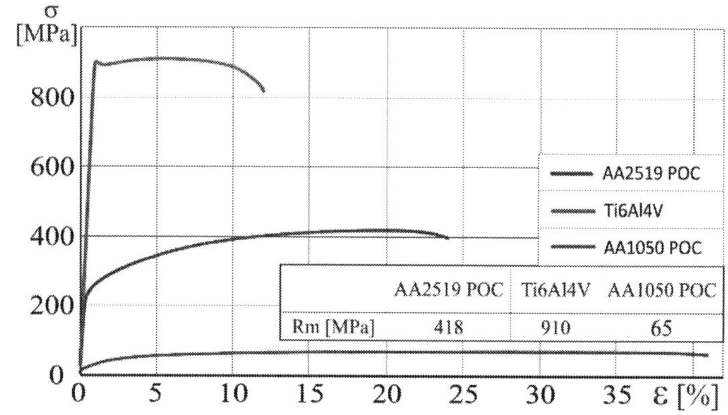

Fig. 4. Monotonic stretching of flat samples from AA2519-AA1050-Ti6Al4V layered composite after heat treatment

The fatigue strength characteristic obtained for the titanium alloy Ti6Al4V did not change, while for the alloy AA2519 the maximum tensile strength increased to Rm=418Mpa with a simultaneous increase in the relative elongation to about 25 %. This difference is mainly due to structural changes in the form of considerable fragmentation of the structure. The test of monotonic stretching of layered composites, whose base materials are characterized by a continuous structure, can be compared to the theoretical composite strength calculated on the basis of the law of compound. On its basis, it is possible to estimate the probable maximum tensile strength R_m in accordance with the law of compound [11]. For the analyzed case, this right is described by the equation:

$$Rmk = Rmw \times Vw + Rmo \times Vo + Rmm \times Vm \qquad (1)$$

where R_{mk} – tensile strength of the composite, Mpa;
R_{mw} – tensile strength of reinforcement, Mpa;
V_w – ratio of the cross-sectional area of the reinforcement to the cross-sectional area of the compound;
R_{mo} – tensile strength of the warp, Mpa;
V_o – ratio of the cross-sectional area of the matrix to the area of the reinforcement section;
R_{mm} – strength on interlayers, Mpa;
V_m – the ratio of the cross-sectional area of the interlayer to the cross-section area of the reinforcement.

Produced laminate in the initial state:

$$Rmk = 910Mpa \times 0{,}47 + 332Mpa \times 0{,}47 + 65Mpa \times 0{,}060;$$

$$Rmk = 587{,}64\ Mpa.$$

Heat treated laminate:

Rmk = 910 Mpa × 0,47 + 418 Mpa × 0,47+ 115 Mpa × 0,060;

Rmk =631,06 Mpa.

The obtained results indicate that the produced laminate in the initial state should have a higher tensile strength at the level of R_m = 588 Mpa, which after the thermal treatment should increase by about 7 % increases to 631 Mpa. It should be remembered that the applied law of mixtures does not take into account the relative deformation for which the maximum tensile strength is obtained. As a result, the highest theoretical tensile strength of the laminate is obtained. The results obtained from the right mixtures have been verified by performing monotonic stretching tests on the produced laminates. The graphs obtained from the uniaxial stretching tests are shown in Figure 5. The blue curve indicates the b-laminate in the initial state, while the red curve is marked with the laminate after heat treatment.

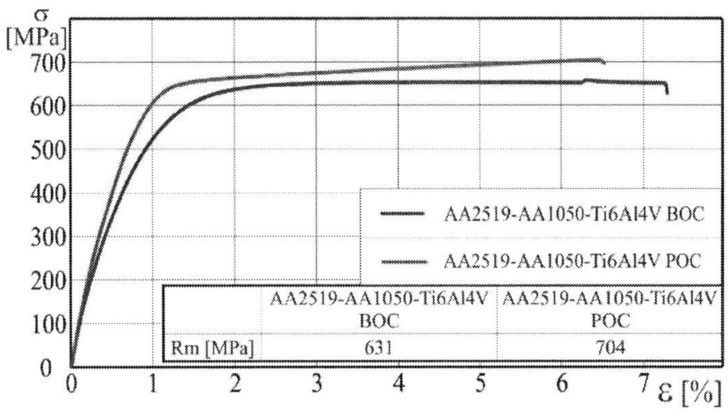

Fig. 5. Monotonic stretching of flat samples from the AA2519-AA1050-Ti6Al4V layered composite before and after heat treatment

The traces of the monotonic stretching graphs for the laminate without heat treatment indicate its stabilization after exceeding the conventional yield stress. For the treated laminate, a uniform increase in stress due to plastic deformation is visible. These trends indicate the continuous strengthening of the material, while narrowing the sample. The comparison of the obtained results of maximum tensile strength based on the law of mixtures and tests performed is shown in Figure 6.

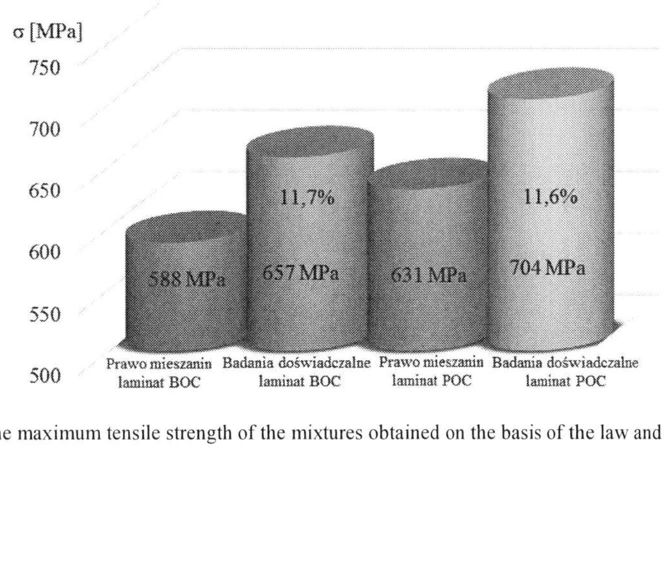

Fig. 6. The results of the maximum tensile strength of the mixtures obtained on the basis of the law and the tests performed

Literature analysis confirmed by own tests indicates the occurrence of material strengthening as a result of crumbling caused by the explosive bonding process. Another factor affecting the increase of tensile strength is the formation of phase materials of Al_3Ti intermetallic compounds at the interface. The diffusion nature of Al-Ti joint formation and thermal-mechanical processes mean that despite a very short diffusion time, in the very narrow area of the transition zone, the elements are segregated and, as a consequence, the mechanical properties change locally [11]. These factors, as well as the layered structure of the produced composite, can have a significant impact on the unevenness of the surface deformation of the material. Verification of the process of deformation of the laminate surface in the monotonic uniaxial stretching test was carried out using the digital image correlation (DIC) method. The observations were carried out on three sides of the laminate: aluminum alloy, titanium alloy and joint. The results of the tests carried out for the laminate in the initial state are shown in Figure 7. The results of the tests carried out for the laminate after heat treatment are shown in Figure 8.

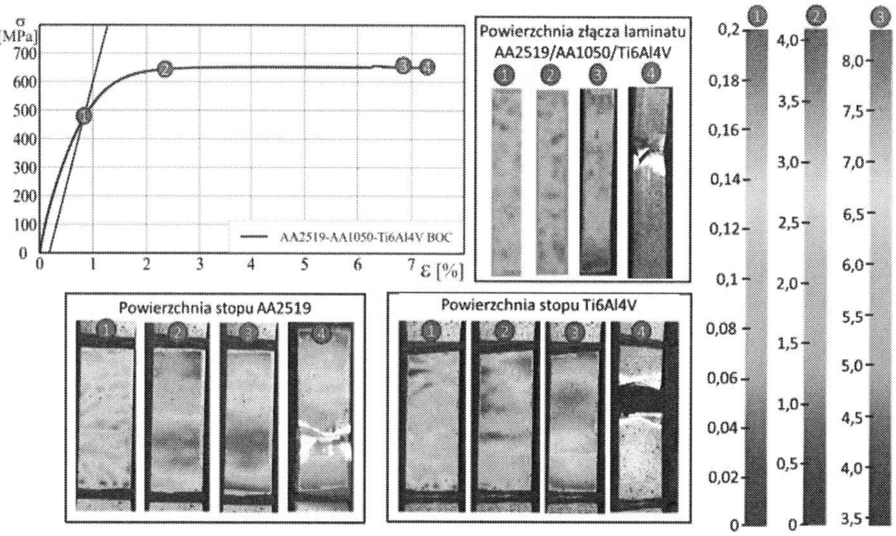

Fig. 7. Analysis of strain distribution by the DIC method for selected points of the monotonic test of stretching a sample of a flat layered composite AA2519-AA1050-Ti6Al4V

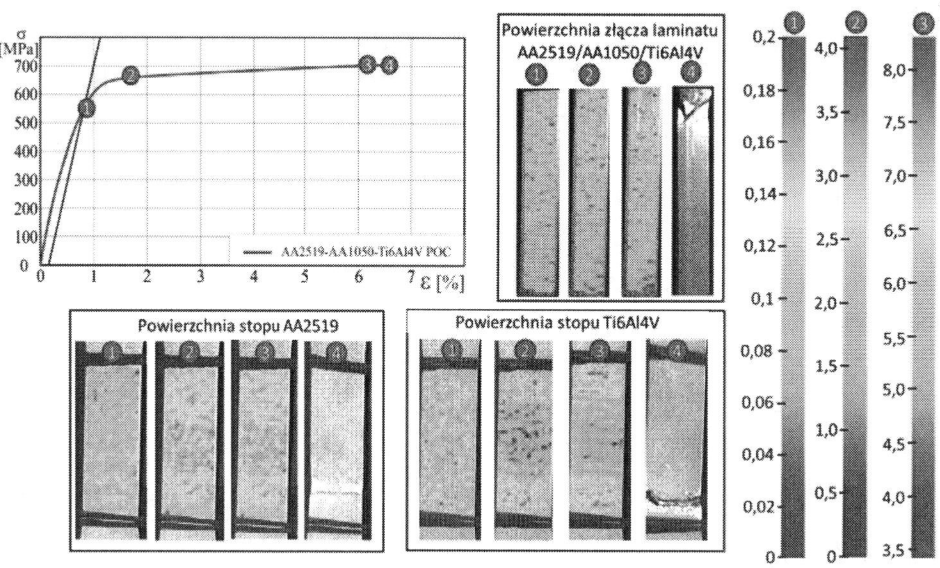

Fig. 8. Analysis of strain distribution by the DIC method for selected points of a monotonic test of stretching a sample of a flat layered composite AA2519-AA1050-Ti6Al4V

The obtained test results for the laminate in the initial state for the point marked with the number 1 show an even level of deformation for each side. Only after exceeding the conventional yield stress, the areas of uneven material displacement are distinguished. It should be noted that the deformations of the side do not show banding that may result from different material properties. Point 3 illustrates the concentration of maximum deformations at the place of sample fracture. The location of maximum deformations and the gradual nature of their formation indicates the continuity of the joint and its high quality. Deformations for the heat treated laminate are characterized by a higher level of uniformity. This is particularly noticeable on the surface of AA2519 alloy for which the parameters of the heat treatment process were selected. Also, the joint surface does not exhibit local and banded differences in strain, which makes it difficult to locate the cracking zone. The deformations on the titanium side start to lose their uniformity at the point marked with number 2. This is caused by the local formation of the narrowing of the AA2519 alloy.

Conclusions

On the basis of the obtained results, it can be stated that the laminate is a perspective material of construction, which can be subjected to modifications aimed at improving its mechanical properties. A large heterogeneity of deformation fields was observed on the surface of AA2519 aluminum alloy samples. The diffusion nature of the Al-Ti joint formation process means that the strength properties of the laminate can be shaped by the appropriate selection of heat treatment parameters carried out after the explosive bonding process. The test results indicate the stability of the laminate produced.

Acknowledgements

The project is carried out under Project PBS2/A5/35/2013 funded by the National Research and Development Centre.

References

[1] Lysak V. I., Kuzmin S.V. Explosive welding of metal layered composite materials / Ed. B. E. Paton. Kiev: E. O. Paton Electric Welding Institute of NASU, 2003, 117 p.

[2] Abrahamson G. R. Residual periodical deformations of surface under action of moving jet. Proc. of ASME, Ser. E, Appl. Mechanics, 1961, Vol. 28, No. 4, pp. 45–55.

[3] Cowan G. R., Holtzman A. H. Flow configuration in colliding plates: explosive banding. J. Appl. Phys. 1963, Vol. 34, No. 4, p. 928–939.

[4] Cowan G. R., Bergmann O. R., Holtzman, A.H. Mechanism of Bond Zone Wave Formation in Explosion Clad Metals. Metallurgical Transactions, 1971, V. 2, No. 11, pp.3145–3155.

[5] Godunov S. K., Kiselev S. P., Kulikov I. M., Mali V. I. Numerical and Experimental Simulation of Wave Formation during Explosion Welding. Proceedings of the Steklov Institute of Mathematics, 2013, Vol. 281, pp. 12–26.

[6] Achenbach J.D. Wave propagation in elastic solids. New York: Amer. Elsievier Publ. Co, 1973, 425 p.

[7] Hutsaylyuk V., Sulym H., Pasternak Ia., Turchyn I. Transient Plane Waves Propagation in Non-homogeneous Elastic Plate. Composite Materials: The Great Advance. Proc. of the 19th Int. Conf. Composite Materials, ICCM19, Montreal, Canada, 28 July–2 August 2013, pp.8890–8897.

[8] Sniezek L., Szachogłuchowicz I., Gocman K. The mechanical properties of composites AA2519-Ti6Al4V obtained by detonation method. Intelligent Technologies in Logistcs and Mechatronics Systems ITELMS'2014, Proc. of The 9TM International Conference edited by Z. Bazaras and V. Kleiza, 2013, pp. 214–219.

[9] Sniezek L., Szachogłuchowicz I., Hutsaylyuk V. Research of Property Fatigue Advanced Al/Ti Laminate. Intelligent Technologies in Logistcs and Mechatronics Systems ITELMS'2014, Proc. of The 9TM International Conference edited by Z. Bazaras and V. Kleiza, 2014, pp. 232–238.

[10] Sulym H., Hutsaylyuk V., Pasternak Ia., Turchyn I. Stress-strain state of an elastic rectangular plate under dynamic load. Mechanika, 2013, Vol. 19, No 6, pp. 620–626.

[11] Gorskov A. G., Medvedsky A. L., Rabinsky L. N., Tarlakovsky D. V. Waves in continuum media. Moscow: Fizmatlit, 2004. 472 p.

ITMS'2018

The 12th International Scientific Conference Intelligent Technologies in Logistics and Mechatronics Systems (ITELMS'2018), 26–27 April 2018, Panevėžys, Lithuania

The Shot Peening Impact Studies on the Fatigue Strength of Carbonated and Tempered 21NiCrMo2 Gears

Tomasz Tchorz[a], Lucjan Sniezek[a], Krzysztof Grzelak[a], Marcin Wachowski[a]

[a]*Military University of Technology, gen. S. Kaliskiego 2 str., 00-908 Warsaw, Poland*

Abstract

The article presents the impact of shot peening gears on their contact fatigue strength. Gears are made of indigenous material stainless 21NiCrMo2. The gears were subjected to a thermo-chemical treatment, which resulted in a surface hardness of 58-60 HRC. A series of gears was subjected to shot peening treatment. The gears were subjected to comparative fatigue tests on the circulating power machine. The fatigue testing programs carried out on the test bench have been developed on the basis of measurements of loads occurring in the driving system of the farm tractor during the performance of heavy agro-technical works.

© 2018 T. Tchórz, L. Śnieżek, K. Grzelak, M. Wachowski
Peer-review under responsibility of the Kaunas University of Technology, Panevėžys Faculty of Technologies and Business

Keywords: mechanics, farm tractors, gears, fatigue tests, shot peening

1. Introduction

In the design and gears construction, the aim is to minimize the mass and their size while increasing the gears strength. Minimizing the gears size is primarily determined by the contact strength of the teeth's lateral surface, as well as the strength of the teeth for bending. The destruction of gears caused by the wear of the teeth lateral surfaces as a result of fatigue crashes has been estimated at 40÷50 % of the total number of gears. Destruction by breaking a tooth affects about 20 % of used gears, while the remaining 30% are other types of wear.

The development of fatigue damage in gears occurs in two areas: on the working surface in the tooth contact zone and at the root of the tooth in the critical section, where the greatest stress caused by bending the tooth occurs. The first case concerns the long-term process of gears wear as a result of high pressures. The result of these wears type are fatigue cracks on the lateral surface of the tooth in the form of pitting. In the second case, the gears are damaged as a result of breaking the tooth at the root [1].

© 2018 T. Tchórz, L. Śnieżek, K. Grzelak, M. Wachowski
Peer-review under responsibility of the Kaunas University of Technology, Panevėžys Faculty of Technologies and Business
B427H0034 . Ed*Learning*.it

2. Object of study

The research topic is a pair of model gears made of steel 21NiCrMo2, which were subjected to carburization and hardening treatment. The gears parameters subjected to fatigue tests are shown in the Table 1.

Table 1. Parameters of model gears

Parameter	Signature	Pinion	Cooperating gear
Number of teeth	z1; z2	25	26
Module	m	3.5 [mm]	3.5 [mm]
Gear width	b1; b2	5 [mm]	15 [mm]
Pitch diameter	d1; d2	87.5 [mm]	91 [mm]
Root diameter	df1; df2	78.82 [mm]	82.18 [mm]
Tip diameter	da1; da2	94.57 [mm]	97.93 [mm]
Correction factor	x1; x2	0.035	-0.01
Shape factor	qr1; qr2	2.87	2.87
Pressure angle	αw	20°	20°
Center distance	aw	89.25 [mm]	89.25 [mm]
Tooth high factor	y	1	1

The model gears were subjected to a thermo-chemical treatment consisting of:

- carburizing at 920 °C for four hours and ten minutes;
- hardening from 850 °C for two hours in oil;
- tempering at 190 °C for one hour and thirty minutes.

The obtained thickness of the carburized and hardened layer is 0.8 mm. Figure 1 (*a*) presents a general view of the tooth's cross-section. As a result of the thermo-chemical treatment, the thickness of the carburized layer about 0.8 mm was obtained with a fine-grained martensitic structure, Figure 1 (*b*), with a hardness 58–60 HRC. The tooth core is characterized by a thick-grained martensitic structure Figure 1 (*c*) with a hardness of 52 HRC.

Courses of microhardness changes measured near the tooth's tip, pitch diameter and tooth root have similar distributions. Microhardness near the tooth's working surface is about 740 HV and decreases with increasing distance from the surface, reaching a hardness of 620–640 HV in the tooth core.

A part of gears was subjected to shot peening in order to introduce additional tooth strengthening by implementation of compressive residual stresses and increasing microhardness of the surface layer [2, 6]. The parameters of this treatment were selected based on the studies of the shot peening effect on flat samples made of steel 21NiCrMo2, which were also subjected to thermo-chemical treatment. The flat samples were subjected to shot peening with different parameters. Based on the measurements analysis of roughness, microhardness and surface residual stress the best shot peening parameters were determinate. They were used during gears machining process. The applied shot peening parameters are shown in Table 2.

As a result of shot peening (achieved increase) microhardness increase in the tooth's lateral surface layer was achieved. It is also noticeable that the value of compressive residual stresses in the surface of samples subjected to shot peening in relation to non-peened samples increases almost five times. The largest increase in microhardness was observed in the area of tooth's tip. The results of microhardness measurements for this area are shown in Figure 2. The measurements were taken for the pinion before and after the shot peening.

The maximum increase in microhardness occurs near the top of surface layer and exceeds 100 HV0.2. Shot peening influence on the growth of microhardness is noticeable to a depth of 0.5 mm. Stabilization of microhardness at the level of 520 HV0.2 occurs at a depth of 1.2 mm, this value corresponds to the hardness occurring in the tooth's core.

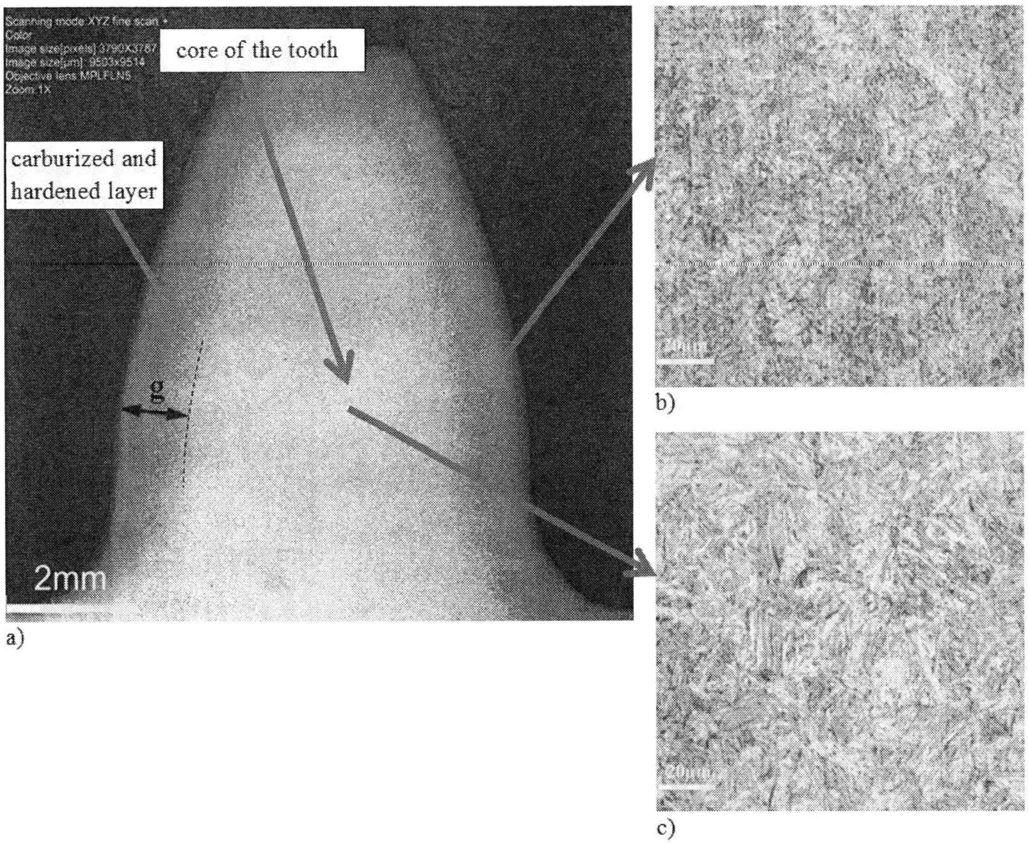

Fig. 1. Structure of the tooth material: a) general view of the tooth with a carburized and hardened layer of thickness g ≈ 0.8 mm; b) microstructure of carburized and hardened layer; c) tooth core microstructure

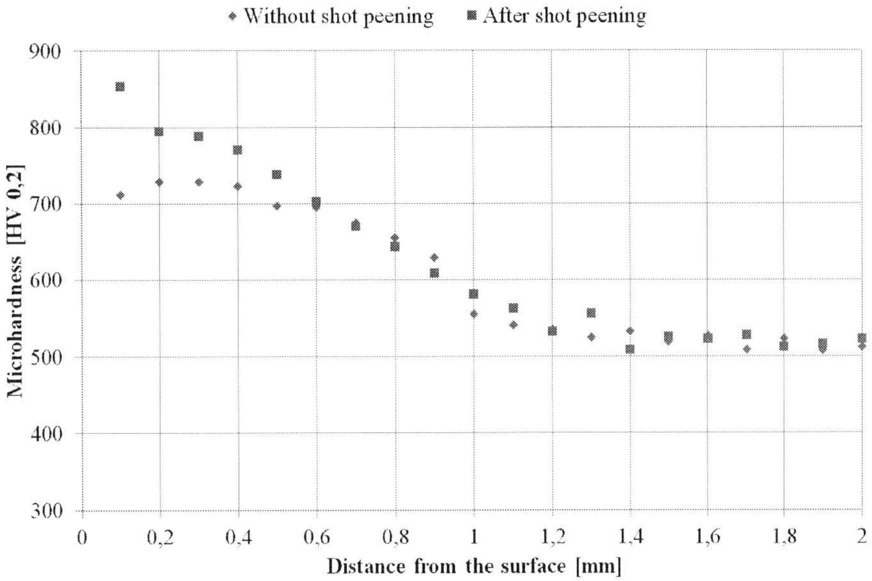

Fig. 2. Distribution of microhardness in the teeth's lateral surface layer before and after shot peening

Table 2. Value of gears shot peening parameters

Parameter	Unit	Value
Diameter	mm	0.8
Hardness	HV	640
Nozzle distance	mm	200
Air pressure	MPa	0.47 ± 0.05
Shot speed	m/s	75
Shot peening time	min	4
Shot peening intensity [f_c]	mm	0.15

3. Testing methodology

Gears fatigue strength tests are used to determine damage, such as: lateral surface fatigue wear, called by pitting and tooth fracture at the root. There is a need to determine when the damage will occur from the first start machine at a given load. By determining the useful transmission life for operation in normal operation conditions, without the need of human intervention in the device, serious damage to the whole device can be avoided. In order to determine the aforementioned damages, tests are carried out on various types of dynamometers for testing entire transmission systems. When the tests are limited to a single pair of gears, circulating power machine are used, on which the gear are tested due to damage to the tooth's lateral surface.

Devices for testing the contact strength of gears enable conducting tests in specific load conditions, taking into account the influence of oil and other factors affecting the tooth's lateral surface contact strength. Circulating power machines are characterized by (the) internal state of gear load performed by elastic deformation of the connector between the test and closing gear. As a result, the engine's power is much lower than the power transmitted by gears because it is used only to overcome the frictional resistance. The test stand is equipped with a hydraulic lubrication system for the head elements. During the tests, Boxol 26 hydraulic-gear oil was used. The test stand is shown in Figures 3 (*a*) and 3 (*b*) shows the construction of the test head.

Fatigue tests on circulating power machine were carried out on the basis of two research programs. The test programs have been developed based on measurement of the agricultural tractor drive system loads. In order to map the actual working conditions, a similar value of contact stresses to the stress occurring in the pair of gears to be tested was applied during the tractor's heavy agrotechnical works.

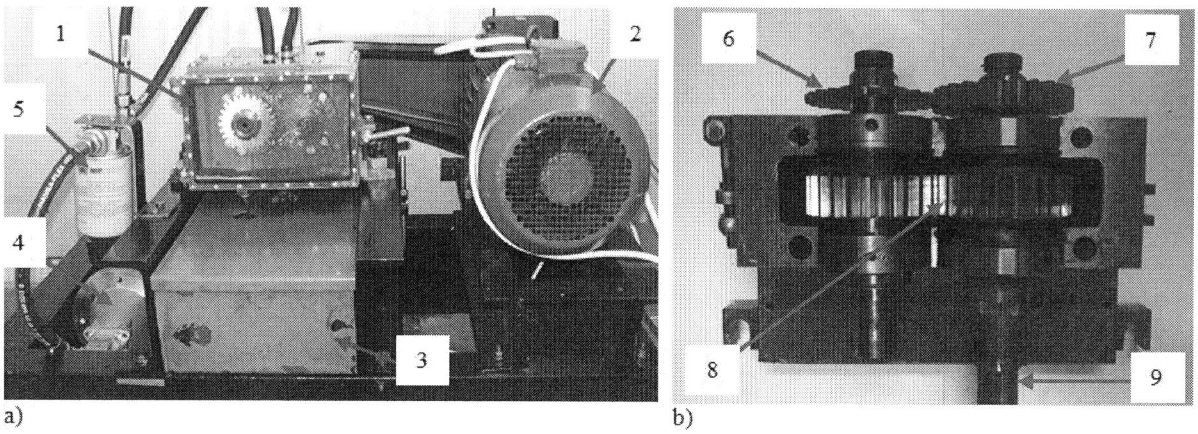

Fig. 3. Circulating power machine : a) general construction: 1 – research head, 2 – engine, 3 – oil tank, 4 – hydraulic pump, 5 – hydraulic filter; b) construction of a research head: 6 – tested gear, 7 – cooperating gear, 8 – closing gear, 9 – torsin shaft

4. Analysis of fatigue test results

As a result of shot peening the gears in addition to the increase in the value of compressive residual stresses and microhardness in the tooth's lateral surface layer, an increase in roughness was also obtained. Observation of the teeth consumption level was carried out using the replicas method, which were removed from the teeths' lateral surfaces. In the next step, replicas were scanned using a confocal microscope. The condition of tooth's lateral surface fragment before and after shot peening are shown in the replicas pictures in Figure 4. Figure 4 (*a*) shows the tooth's lateral surface fragment in the initial state which has not been shot peened. Figure 4 (*b*) shows a lateral surface fragment after shot peening.

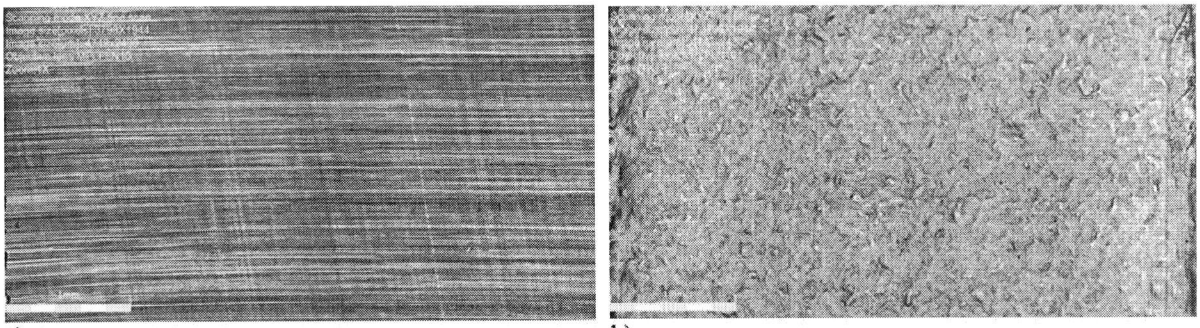

a) b)

Fig. 4. Tooth's lateral surface fragment in the initial state: a) non-peened surface; b) shot peened surface

In the first stage fatigue tests were carried out at a contact stress value equal to 1610 Mpa. In the course of fatigue tests the wear development in the form of pitting and micropitting on the lateral teeth's surfaces have been observed. Its progressing is registered every 10 million cycles. Numerous research works, devoted to the subject of contact fatigue testing of gears, determine the criteria for evaluation of test results [8, 9, 10]. The limit value, which excludes a pair of gears from further research it is achieving a 0.5 % share of wear on all teeth's surfaces. Another criterion for research completion is to exceed 4 % share of the area for (wear) a single tooth wear. The results of fatigue tests carried out according to the first test program is shown in Figure 5.

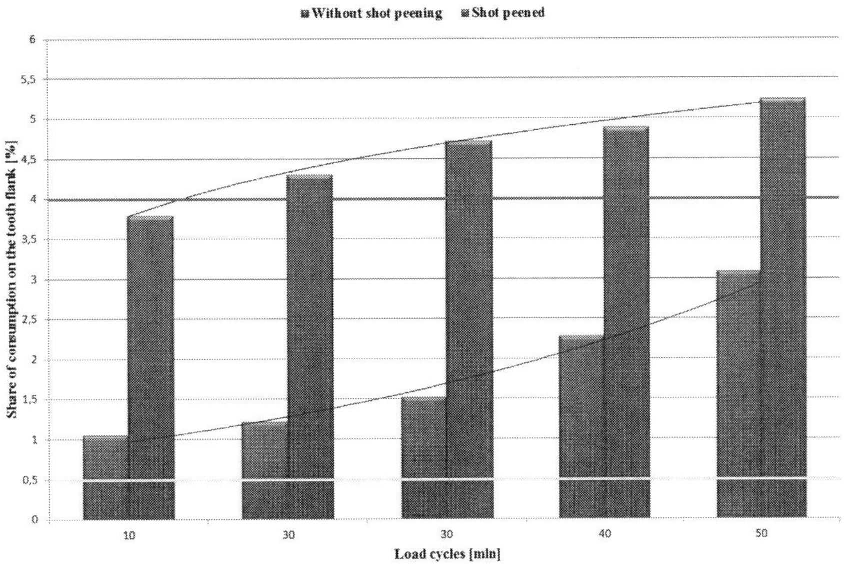

Fig. 5. Fatigue wear development on the pinion lateral surface

From the first measurement, a significantly higher share of wear is observed for the teeth of shot peened gear in relation to the non-peened gear. Development of gear tooth wear without shot peening acquires exponential character, while in the case of gear subjected to shot peening, the consumption development obtains impressiveness character. The observed phenomenon is caused by the large roughness of the tooth's surface after shot peening, which increases the contact stress at the surface unevenness tops, which causes intensive development of wear in the initial stage of new pair of gears cooperation. The micropitting phenomenon appears here. After reaching a certain level it does not develop.

The surface condition of selected teeth after reached number of basic cycles, which is 5×10^7, show pictures of replicas in Figure 6. Figure 6 (*a*) shows a fragment of non-peened lateral tooth's surface, Figure 6 (*b*) shows the fragment of tooth's flank after shot peening. Fatigue cracks develop around the pitch diameter, which is marked by a blue line.

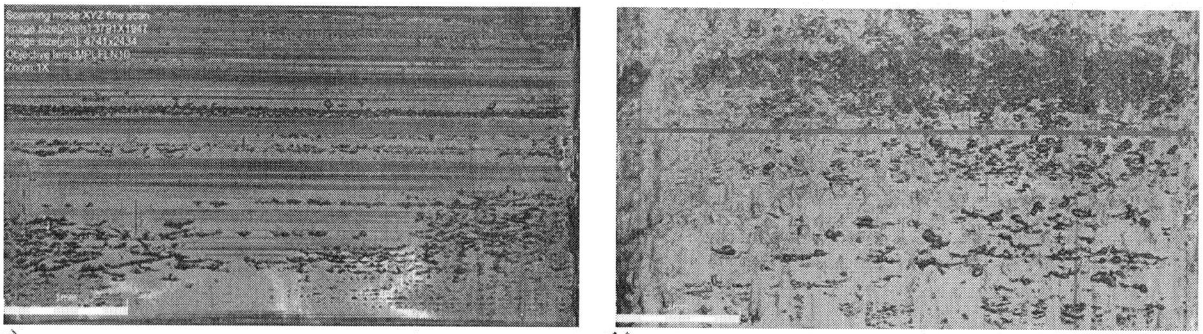

a) b)

Fig. 6. Fragments of tooth's flanks after reaching number of basic cycles at the level contact stress of 1610 MPa: a) non-peened surface; b) shot peened surface

On images obtained TEM (Fig. 7), the pitting geometry difference is clearly noticeable. In the case of carburized and hardened gears, the intensity of wear development increases with increasing number of load cycles, which is the opposite of shot peened gears. It is also considerable that larger fragmentation of pitting do not connect with each other (Fig. 7, *b*), there is also the case for non-peened gears (Fig. 7, *a*).

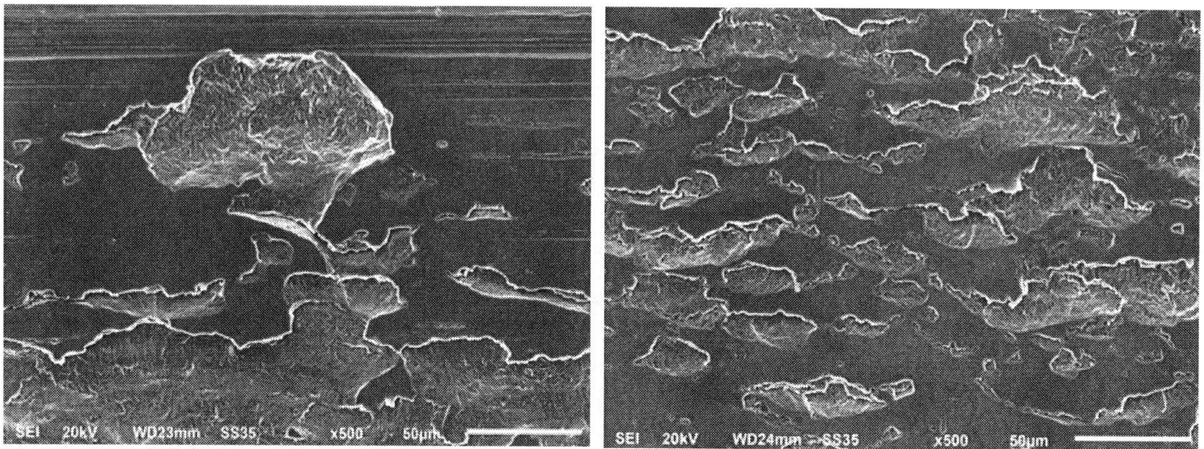

a) b)

Fig. 7. Enlarged tooth's flanks fragments after reaching number of basic cycles at the level contact stress of 1610 MPa: a) non-peened surface, b) shot peened surface

The fatigue wear process was also observed in the tooth's cross-sections after tests were completed. It is shown in Figure 8. Figure 8 (*a*) shows the visible crack propagation in a circle of non-peened gear in the material at an angle of 38°, at the depth of about 15 μm. Figure 8 (*b*) shows the crack propagation for the gear after the after shot peening, in which the propagation progressed at a maximum angle of 17° and after it reached a depth of about 3 μm. In both cases, the depth of fatigue cracks did not exceed 20 μm, which indicates the presence of micropitting. In addition, in the shot peened tooth's section a loss of material was observed, in which no further cracks develop.

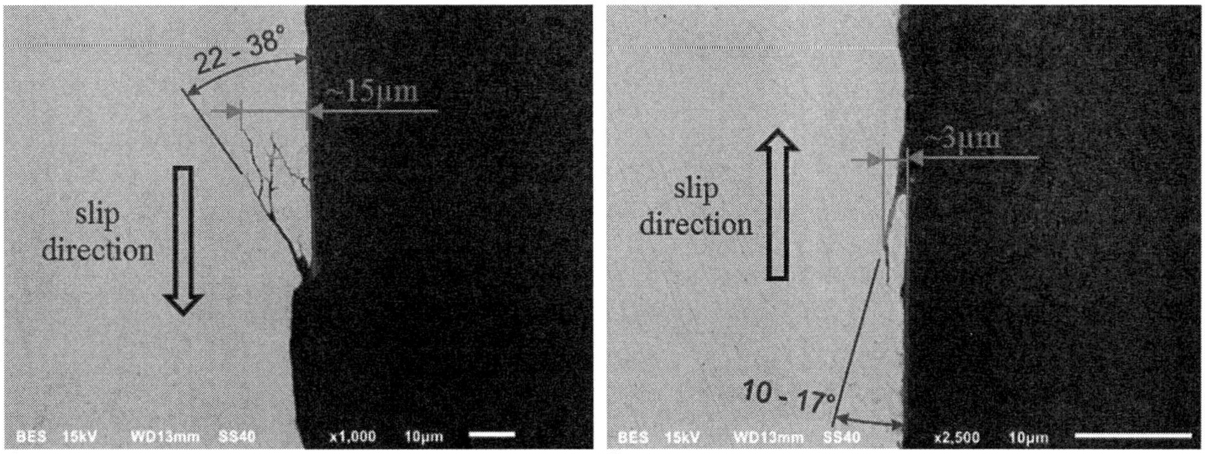

a) b)

Fig. 8. Tooth's cross-sections after reaching number of basic cycles at the level contact stress of 1610 MPa: a) non-peened surface, b) shot peened surface

In the next stage, the test was carried out at the increased contact stress value up to 1906 MPa. Based on the photographs of selected teeth's surfaces shown in Figure 9, after reaching the base number of load cycles, a clear difference in wear can be observed. In the case of a non-peened gear, it is approximately 11.5 %, while for a shot peened gear this value is 6.8 %. The nature of fatigue cracks development is analogous to that discussed at lower contact stresses, but in this case the pitting cracks are much more equally for the gear without shot peening (Fig. 9, *a*) and after shot peening (Fig. 9, *b*).

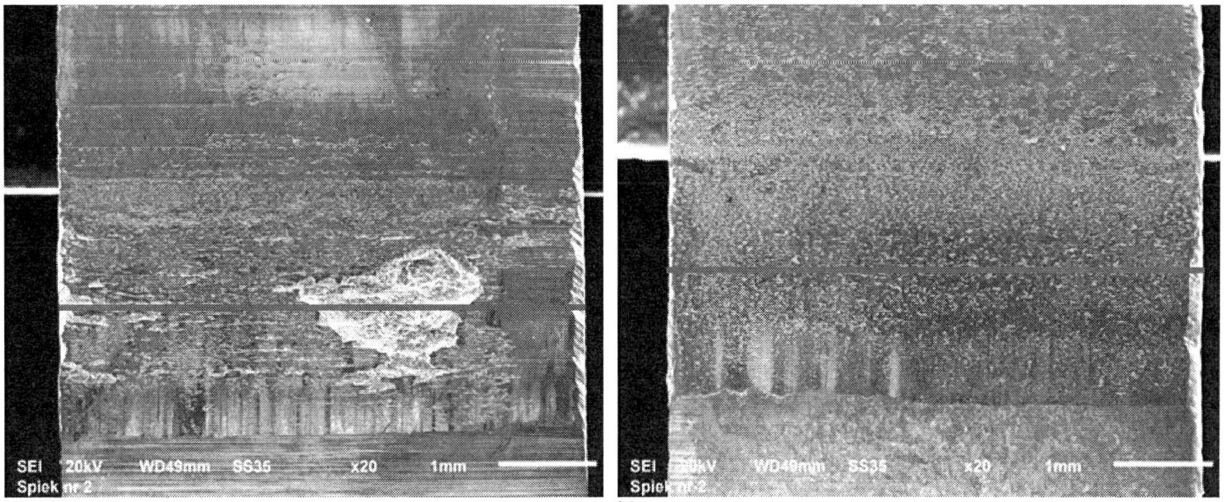

a) b)

Fig. 9. Fragments of tooth's flanks after reaching number of basic cycles at the level contact stress of 1906 MPa: a) non-peened surface, b) shot peened surface

Analysis of teeth's cross-sections showed significant disproportions in the state of the lateral surfaces (Fig. 10). In the non-peened gear, in Figure 10 (*a*), a deep chipping and further propagation of cracks into the material in the maximum tangential stress planes is visible at the angle reaching 45°. The depth of cracks exceeds 200 μm, further cracks propagating towards the core are noticeable. Figure 10 (*b*) shows the cross-section of shot peened tooth, in which cracks are also noticeable, but their depth does not exceed 10 μm.

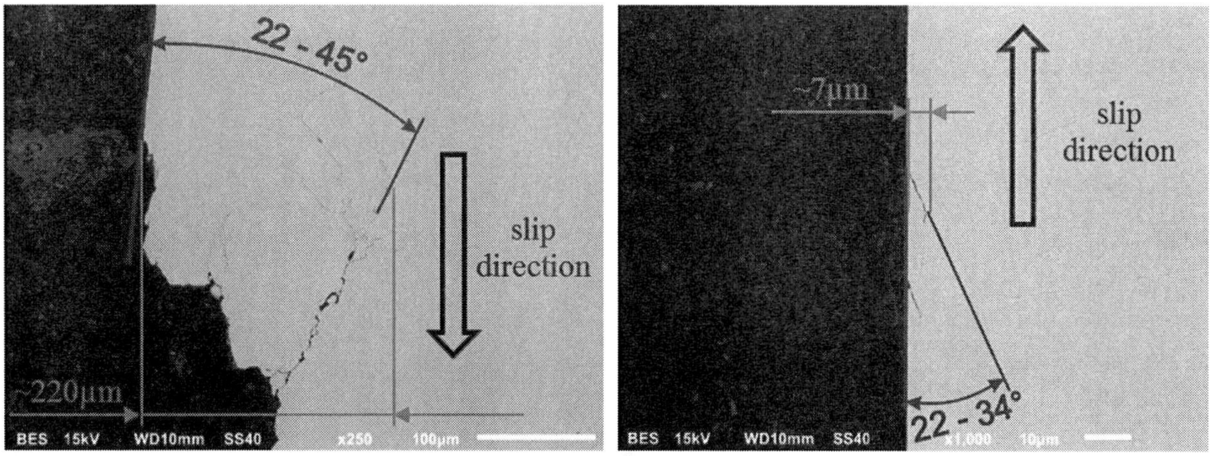

a) b)

Fig. 10. Tooth's cross-sections after reaching number of basic cycles at the level contact stress of 1906 MPa: a) non-peened surface; b) shot peened surface

In order to determine the effect of shot peening on fatigue strength of the teeth's lateral surface with a high level value of contact stress, mixed state of loads tests have been carried out. Initially there was done lapping process on the gears subjected to load as a result of which 1440 MPa of contact stresses were obtained. The time needed took 95 hours, equivalent to 10 million load cycles. Photos of the selected teeth's lateral surface replicas subjected to the lapping process is shown in the Figure 11. There is visible a significant increase in consumption in comparison to the surface without shot peening on the tooth surface subject to shot peening. The share of non-peened surface wear has reached 0.2 % (Fig. 11, *a*), while for the shot peened one it is 2.9 % (Fig. 11, *b*).

a) b)

Fig. 11. Photos of the teeth's replicas previously subjected to lapping process: a) tooth without shot peening; b) shot peened tooth

In the second stage of this research the level of contact stresses was increased to the value of 2280 MPa in order to speed up the consumption. After 14 hours test performed, which corresponds to 1,5 million load cycles, there was observed very intensive growth of the teeth's lateral surface consumption for gear not subjected to shot peening. The share of the pitting surface for the most worn tooth is about 16 % (Fig. 12, *a*). With the same stress level and the same testing time on the teeth's surface subjected to shot peening, there were observed smaller development of pitting, which share is about 8.6 % (Fig. 12, *b*). The biggest wear of lateral surface replicas teeth photos are shown in the Figure 12.

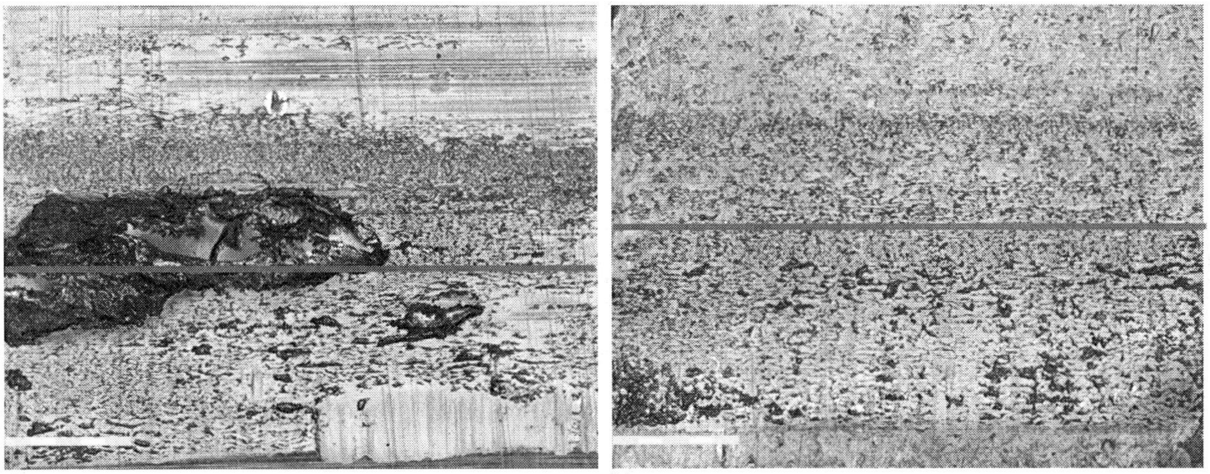

a) b)

Fig. 12. The teeth's replicas photos after reaching 1.5x106 load cycles at the level contact stress of 2280 MPa: a) tooth not subjected to shot peening; b) shot peened tooth

Conclusions

1. The results of microhardness measurements in the teeth's cross-section after shot peening indicate (an) the increase from the value of 850 HV0.2 at the lateral surface to 770 HV0.2 at a depth of 0.4 mm and then, (along) with the increase in distance, regular fall to 520 HV0.2 in the tooth's core.
2. The results of previous strength tests of gears on the circulating power machine indicate the fatigue character of the formation and development of lateral surface damage in the form of pitting and micro pitting.
3. There is a noticeable reduction in the total area of losses caused by the phenomenon of pitting on the teeth's lateral surface of subjected to shot peening reaching 7.4 % at the highest level of contact stresses reaching 2280 MPa.
4. A differentiated geometry of cracks propagating from pits after pitting cracks in the tested gears were noticed. In the case of carburized and hardened gears, these cracks developed at (an) the angle of 22–45° to the tooth's side surface, and in the case of shot peened gears the angle value was 10–34°.

References

[1] Weresa E., Rak Z., Seweryn A. Metodyka prowadzenia badań wytrzymałości zmęczeniowej kół zębatych, Modelowanie inżynierskie Nr. 45, t. 14/2012, pp. 241–248.

[2] Kirk D. Development of Shot Peening Control and Residual Stress Patterns. Europe- USA seminar on Shot Peening 29–30 sept. 1992, Cincinati, Ohio, USA–14 oct. 1992, Grenoble, France.

[3] Majzoobi G. H., Azizi R., Alvi Nia A. A three-dimensional Simulation of Shot Peening Process Using Multiple Shot Impacts. Journal of Materials Processing Technology, 15 May 2005, pp. 164–165.

[4] Nakonieczny A. Dynamiczna powierzchniowa obróbka plastyczna kulowanie. Instytut Mechaniki Precyzyjnej, Warszawa 2002.

[5] Schulze V. Modern Mechanical Surface Treatment. States, Stability, Effects. Wiley-VCH 2006.

[6] Shot Peening Applications. Metal Improvement Company, Inc. New Jersey 1992.
[7] Tchórz T., Śnieżek L., Grzelak K., Mierzyński J. Parameters selection of shot peening gears of carburized and hardened steel 21NiCrMo2. Journal of Kones Powertrain and Transport Vol. 23, No. 2, 2016, pp. 389–395.
[8] Zwolak J., Palczak A. Effect of gear teeth finishing method on properties of teeth surface layer and its resistance to pitting wear creation. J. Cent. South Univ. 2016, 23, pp. 68–76.
[10] Zwolak J., Witek M. Kryteria wyboru materiału na koła zębate w odniesieniu do stosowanej technologii obróbki wykończeniowej zębów nawęglanych i hartowanych. Tribologia 6/2016, pp. 293–306.
[10] Rysiński J., Wróbel I. Nowoczesne metody oceny stanu powierzchni części maszyn. PAK, Vol. 57, 12/2011, pp. 1595–1597.

ITMS'2018

ktu 1922 panevėžys faculty of technologies and business

The 12th International Scientific Conference Intelligent Technologies in Logistics and Mechatronics Systems (ITELMS'2018), 26–27 April 2018, Panevėžys, Lithuania

Possibilities of Micromachined Ultrasound Transducer Technology for Matrix Atomic Force Microscopy

Sergejus Volosevičius[a*]

[a]*Kaunas University of Technology, Nemuno g. 33, LT–37164 Panevėžys, Lithuania*

Abstract

The matrix atomic force microscopy (AFM) is particularily interesting where large field of views and high imaging speeds are required. One of the possible application areas is the industry of optical elements, which require surface examination of the optical coatings. The alternative for AFM inspection, which is also widely used, white light interferometry, have very limited application for the optical coatings due the parasitic interferometry effects. Therefore we reviewed in this article several possibilities of applying microelectromechanical structures to produce matrix AFMs. While one of the first MEMS designed for this purpose were microcantilever arrays, capacitive micromachined transducer technology (CMUT) appear to us as a better alternative due the well established fabrication technology and good knowledge base for fully sampled matrix operation. The main limiting factor for CMUT AFMs is the limited travel range of the CMUT membrane. We believe that this can be solved by using the elements of non-linear mechanics in CMUT structures, such as multiple membranes or V-shaped structures. However, since these structures are still under the research, significant scientific efforts are still to be put in it.

© 2018 S. Volosevicius
Peer-review under responsibility of the Kaunas University of Technology, Panevėžys Faculty of Technologies and Business

Keywords: atomic force microscopy, cmut structure, surface roughness, transducer technology

1. Introduction

One of the most demanding fields requiring sophisticated means for the surface quality inspection are the optical coatings. Special demands and importance for the optical coatings quality are raised by the manufacturers of the optical elements for the laser systems. There are multiple kinds of optical coatings, which are determined by the characteristics of the optical basis, design of the coating, technical conditions for exploation, working range of the optical element in terms of wavelengths, optical breakthrough threshold, permittable surface roughness and

* Corresponding author. Tel.:+370-67-232-936
E-mail address: volosevicius@gmail.com
© 2018 S. Volosevicius

B427H0035 . EdLearning.it
Peer-review under responsibility of the Kaunas University of Technology, Panevėžys Faculty of Technologies and Business

mechanical strength. Main technology for optical coatings is the physical vapor deposition in vacuum. For evaporation of the coating precursors electron beam, ion beam and magnetron sputtering are used [1, 2]. Atomic force microscopy (AFM) is de facto golden standard for the surface inspection with resolutions down to atomic level, and it is extensively used during to examine the quality of the optical coatings [3]. The AFM technology is based on the sensing of the inter-atomic forces, widely known as Van der Waals forces. It is made by scanning of the surface with the cantilever tips of few nanometer radius of curvature. The cantilever is deflected due the inter-atomic interaction, and this deflection is converted to the digital signal. The signal is processed and the information about three-dimensional topography and other physical properties of the surface at very high resolution is revealed. Usually AFM's have different working regimes: contact, non-contact, tapping and others [4]. Also, this imaging and measurement technology is known as being slow. For example, when the field of view is 100 x 100 μm, it is usually divided in to 256 or 512 scan lines. For adequate resolution, one line takes from 1 s to 10 s to scan. Slower scanning will give higher resolution and higher quality of the data. Apparently, it will take hours to scan even comparatively small area of 100 x 100 μm, which is not quite appropriate in many cases of optical elements inspections.

For rapid inspection of optical elements, white light interferometry, for example narrow beam Michelson type interferometry ("Fizeau") is used [8,9]. Entire area of the optical element can be examined within several minutes with quite high resolution [8]. However, interferometry techniques have limitations while expecting optical elements with optically transparent coatings, because of the parasitic interference produced by it. Another limitation is the resolution working range in the vertical direction, which cannot be more than one micrometer or so.

2. Approaches to increase the productivity of AFM

There are several attempts made to increase the AFM imaging speed. One of them is matrix AFM, which employs not a single cantilever probe, but a matrix of them. However, additional problems related with accessing of the single cantilever and acquiring the data from them arise. Together with the increase of the field of view, the mechanical design complicates, dimensions of actuators and moving mass increase, which decreases the scanning speed. Therefore overall costs of AFM inspection remains quite high. These deficiencies can be solved by using microelectromechanical systems (MEMS) approach. It is widely known that MEMS technology allows fabrication of a matrix of the cantilevers in a single chip and integrate it with the microelectronic circuits [5]. Previously published research about MEMS matrix cantilever design is schematically shown in Figure 1. It is based on thermo mechanical detection instead of Van der Waals forces. In the cantilever matrix there are resistive elements integrated, which are used as local heaters of the probe. As the warm cantilever approaches the surface, the change in the thermal impedance between the cantilever and the surface produces a measurable change in the temperature of the cantilever. This approach makes the overall design very much compact, simple and easy to operate. However, chosen detection method cannot compete with the AFM in terms of the resolution.

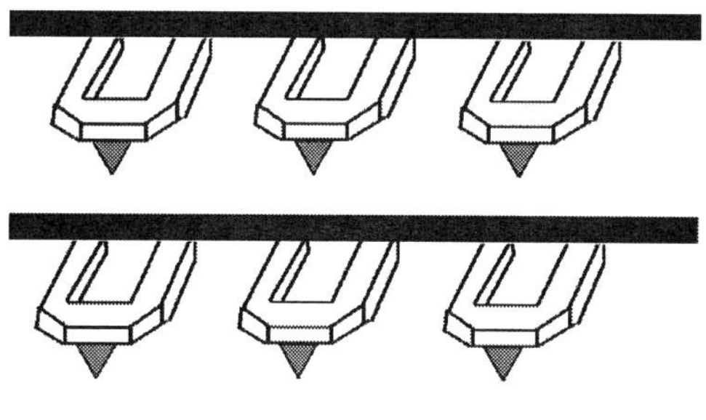

Fig. 1. Millipede cantilever array [5]

Another MEMS technology, which has well established methods for fabrication of fully sampled matrixes of transducers, is capacitive micromachined ultrasound transducers (CMUT). One of possible applications of CMUTs is transduction of the electrical signal to mechanical with comparatively high resonance quality if operated in the air. Also, these devices are highly sensitive to the load, which makes them excellent candidates for sensing of the interatomic forces. Application of CMUT concept for matrix AFM is already known under the brand name of FIRAT probe-AFM [6]. The CMUT cells are built over the optically transparent substrate. The metal electrode on the substrate is shaped as the optical diffraction grating. Laser beam, partially reflected by the optical grating and partially by the moving membrane, is detected by the photodetector (Fig. 2). The nanometer-sharpened tip in this design is built over the CMUT membrane, as contrast to the cantilever-fixed tip in the classical AFM. As CMUTs are actuated by the electrostatic force and have sealed design, they enable high frequencies, energy effectiveness, and better robustness and mechanical stability than cantilevers [6, 7]. Still the vertical range of FIRAT probe AFM is limited to the maximum displacement of the CMUT membrane, which is practically limited to less than one micrometer. This is one of the reasons why this and similar applications did not yet succeeded in the market. However, if there would be a ways of increasing the possible displacement of the CMUT membrane to the range of tens of microns, it would bring the idea of the matrix AFM based on CMUT structure to the new level of possibilities. Recent research of MEMS technologies investigated several approaches of non-linear mechanics that can be used to change the mode of operation of the electrostatic transducers and increase the displacement range without actually increasing the gap between electrodes [15–19].

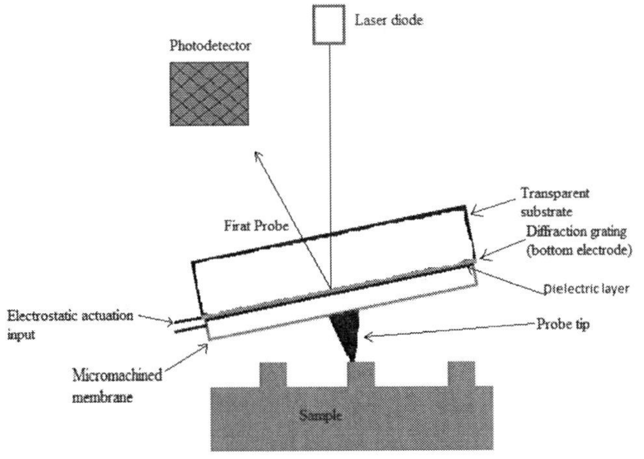

Fig. 2. FIRAT probe AFM with prototype transducer [6]

3. Electrostatic MEMS structures with the elements of non-linear mechanics

Kinematics and electromechanics of the electrostatic MEMS can be simplified to the well known single degree of freedom system of three mechanical components mass, spring and damper, as shown in the Figure 3, *m*, *k* and *b*, correspondingly [23]. As for electrical elements, voltage source (*V*) is used to provide the device with the charge. The capacitance of the device is represented by the capacitor *C*, and the losses at electrical part are represented by the resistor *R*. The voltage is applied between the electrodes; one of them is fixed and one can move. The moving electrode is usually integrated to the spring-like mechanical structure: membrane or cantilever. If the charge is accumulated between the electrodes, moveable electrode is deflected due the electrostatic force. When the device is discharged, the moveable electrode is released. Generally, the displacement of the moveable electrode is limited by the initial gap between the moveable element and the substrate, which is determined by the design of the device. The gap is usually kept less than one micrometer to have the reasonable control voltage, usually below 200 V [23].

In the Figure 4, basic design of the single CMUT cell is shown. The bacground of the device is made of highly doped silicon. Isolation posts usually are fabricated out of the silicon dioxide, and the membrane is made of the silicon nitride or other material with high elasticity modulus [11, 12]. If the membrane is made of the isolating material, the top electrode is made by depositing and patterning the metal film over the membrane. The operation frequency and voltage, displacement and other properties of the device are the functions of the vertical and horizontal dimensions of the cell and mechanical properties of the materials used in the structure.

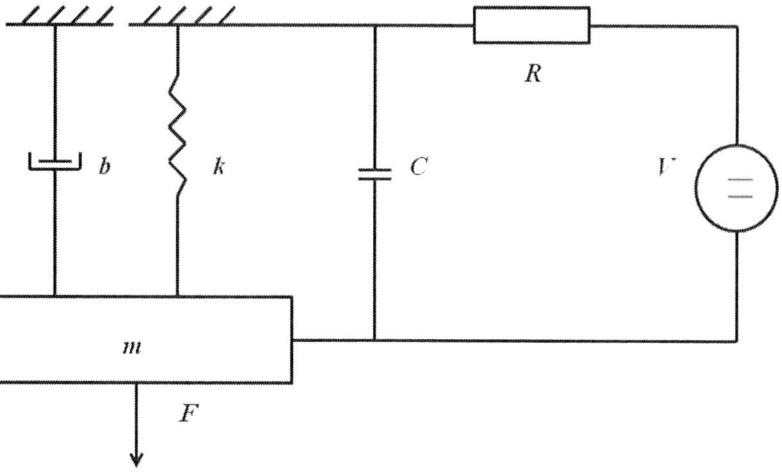

Fig. 3. Simplified explanation of CMUT operation [23]

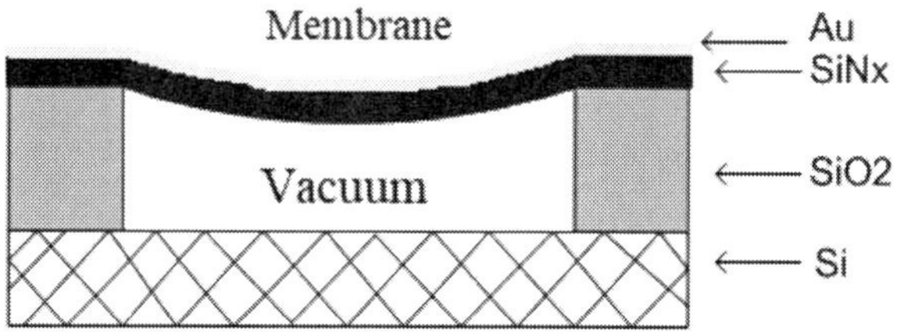

Fig. 4. Basic structure of a CMUT cell

If one desires to change the mode of CMUT operation, usually the membrane is modified by introducing the elements of non-linear mechanics. In one of the simplest cases, when the fabrication process of the device is kept without essential changes, additional mass is formed at the internal part of the membrane (Fig. 5), and therefore larger area at the center of the membrane is maintained undeformed during CMUT operation. This modification can increase the receive sensitivity and acoustical output of the device [15, 16]. Added mass increases the quality of the resonance and allows decrease the effective area of the electrodes, reducing the parasitic capacitance while maintaining the same effectiveness [15].

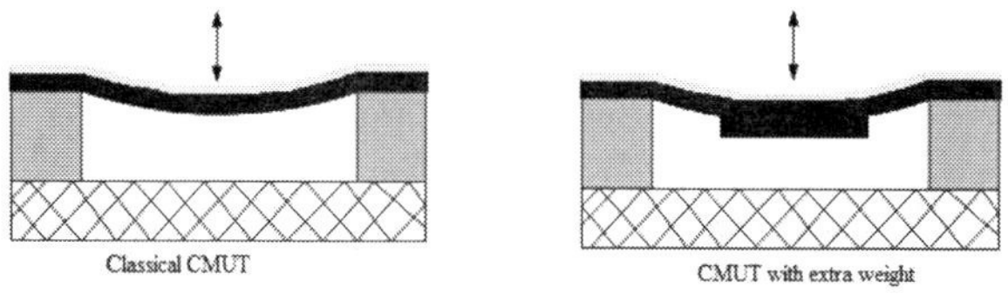

Fig. 5. Modification of the CMUT by adding the mass to the center of membrane

Effects of increased mass at the membrane center is also discussed in another recent publications [11–17]. By applying wafer bonding technology and silicon on isolator wafers, π shaped structures can be fabricated with the membrane (Fig. 6). Using of the basic CMUT design (Fig. 4) is usually related with the non-linear mechanical effects, when the output pressure is larger than 0.5 MPa. As a contrast, π shaped design will provide with the 1.6–2.46 times better linearity [18].

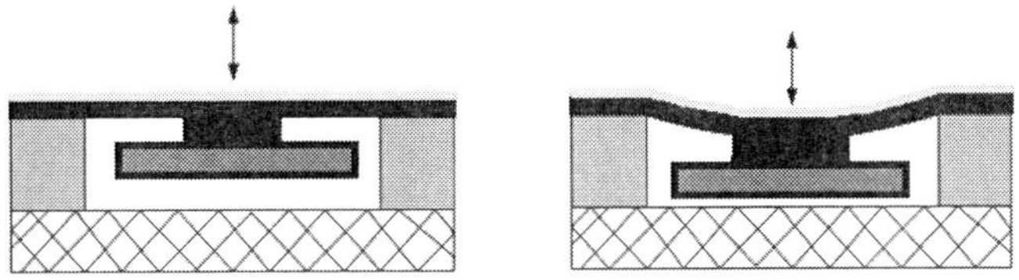

Fig. 6. Modified π- shape CMUT

Another trend of complex mechanical structures in CMUT is multimembrane designs. The fabrication technology and operation of two-membrane CMUT is recently demonstrated under the brand name of M^3-CMUT [19]. The idea is illustrated in Figure 7. When a dc voltage is applied across the two membranes of the M^3-CMUT, the effective cavity height is determined by the displacement of both deflectable membranes. This idea allows to double overall displacement without increasing the voltage. However, due the complexity of the structure, very high requirements for the fabrication technology are to be maintained. Also, the overall displacement demonstrated so far is still below one micrometer [19].

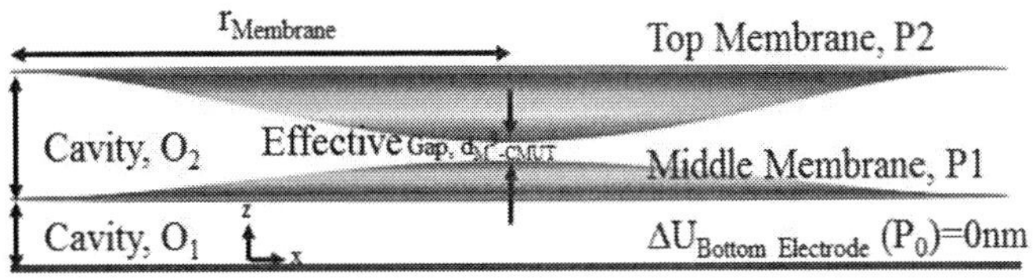

Fig. 7. Example of membrane displacement map of a M3-CMUT cell [19]

Yet another example of using the non-linear mechanics for increasing the displacement of the electrostatic transducer is so-called quasi-static transducer featuring the elements of V or Λ shape [20]. While there is the same basic structure of the electrostatically actuated MEMS with comparatively small gap (200 nm in the present embodiment [20]) inside of the device, the V shape will convert the large force acting inside of the electromechanical structure to the smaller force, but larger displacement deflecting the cantilever. The idea of the small gap and large displacement actuator is shown in the Figure 8. At the present embodiment, it was demonstrated to have the travel range, which is 138 % of the gap in the electrostatic structure. This kind of the electrostatic actuators is still under development.

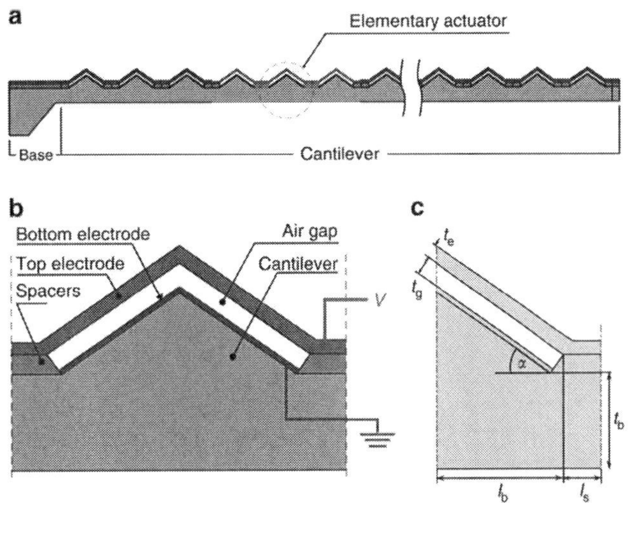

Fig 8. Schematic design of electrostatically actuated cantilever with V-shaped structures [20]

Conclusion

By reviewing several possibilities of matrix AFM, we found that CMUT technology is among of the best suitable, because of well established fabrication technology and well known methods of fully sampled matrix operation. However, the drawback of the limited travel range of the CMUT membrane reduces the applicability of such approach in the research fields, where it can be useful, for example in the inspections of the optical elements surfaces. At the same time, it is apparent that there are several examples of using non-linear mechanics in electrostatical MEMS with the goal to increase the useful travel range. Particularly interesting are multimembrane CMUTs and V shaped structures demonstrated in microcantilevers. Therefore the research field of application of non-linear mechanics in electrostatic actuators provides very interesting topic for the further research.

References

[1] Belov I. M., Manyakin S. M. and M.P. Volkov. Application of electron beam evaporation method for making coatings on the elements of thermoelectric materials based on Bi2Te3. Xvii International Conference on Thermoelectrics, Proceedings Ict 98, 1998: pp. 145–147.

[2] Gavrilov N.V., et al. Magnetron sputtering system for coatings deposition with activation of working gas mixture by low-energy high-current electron beam. 12th International Conference on Gas Discharge Plasmas and Their Applications, 2015. 652 p.

[3] Salapaka S. M., Ramamoorthy A. and M. V. Salapaka, AFM Imaging-Reliable or Not? VALIDATION AND VERIFICATION OF IMAGES IN ATOMIC FORCE MICROSCOPY. Ieee Control Systems Magazine, 2013. 33(6): 106–115

[4] Tortonese M. Cantilevers and tips for atomic force microscopy. Ieee Engineering in Medicine and Biology Magazine, 1997. 16(2): 28–33.

[5] King W.P., et al. Design of atomic force microscope cantilevers for combined thermomechanical writing and thermal reading in array operation. Journal of Microelectromechanical Systems, 2002. 11(6): 765–774.

[6] Onaran A.G., et al. A new atomic force microscope probe with force sensing integrated readout and active tip. Review of Scientific Instruments, 2006. 77(2).

[7] Daeinabi K. and M. Teshnehlab. Principles of Nano-Robotics Based on Atomic Force Microscopy. 2009 IEEE International Conference on Mechatronics and Automation, Vols 1–7, Conference Proceedings, 2009: 1589–1595.

[8] Sampathkumar A. and J. A. Ketterling. Optical characterization of surface-displacement fields of high-frequency transducers. 2014 IEEE International Ultrasonics Symposium (Ius), 2014: 1968–1970.

[9] Romanov P., et al. Modified Interferometer Fizeau for Diagnostics of Wide Aperture Optical Elements. Caol 2008: Proceedings of the 4th International Conference on Advanced Optoelectronics and Lasers, 2008: 393–394.

[10] Ladabaum I., et al. Surface micromachined capacitive ultrasonic transducers. Ieee Transactions on Ultrasonics Ferroelectrics and Frequency Control, 1998. 45(3): 678–690.

[11] Zhou S. W., Reynolds P. and J. A. Hossack. Improving the performance of capacitive micromachined ultrasound transducers using modified membrane and support structures. 2005 IEEE Ultrasonics Symposium, Vols 1–4, 2005: 1925–1928.

[12] Logan A. and J. T. W. Yeow. Fabricating capacitive micromachined ultrasonic transducers with a novel silicon-nitride-Based wafer bonding process. IEEE Transactions on Ultrasonics, Ferroelectrics, and Frequency Control, 2009. 56(5): 1074–1084.

[13] Ahrens O., et al. Fabrication of gap-optimized cMUT. IEEE Transactions on Ultrasonics Ferroelectrics and Frequency Control, 2002. 49(9): 1321–1329.

[14] Maity R., et al. Lumped Electromechanical Modeling of Capacitive Micromachined Ultrasonic Transducers. Materials Today-Proceedings, 2016. 3(6): 2289–2294.

[15] Huang Y. L., et al., Capacitive Micromachined Ultrasonic Transducers with Piston-Shaped Membranes: Fabrication and Experimental Characterization. IEEE Transactions on Ultrasonics Ferroelectrics and Frequency Control, 2009. 56(1): 136–145.

[16] Huang Y. L., et al., Capacitive micromachined ultrasonic transducers (CMUTs) with piston-shaped membranes. 2005 IEEE Ultrasonics Symposium, Vols 1–4, 2005: 589–592.

[17] Senlik M. N., Olcum S. and A. Atalar. Improved performance of cMUT with nonuniform membranes. 2005 IEEE Ultrasonics Symposium, Vols 1–4, 2005: 597–600.

[18] Cheng, T.C. and T.H. Tsai, Differential pi-Shaped CMUTs with Improved Linearity and Sensitivity. 2016 IEEE International Conference on Consumer Electronics-Taiwan (Icce-Tw), 2016: 197–198.

[19] Emadi T. A. and D. A. Buchanan, Multiple Moving Membrane CMUT With Enlarged Membrane Displacement and Low Pull-Down Voltage. IEEE Electron Device Letters, 2013. 34(12): 1578–1580.

[20] Conrad H., et al. A small-gap electrostatic micro-actuator for large deflections. Nature Communications, 2015. 6, 10078 p.

[21] Sulchek T., et al. Parallel atomic force microscopy with optical interferometric detection. Applied Physics Letters, 2001. 78(12): 1787–1789.

[22] Caliano G., et al. Design, fabrication and characterization of a capacitive micromachined ultrasonic probe for medical imaging. IEEE Transactions on Ultrasonics Ferroelectrics and Frequency Control, 2005. 52(12): 2259–2269.

[23] Viržonis D., Kodzius R., Vanagas G. Integration of capacitive micromachined ultrasound transducers to microfluidic devices. Microfluidics: Control, Manipulation and Behavioral Applications. Nova Science Publishers 2013/10/22, 127–149.

The 12th International Scientific Conference Intelligent Technologies in Logistics and Mechatronics Systems (ITELMS'2018), 26–27 April 2018, Panevėžys, Lithuania

The Influence of the Post-Weld Heat Treatment on the Microstructure and Mechanical Properties of Niobium/Non-Alloy Steel Clad Plate Obtained by Explosive Welding

Marcin Wachowski[a*], Robert Kosturek[a], Aleksander Gałka[b], Marcin Małek[c]

[a]Military University of Technology, Faculty of Mechanical Engineering, 2 Gen.W.Urbanowicza str., 00-908 Warsaw, Poland
[b]ZTW EXPLOMET, Oświęcimska 100H, 45-641 Opole, Poland
[c]Military University of Technology, Faculty of Civil Engineering and Geodesy, 2 Gen.W.Urbanowicza str., 00-908 Warsaw, Poland

Abstract

In this investigation niobium and non-alloy steel (S355) have been successfully bonded through the explosive welding technique. The obtained bimetal clad plate was subjected to two different types of post-weld heat treatment: stress-relief annealing (610 °C/2 h) and normalizing (910 °C/40 min). In order to identify the changes of its properties the microstructure analysis (with light and scanning electron microscope) and the mechanical testing (peel test) have been performed. The diffusion changes in the joint zone after normalizing and its impact on the joint quality were identified.

© 2018 M. Wachowski, R. Kosturek, A. Gałka, M. Małek
Peer-review under responsibility of the Kaunas University of Technology, Panevėžys Faculty of Technologies and Business

Keywords: non-alloy steel, post-weld heat treatment, explosive welding

* Corresponding author. Tel.:+48-261-839-245.
E-mail address: marcin.wachowski@wat.edu.pl

© 2018 M. Wachowski, R. Kosturek, A. Gałka, M. Małek
Peer-review under responsibility of the Kaunas University of Technology, Panevėžys Faculty of Technologies and Business

B427H0036 . Ed*Learning*.it

1. Introduction

One of the most promising engineering materials in terms of the improvement of the industrial installation's corrosion resistance (especially in the elevated temperature) are refractory metals and its alloys. The resistance to the aggressive environment of this material group (e. g. tantalum, zirconium and niobium) exceeds the resistance of some currently used corrosion-resistance alloys (e. g. Hastelloy) [1–2]. Despite the high melting points of the refractory metals, along with increasing of the temperature their mechanical properties (such as tensile strength) severely decreases 0. The approach that is worth consideration is using the refractory metals as layers or clad materials to provide the necessary corrosion resistance to the common construction materials such as non-alloy steel.

One of the most interesting refractory metals in terms of application in the chemical industry is niobium, which has excellent corrosion resistance to most of aggressive environments. The appropriate technology to clad non-alloy steel with niobium is the explosive welding [4, 5]. This solid-state welding technology uses the energy of detonation of high explosive material to create a metallic bond between two metal surfaces during high velocity collision [6–11]. Severe plastic deformation of welding metals in the joint zone causes their hardening and present of residual stress in the joint. For this reasons the explosively welded joints are subjected to the heat treatment: stress-relief annealing and normalizing [12–14]. Both of this heat treatments could have significant influence of the microstructure of the joint as well as its mechanical properties [15, 16].

2. Experimental

The non-alloy steel (S355) has been successfully cladded with niobium by using explosive welding process. As an explosive material, the mixture of ammonium nitrate fuel oil (ANFO) was used. The scheme of the explosive welding system is shown in Figure 1. The thickness of the welded elements was equal to 2 mm for niobium and 3 mm for S355 steel.

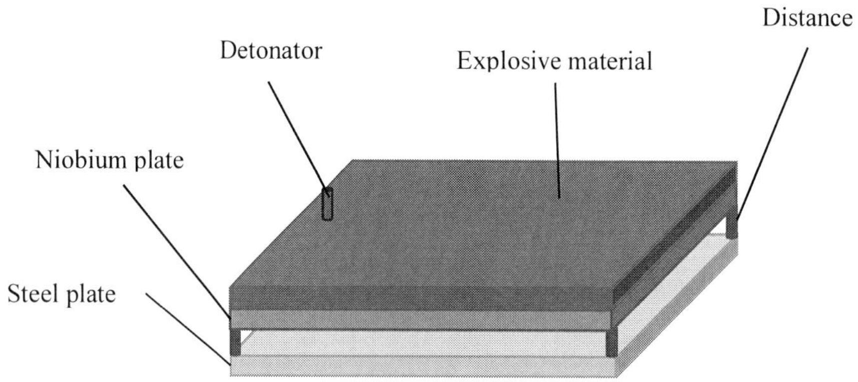

Fig. 1. Scheme of the explosive welding system

The obtained explosively welded clad plate has been cut and subjected to the two separated heat treatment (stress relief annealing and normalizing) in order to establish the influence of the heat treatment on the properties of the joint. The both of this operations, the explosive welding process and the heat treatments have been performed in EXPLOMET High-Energy Techniques Works, the polish company concerned with explosive metalworks. The parameters of the heat treatments (which have been developed by Explomet) and the samples designations are presented in Table 1 and Table 2 respectively.

Table 1. Parameters of the heat treatments

Heat Treatment	Temperature [°C]	Time [min]
Stress relief annealing	610	120
Normalizing	910	40

Table 2. Samples designation

Sample	State
Nb/CS EXW	As welded
Nb/CS OCO	After stress relief annealing
Nb/CS OCN	After normalizing

In order to examine the joint microstructure and changes caused by heat treatments the samples were examined by using light microscope (Olympus LEXT OLS 4100) and scanning electron microscope (Jeol JSM 6610) with energy-dispersive x-ray spectroscopy (EDS) and back-scattered electron (BSE) detector. Metallographic observations were carried out using samples cut from plate in the direction perpendicular to the welding direction. The samples were ground, polished using 0.6 µm Al_2O_3 powder and etched in mixture of hydrofluoric acid (40 %) and nitric acid (65 %) with proportions 1:1.2 to revel the microstructure of niobium. The microstructure of steel was revealed using nital. The quality of joint have been performed in the peel test (Fig. 2).

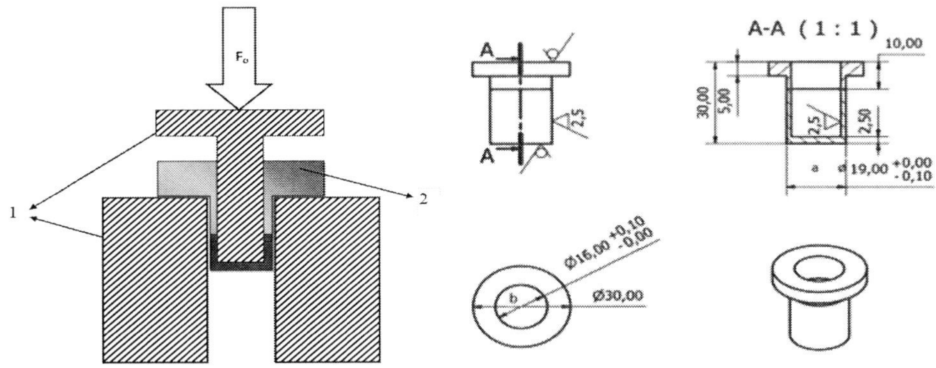

Fig. 2. Scheme of the peel test performance. 1 – elements of the testing machine, 2 – sample (left) and the scheme of the sample (right) 0

3. Results

The microstructure of non-alloy steel (S355) and niobium in the as-received state was observed with use of light microscope and its pictures are shown at Figure 3.

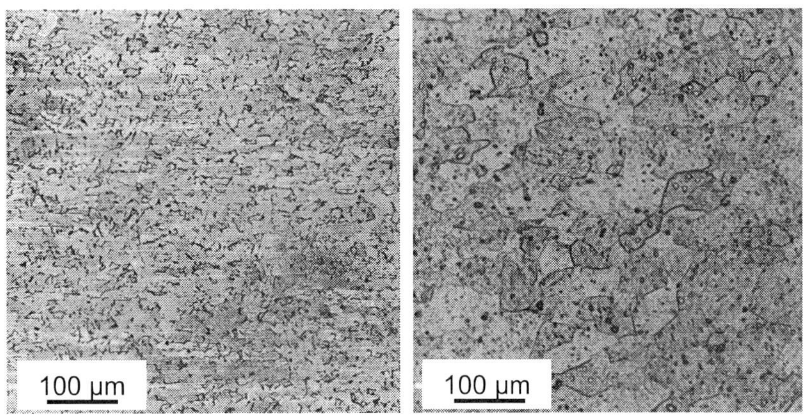

Fig. 3. The microstructure of non-alloy steel (S355) (left) and niobium (right) in the as-received state (after etching)

The light microscope observations allowed to estimate the grain size of the material to be joined. Grain size estimation was performed using Olympus microscope software. Steel S355 has a ferrite-pearlite microstructure with average grain size equal to 24.8 ± 8.6 µm the average grain size of niobium is equal to 94.1 ± 35.3 µm. The bimetal joint obtained in the explosive welding process was also subjected to the light microscope observations, what allows to observe the characteristic wavy-shape joint interface (Fig. 4–5).

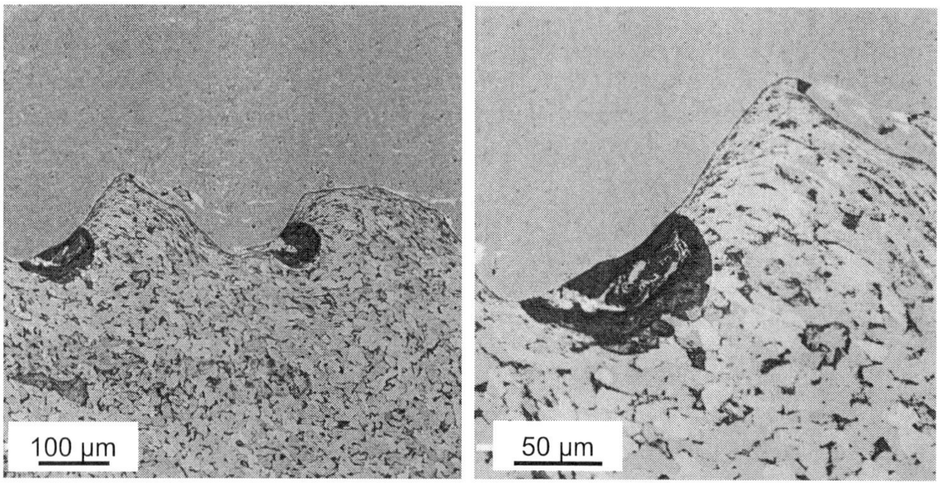

Fig. 4. The microstructure of joint in Nb/CS EXW (sample after etching steel S355). Niobium above the join line, steel below

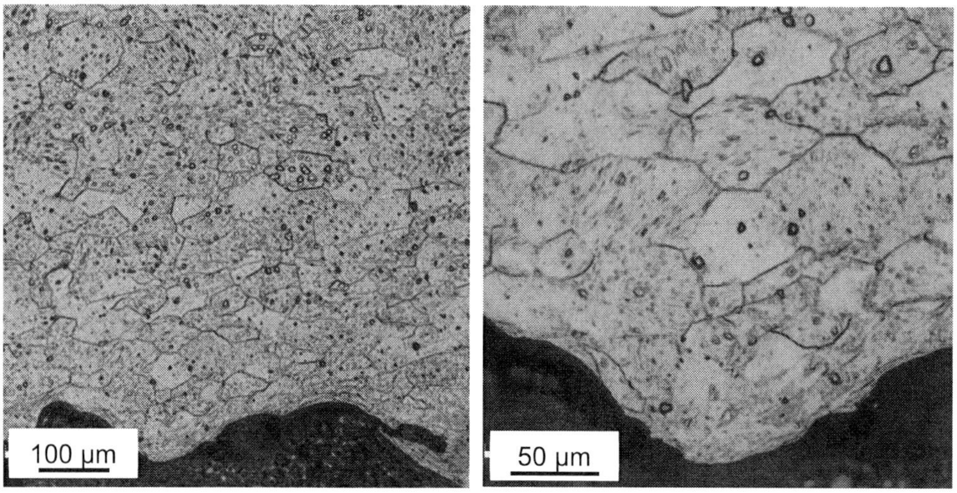

Fig. 5. The microstructure of joint in Nb/CS EXW (sample after etching niobium). Niobium above the join line, steel below

Figures 4–5 show severe plastic deformation of grains of both materials in the joint zone as the result of the high velocity collision during explosive welding process. The largest deformation has taken place in the waves, where compressed elongated grains occurs. Scanning electron microscope observation of the weld interface revealed the present of the melted zones in the joint line. Melted zones create during welding process as the result of the intense heat-generated deformation and the friction between joining materials. Both scanning electron microscope observation and spot chemical analysis using EDS indicate mixing of niobium and S355 steel in the melted zones (Fig. 6, Table 3).

Fig. 6. SEM picture of the melted zone in Nb/CS EXW

Table 3. Results of spot chemical analysis

Spot	Niobium [% mass]	Iron [% mass]
1	36.10	63.90
2	8.06	91.94
3	16.52	83.48

In order to obtain more accurate results of the chemical elements concentrations in the melted zone the element mapping has been performed (Fig. 7).

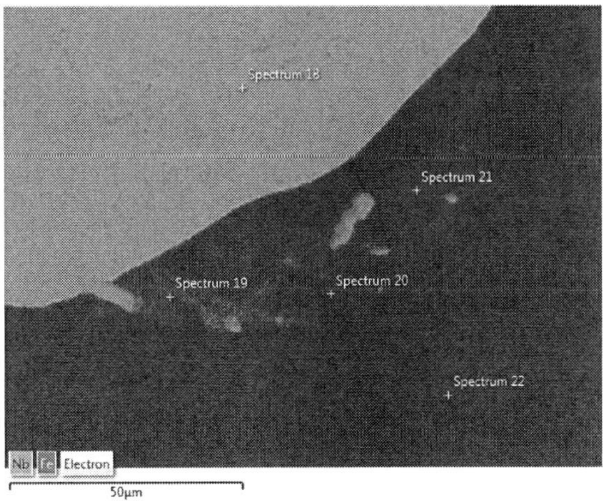

Fig. 7. The element mapping of melted zone in Nb/CS EXW

Similar investigations have been performed for samples after heat treatments. The results of light microscope observations of the sample after stress relief annealing do not show any significant changes in the grain sizes and deformation of textures in the joint zone (Fig. 8). However, occurring of the decarburization areas (pearlite free zones) in the wavy regions of joint was reported.

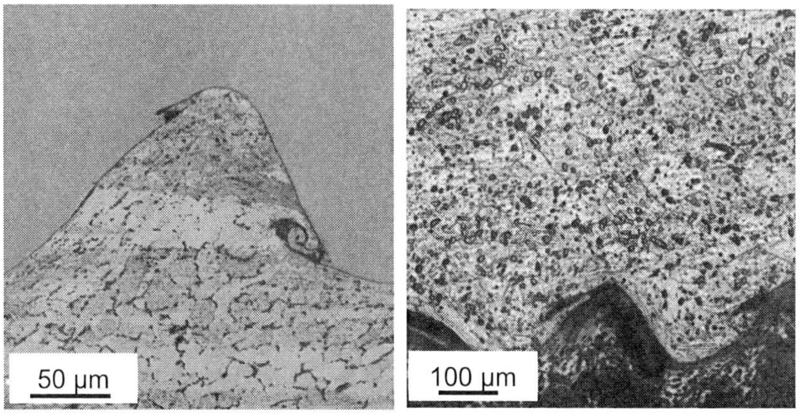

Fig. 8. The microstructure of joint in Nb/CS OCO after etching steel S355 (left) and niobium (right)

Scanning electron microscope observation and element mapping results indicates lack of any significant changes in the elements concentrations in the joint zone due to stress relief annealing heat treatment (Fig. 9).

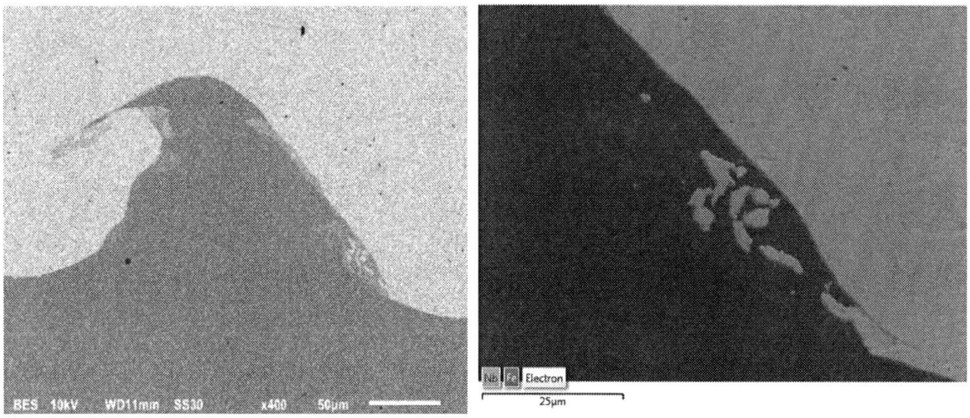

Fig. 9. SEM picture of the joint interface in Nb/CS OCO (left) and mapping of the melted zone (right)

The present of the decarburization area in steel S355 was also reported in the sample after normalizing (Fig. 10). The width of this area was estimated at around 200 µm. The another change in the microstructure is disappearance of the deformation texture of steel. In the joint zone the grains of steel S355 are equiaxed, what indicates on the occurring of recrystallization process. As has been reported, some grains suffer significant overgrowth near the joint zone with their sizes equal to 80 µm. Microstructure of niobium (Fig. 10), did not subject any significant evolutions. The severe plastic deformation of its grains in the joint zone is still noticeable.

The analysis of chemical elements concentration allowed to state that the joint zone suffers slightly change in chemical composition due to normalization heat treatment. Figure 11 shows diffusion zone with its size equal to about 1–2 µm and results of the chemical elements line scan.

The results of the peel test show very good quality of explosively welded joint with its strength in as-welded state equal to 380 MPa and the failure place located in niobium. Stress relief annealing causes decreasing the joint strength by about 45 MPa to the value of 335 MPa. The peel off strength of sample after normalization is equal to the 260 MPa, and it differs from strength in as-welded state by about 120 MPa. The obtained results are given in Table 4.

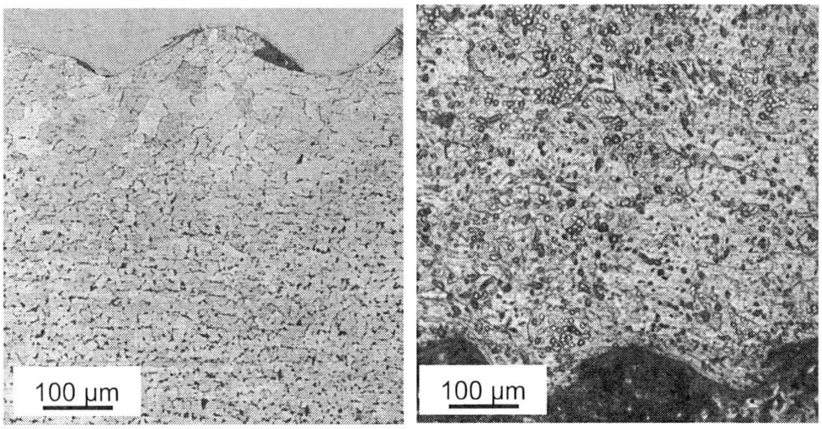

Fig. 10. The microstructure of joint in Nb/CS OCN after etching steel S355 (left) and niobium (right)

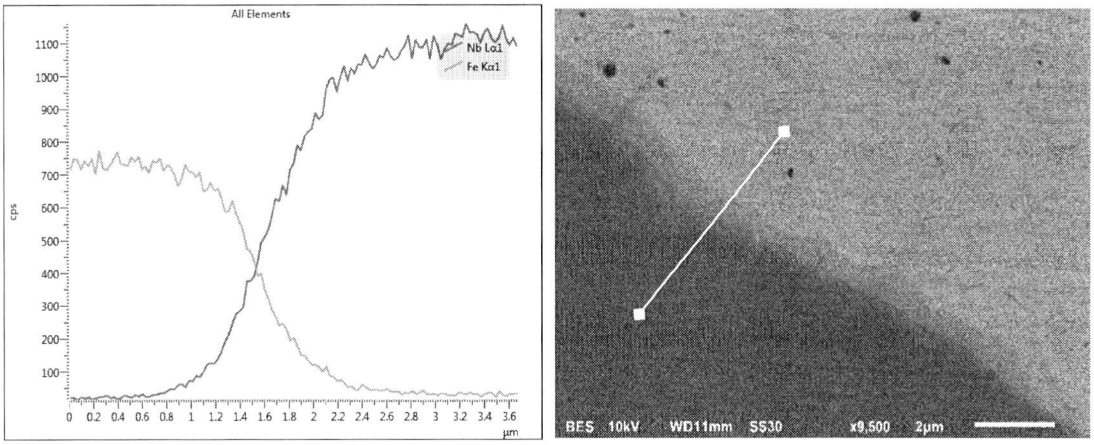

Fig. 11. The results of the line scan of the chemical elements through the joint interface (left) and SEM image of diffusion zone with the line scan marker (right)

Table 4. Results of peel test

Sample	Peel off strength [MPa]	Place
Nb/CS EXW	380	In niobium
Nb/CS OCO	335	In joint
Nb/CS OCN	260	In joint

4. Discussion

The explosive welding process allowed to obtain the very good quality niobium/non-alloy steel bimetallic joint. Light microscope and scanning microscope observation revealed characteristic wavy-shape joint with no imperfections, such as cracks and voids. The occurrence of melted zones with predominating concentration of iron has been reported in the joint. The grain structure of joined materials has noticeable deformation texture with severe plastic deformation of compressed elongated grains directly next to the joint line. The results of the peel test indicates high strength of the explosively welded joint (380 MPa) and what is important, the failure location occur in niobium and not in the joint. Stress relief annealing heat treatment (610 °C/120 min) did not affect the concentration of the chemical elements in the joint zone in a significant way. Both materials still have deformation texture, but the pearlite-free zones appear in steel layer next to the joint line as the result of the decarburization. The cause of decarburization can be explain by high affinity of niobium for carbon, what can result in the creation of the niobium carbides along the joint line. The supposed small size of this carbides makes them difficult to observe in the light and scanning electron microscope observations. The presence of carbides in the joint leads to the decreasing of the joint strength, what is confirmed by the results of the peel test, which indicates that as the result of stress relief annealing the strength of the joint decreases by 45 MPa and the failure location is in the joint. Normalizing heat treatment (910 °C/40 min) causes further decarburization of steel, as well as its recrystallization and the vanishing of the deformation texture. The width of decarburization area is equal to 200 μm. Some of steel grains suffers significant overgrowth. In the same time the grain structure of niobium does not change in a noticeable way, due to high recrystallization temperature of niobium (about 900–1300 °C) and the plastic deformation of grains is still observed. The line scan analysis shows occurring of the diffusion zone in the joint with 1–2 μm width. The shape of the line scan does not suggest presence of intermetallic phases in the diffusion zone. The significant decreasing in the joint strength elaborated in the peel test of the normalizing sample is about 120 MPa compared to the sample in as-welded state. This phenomenon is probably concerned with formation of niobium carbides on the joint line.

Conclusions

1. Explosive welding technology allows to obtain a good quality niobium/non-alloy steel joint, free of cracks, voids with low participation of melted zones and the strength elaborated in the peel test equal to 380 MPa.
2. Stress relief annealing causes decarburization of S355 steel, as well as decreasing of joint strength by about 45 MPa compared to the as-welded state.
3. Normalization resulted in the further decarburization of S355 steel, as well as further decreasing of joint strength. Also the significant changes in microstructure and chemical composition of joint have been observed. The grains of non-alloy steel suffers recrystallization and overgrowth. Additionally the diffusion zone between steel and niobium has been formed with 1–2 μm width.
4. The main reason of decreasing in joint strength due to heat treatments is probably the formation of niobium carbides along the joint line, what is the result of high niobium affiliation for carbon, and finds its confirmation in decarburization phenomenon observed in the samples subjected to the heat treatments.

References

[1] Webster RT. Refractory Metals and Their Industrial Applications. In: Smallwood RE editors. *The American Society For Testing and Materials*, Baltimore, Md.(b); 1984. 18 p.

[2] Briant CL., Banerjee MK. Refractory Metals and Alloys, *Reference Module in Materials Science and Materials Engineering*, Elsevier; 2016. Current as of 30 March 2016.

[3] Aimone P., Mei Y. Niobium alloys for the chemical process industry. *International Journal of Refractory Metals and Hard Materials* 2018. 71: 335–339.

[4] Maciejny A., Moskal G. Metale trudnotopliwe we współczesnej technice, *Inżynieria materiałowa*. 2009, 30, 3: 137–142.

[5] Elmer J. W, Terrill P., Brasher D., Butler D. Joining depleted uranium to high-strength aluminum using an explosively clad niobium interlayer. *Welding Journal* 2002. 81, 8: 67–173.

[6] Palmer T., Elmer J., Brasher D., Butler D., Riddle R. Development of an Explosive Bonding Process for Producing High Strength Bonds between Niobium and 6061-T651 Aluminum. 85. 2005

[7] Blazynski T.Z. Explosive Welding, Forming and Compaction. *Applied Science* 1983, London.

[8] Findik F. Recent developments in explosive welding. *Materials and Design* 2011. 32 p.

[9] Walczak W. Zgrzewanie wybuchowe metali i jego zastosowania. *Wydawnictwa Naukowo-Techniczne* 1989. Warszawa.

[10] Szachogluchowicz I., Sniezek L., Sulym H., et al. Testing and verification modeling of wave-shape formation under explosion welding to laminate AA 2519-Ti6Al4V. *Procedia Structural Integrity* 2016. 2: 2375–2380.

[11] Sniezek L., Szachogluchowicz I., Sulym H., et al. Elastodynamic modeling of wave initiation processes during the explosion welding. *Intelligent Technologies in Logistics and Mechatronics Systems* 2015. pp. 245–250.

[12] Kosturek R., Najwer M., Nieslony P., Wachowski M. Effect of Heat Treatment on Mechanical Properties of Inconel 625/Steel P355NH

Bimetal Clad Plate Manufactured by Explosive Welding, *Advances in Manufacturing, Lecture Notes in Mechanical Engineering* 2018. pp. 681–686.

[13] Findik F., Yilmaz R., Somyurek T. The effects of heat treatment on the microstructure and microhardness of explosive welding. *Sci. Res. Essays* 2011. 6: 4141–4151.

[14] Manesh H. D., Taheri A. K. The effect of annealing treatment on mechanical properties of aluminum clad steel sheet. *Mater. Des.* 2003. 24:617–622.

[15] Prażmowski M., Paul H., Żok F. The effect of heat treatment on the properties of zirconium – carbon steel bimetal produced by explosion welding. *Arch. Metall. Mater.* 2014. 59; 3: 1143–1149.

[16] Xia C. Q., Jin Z. P. Examination of the diffusion path in a niobium-steel explosion welding weld interface during heat treatment, *Materials Science and Engineering A* 1996. 221: 173–178.

[17] Gałka A., Najwer M. Explosive cladding of titanium and aluminium alloys on the example of Ti6Al4V – AA2519, *Arch. Metall. Mater.* 2015. 60: 2985–2991.

ITMS'2018

The 12th International Scientific Conference Intelligent Technologies in Logistics and Mechatronics Systems (ITELMS'2018), 26–27 April 2018, Panevėžys, Lithuania

The Influence of the Post-Weld Heat Treatment on the Microstructure and Mechanical Properties of Niobium / Stainless Steel Clad Plate Obtained by Explosive Welding

Marcin Wachowski[a*], Robert Kosturek[a], Aleksander Gałka[b], Marcin Małek[c]

[a]Militrary University of Technology, Faculty of Mechanical Engineering, 2 Gen.W.Urbanowicza str., 00-908 Warsaw, Poland
[b]ZTW EXPLOMET, Oświęcimska 100H, 45-641 Opole, Poland
[c]Militrary University of Technology, Faculty of Civil Engineering and Geodesy, 2 Gen.W.Urbanowicza str., 00-908 Warsaw, Poland

Abstract

In this investigation niobium and high temperature stainless steel (253 MA) have been successfully bonded through the explosive welding technique. The obtained bimetal clad plate was subjected to two different types of post-weld heat treatment: stress-relief annealing (610 °C/2 h) and normalizing (910 °C/40 min). Also the clad-plate have been subjected to heat load during utilization process (20 cycles of 400 °C temperature for 12h per cycle). In order to identify the changes of bimetal properties the microstructure analysis (with light and scanning electron microscope) and the mechanical testing (peel test) have been performed. The diffusion changes in the joint zone after normalizing as well as after utilization and its impact on the joint quality were identified.

© 2018 M. Wachowski, R. Kosturek, A. Gałka, M. Małek
Peer-review under responsibility of the Kaunas University of Technology, Panevėžys Faculty of Technologies and Business

Keywords: stainless steel, post-weld heat treatment, explosive welding

* Corresponding author. Tel.: +48-261-839-245
E-mail address: marcin.wachowski@wat.edu.pl

© 2018 M. Wachowski, R. Kosturek, A. Gałka, M. Małek
Peer-review under responsibility of the Kaunas University of Technology, Panevėžys Faculty of Technologies and Business

B427H0037 . EdLearning.it

1. Introduction

Refractory metals and its alloys are very interesting engineering materials in terms of its corrosion resistance in the elevated temperatures [1, 2]. The explosive welding technology allows to manufacture a clad-plates, consist of highquality joint between most of refractory metals and other engineering materials such as non-alloy steel or stainless steel [3–5]. In this technology, the energy detonation of high explosive material is used to create the metallic bound between surfaces of joined materials during high velocity collision [6–10]. The important aspect of this technology in terms of welding refractory metals is that the difference in melting point of welded materials does not have an influence on the quality of the obtained joint [11, 12]. The subject of this investigation was the bimetal niobium/stainless steel (253 MA) clad-plate manufactured by explosive welding process. This bimetal clad-plate is used in the patent protected industry installation. The very important matter is to establish the influence of the heat load on the properties of the joint. Due to high velocity collision in the explosive welding process the joined materials suffer severe plastic deformation, what causes their hardening, as well as occurring the residual stress in the joint [13]. In order to decrease those effects, the clad-plate is subjected to heat treatment, such as stress-relief annealing and normalizing. However, the temperature can have a significant impact on the microstructure of the joint (and also on its mechanical properties), due to diffusion changes in the bound zone [14, 15]. The aim of this research is to establish the influence of the heat treatments on the joint of niobium/stainless steel (253 MA). Additionally, the impact of the utilization (20 cycles of 400°C temperature for 12h per cycle) on the joint have been taken under investigation.

2. Experimental

The austenitic chromium-nickel high temperature stainless steel (253MA) has been successfully cladded with niobium by using explosive welding process. As an explosive material, the mixture of ammonium nitrate fuel oil (ANFO) was used. The scheme of the explosive welding system is shown in Figure 1. The thickness of the welded elements was equal to 2 mm for niobium and 8 mm for 253MA steel.

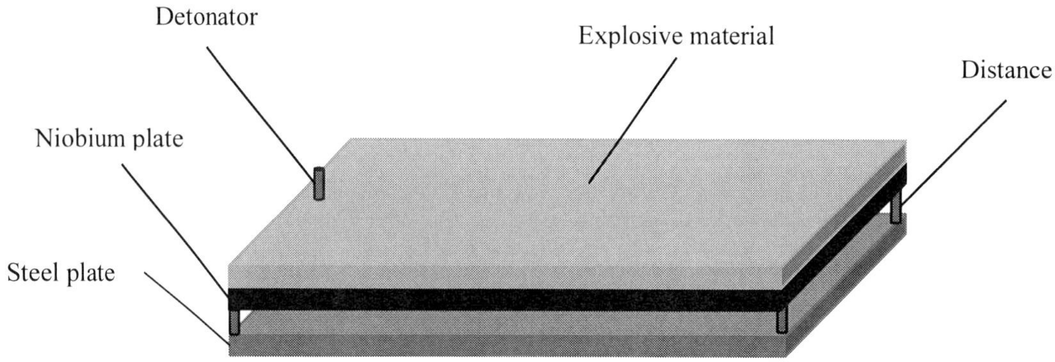

Fig. 1. Scheme of the explosive welding system

The obtained explosively welded clad plate has been cut and subjected to the two separated heat treatment (stress relief annealing and normalizing) in order to establish the influence of the heat treatment on the properties of the joint. The both of this operations, the explosive welding process and the heat treatments have been performed in EXPLOMET High-Energy Techniques Works, the polish company concerned with explosive metalworks. The parameters of the heat treatments (which have been developed by Explomet), the utilization conditions and the samples designations are presented in Tables 1–3, respectively.

Table 1. Parameters of the heat treatments

Heat Treatment	Temperature [°C]	Time [min]
Stress relief annealing	610	120
Normalizing	910	40

Table 2. Estimated parameters of the utilization

Number of cycles	Temperature of one cycle [°C]	Time of one cycle [min]
20	400	720

Table 3. Samples designation

Sample	State
Nb/SS EXW	As welded
Nb/SS OCO	After stress relief annealing
Nb/SS OCN	After normalizing
Nb/SS UTZ	After utilization

In order to examine the joint microstructure and changes caused by heat treatment the samples were examined by using light microscope (Olympus LEXT OLS 4100) and scanning electron microscope (Jeol JSM 6610) with energy-dispersive x-ray spectroscopy (EDS) and back-scattered electron (BSE) detector. Metallographic observations were carried out using samples cut from plate in the direction perpendicular to the welding direction. The samples were ground, polished using 0.6 µm Al_2O_3 powder and etched in mixture of hydrofluoric acid (40%) and nitric acid (65%) with proportions 1:1.2 to revel the microstructure of niobium. The microstructure of 253MA steel was revealed using Keller's reagent. The quality of joint has been performed in the peel test (Fig. 2).

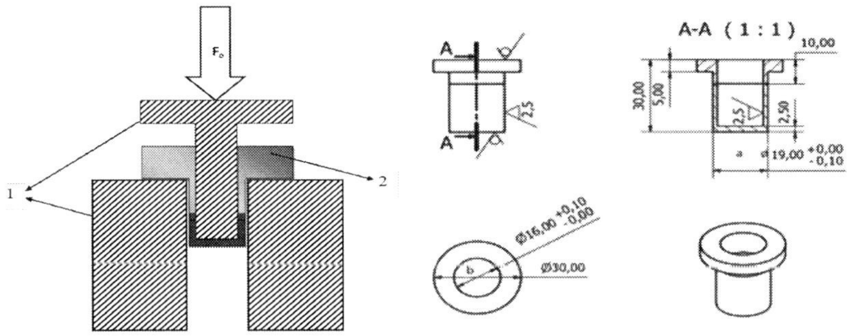

Fig. 2. Scheme of the peel test performance: 1 – elements of the testing machine, 2 – sample (left) and the scheme of the sample (right) [16]

3. Results

The microstructure of stainless steel (253MA) and niobium in the as-received state was observed with use of light microscope and its pictures are shown at Figure 3.

The light microscope observations allowed to estimate the grain size of the material to by joined. Steel 253MA has an austenitic microstructure with average grain size equal to 45.3 ± 17.9 µm the average grain size of niobium is equal to 111.9 ± 40.4 µm. The bimetal joint obtained in the explosive welding process also was subjected to the light microscope observations, what allows to observe the straight shape of the joint interface (Fig. 4).

Figure 4 shows plastic deformation of grains of both materials in the joint zone as the result of the high velocity collision during explosive welding process. The grains near the joint interface show elongation in the welding direction. Observation of the joint does not show any imperfections; such as voids or cracks. Scanning electron microscope observation of the weld interface revealed the rare occurrence of the melted zones in the joint line. Melted zones create during welding process as the result of the intense heat-generated deformation and the friction between joining materials. Both scanning electron microscope observation and spot chemical analysis using EDS indicate mixing of niobium and 253MA steel in the melted zones (Fig. 5, Table 4).

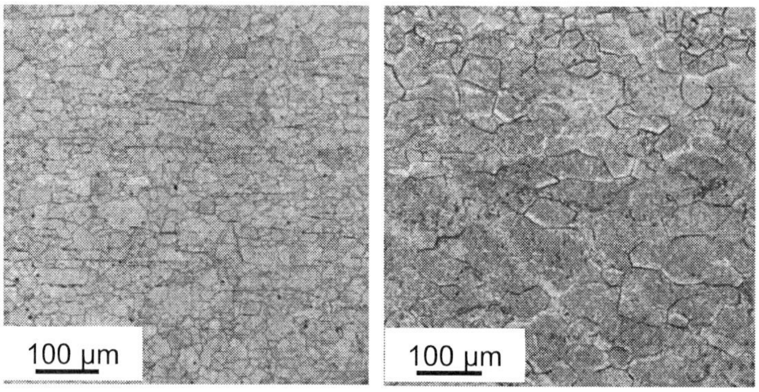

Fig. 3. The microstructure of stainless steel (253MA) (left) and niobium (right) in the as-received state (after etching)

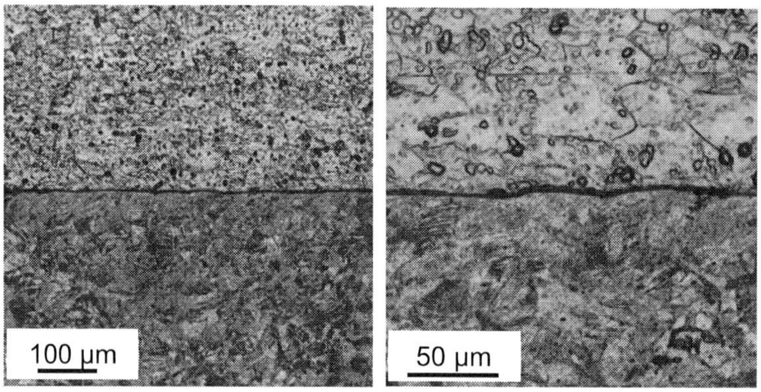

Fig. 4. The microstructure of joint in Nb/SS EXW (sample after etching steel 253MA and niobium). Niobium above the join line, steel below

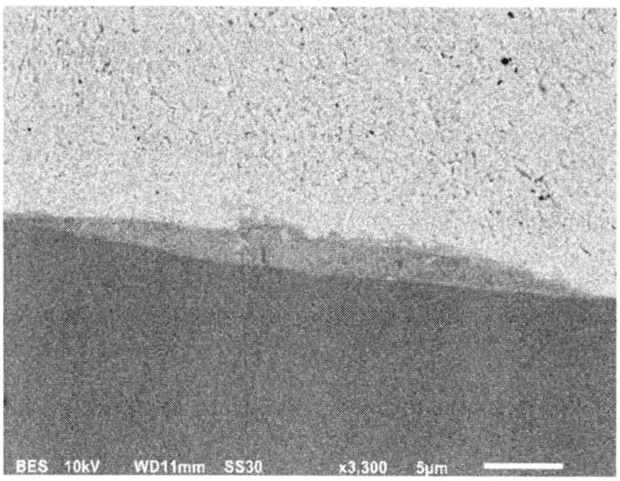

Fig. 5. SEM picture of the melted zone in Nb/SS EXW

Table 4. Results of spot chemical analysis

Spot	Niobium [% mass]	Iron [% mass]	Chromium [% mass]	Nickel [% mass]	Silicon [% mass]
1	32.8	40.2	15.6	10.3	1.1
2	30.9	41.5	15.7	11.0	1.0

Analogous researches have been performed for samples after heat treatments and sample after utilization. The results of the light microscope observation of the sample after stress relief annealing (Nb/SS OCO) shows the occurrence of the recrystallized microstructure near to the joint line (Fig. 6). In the same time, the microstructure of niobium does not suffer any changes in the grain size and in the deformation texture.

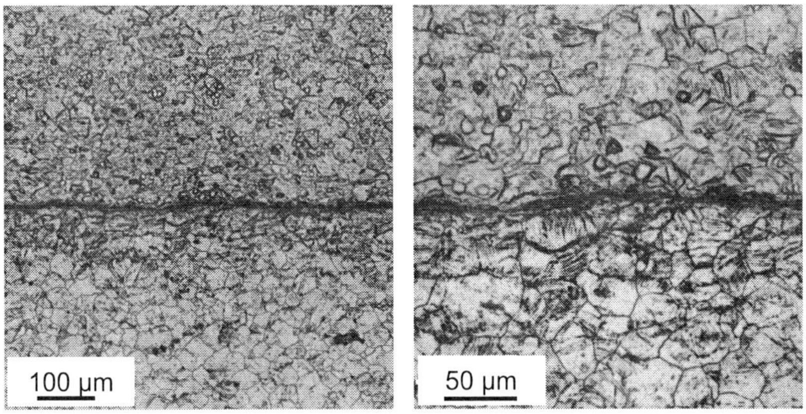

Fig. 6. The microstructure of joint in Nb/SS OCO sample after etching steel 253MA and niobium

Scanning electron microscope observation indicates lack of any significant changes in the elements concentrations in the joint zone after stress relief annealing heat treatment (Fig. 7).

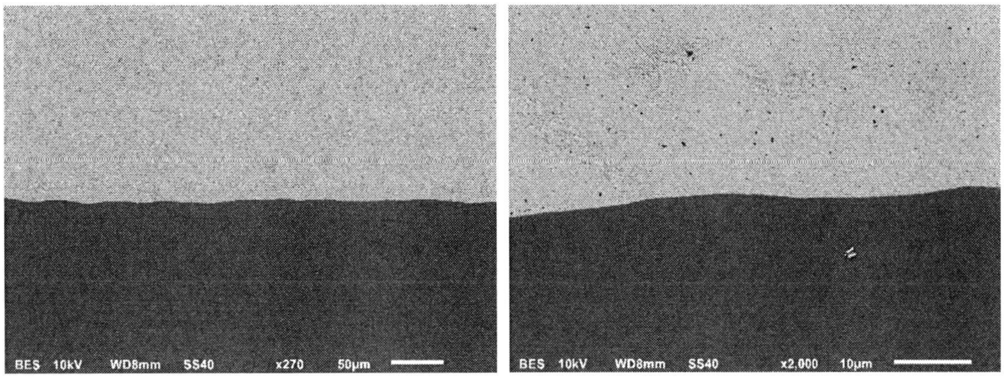

Fig. 7. SEM pictures of the joint interface in Nb/SS OCO

The recrystallization of the 253MA steel near to the joint line was also reported in the sample after normalizing (Nb/SS OCN) (Fig. 8). Microstructure of niobium (Fig. 8), did not subject any significant evolutions. The plastic deformation due to explosive welding process of its niobium grains in the joint zone is still noticeable.

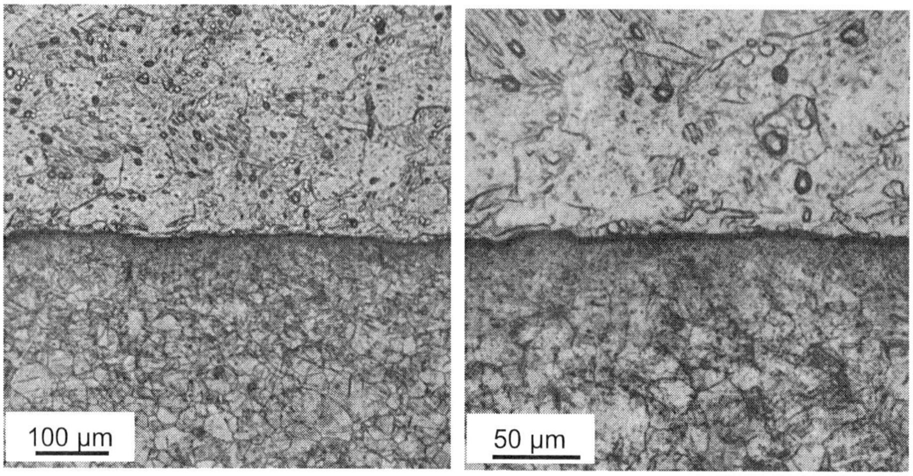

Fig. 8. The microstructure of joint in Nb/SS OCN sample after etching steel 253MA and niobium

The analysis of chemical elements concentration allowed to state that the joint zone suffers slightly change in chemical composition due to normalization heat treatment (Fig. 9). Figure 10 shows the results of the chemical elements line scan, which indicates on occurring of the diffusion zone with its size equal to about 1 µm.

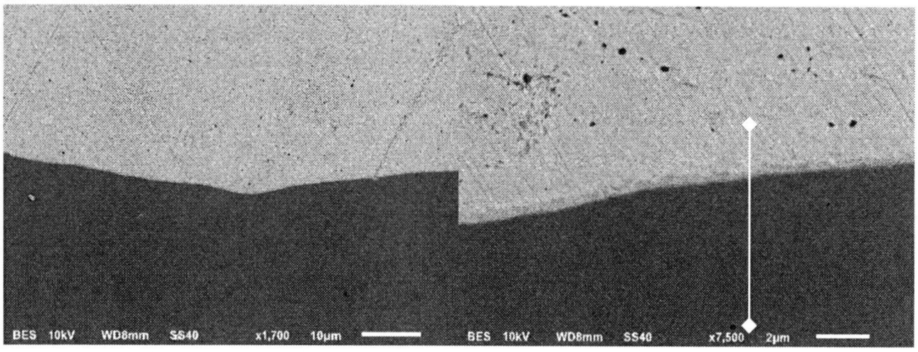

Fig. 9. SEM pictures of the joint interface in Nb/SS OCN with the line scan marker

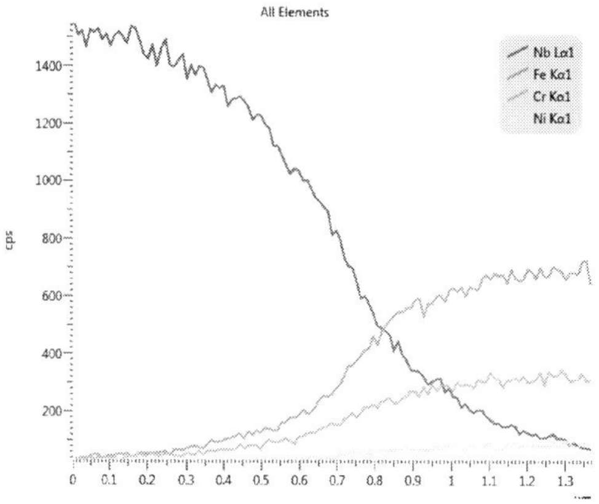

Fig. 10. The results of the line scan of the chemical elements through the joint interface in the sample Nb/SS OCN

The microstructure investigation of the sample after utilization shows significant grain fragmentation, as the result of the recrystallization process in the 253MA steel (Fig. 11). No evolution of niobium grain microstructure has been reported. The scanning electron microscope observation and the line scan of the chemical elements reveals the formation of the diffusion zone with its width about 1 μm (Fig. 12–13).

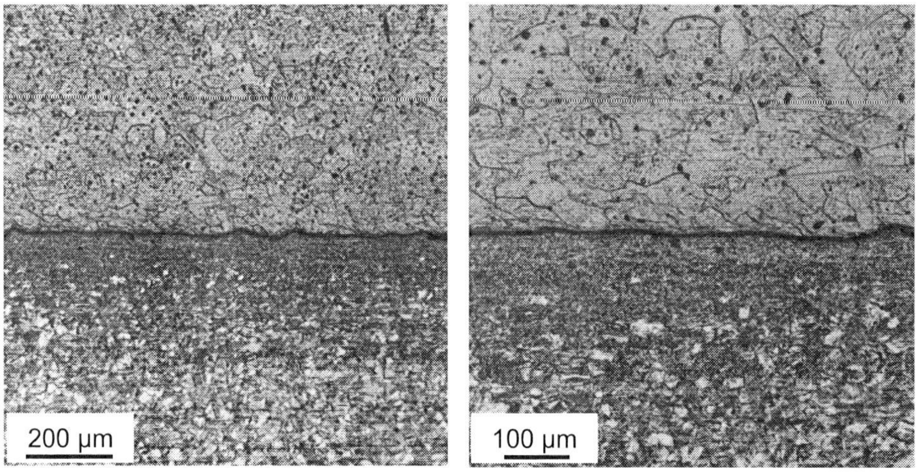

Fig. 11. The microstructure of joint in Nb/SS UTZ sample after etching steel 253MA and niobium

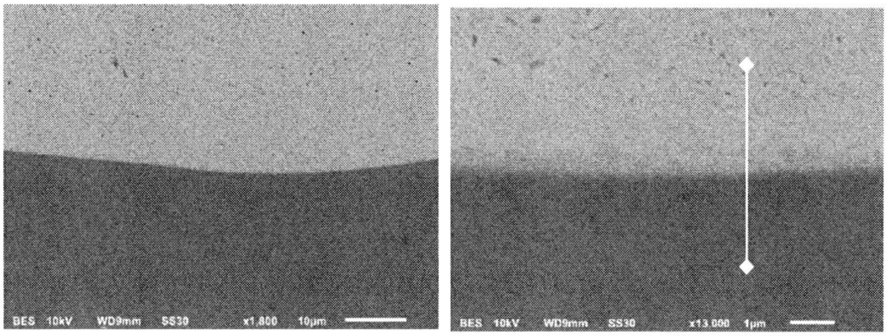

Fig. 12. SEM pictures of the joint interface in Nb/SS UTZ with the line scan marker

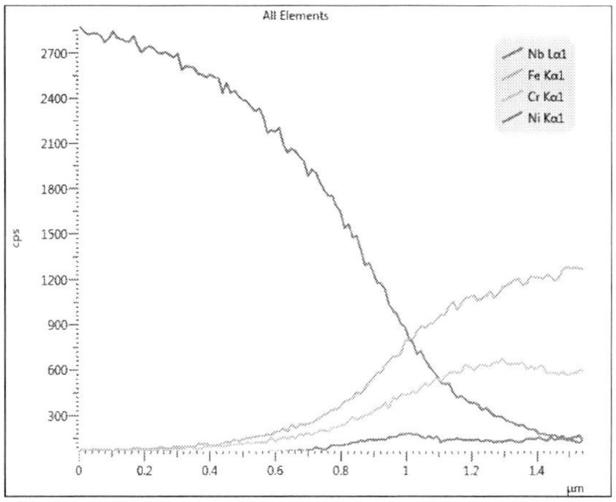

Fig. 13. The results of the line scan of the chemical elements through the joint interface in the sample Nb/SS UTZ

The results of the peel test indicate the good quality of explosively welded joint with its strength in as-welded state equal to 296 MPa and the failure place located in the joint. Stress relief annealing causes decreasing in the joint strength by 34 MPa to the value of 262 MPa. Normalizing results in further decreasing this value (231 MPa) as well as utilization (168 MPa). The obtained results are given in Table 5.

Table 5. Results of peel test

Sample	Peel off strength [MPa]	Place
Nb/SS EXW	296	In joint
Nb/SS OCO	262	In joint
Nb/SS OCN	231	In joint
Nb/SS UTZ	168	In joint

4. Discussion

The explosive welding process allowed to obtain the very good quality niobium/stainless steel bimetallic straight-shape joint. Light microscope and scanning microscope observation did not reveal any imperfections in the obtained joint. The very low participation of the melted zones in the joint line has been reported. The grain structure of the joined materials has deformation texture with elongation it in welding direction. The results of the peel test indicate high strength of the explosively welded joint (296 MPa) with failure location in the joint. Although stress relief annealing did not cause any noticeable changes in the chemical composition of the joint, and the niobium grain structure, the structure of 253 MA steel is partly recrystallized. At the same time the strength of the joint due to stress relief annealing decreases by 34 MPa. The normalizing causes significant grain fragmentation in 253MA steel near to the joint line, as well as formation of the diffusion zone with 1 μm width. The shape of line scan curve does not suggest presence of the significant intermetallic compounds layers in the diffusion zone. Also, in this case the niobium microstructure still has after-welded deformation texture. Decreasing in the peel off strength compared to the as-welded state is equal to 65 MPa. Utilization also causes formation of the 1 μm width diffusion zone and the significant fragmentation of the 253 MA steel grains due to recrystallization process. The occurring of the recrystallization near to the joint is connected with the phenomena that plastic deformation decreases the recrystallization temperature. Although, the estimated parameters of the utilization have lower temperature than the performed heat treatments, it is difficult the state that utilization conditions are strictly set and no overheating occurs. The microstructure of niobium in all samples subjected to the heat load still has deformation texture due to high recrystallization temperature of this metal (about 900–1300°C). The severe decreasing of the joint strength elaborated in the peel test of the sample after utilization (168 MPa) will be taken under further investigations. It is possible that formation of the intermetallic compounds in the joint line, as well as formation of niobium carbides and nitrides are responsible for deterioration of the joint quality. The further research will be undertaken in order to confirm presence of mentioned compounds in the joints subjected to the heat load.

Conclusions

1. Explosive welding of niobium with stainless steel allows to obtain a good quality straight-shape joint free of cracks and voids, with the very low participation of melted zones.
2. As the result of the normalizing and the utilization, the diffusion zone has been formed with its width equal to 1 μm. The shape of line scan curve does not indicate the presence of the significant intermetallic compound layers.
3. The steel 253MA grains fragmentation point to the recrystallization of steel grains due to the heat treatment processes and utilization
4. Niobium microstructure did not suffer any noticeable changes during heat treatment and utilization. The deformation texture is still visible, with grain elongated in welding direction.

Acknowledgements

This work was supported by the university grant RMN 726/2017.

References

[1]. Briant CL., Banerjee MK. Refractory Metals and Alloys, *Reference Module in Materials Science and Materials Engineering*, Elsevier; 2016. Current as of 30 March 2016.
[2]. Aimone P., Mei Y. Niobium alloys for the chemical process industry. *International Journal of Refractory Metals and Hard Materials* 2018. 71: 335–339.
[3]. Elmer, J. W., Terrill P., Brasher D., Butler D. Joining depleted uranium to high-strength aluminum using an explosively clad niobium interlayer. *Welding Journal* 2002. 81; 8: 67–173.
[4]. Palmer T., Elmer J., Brasher D., Butler D., Riddle R. Development of an Explosive Bonding Process for Producing High Strength Bonds between Niobium and 6061-T651 Aluminum. 2005. 85 p.
[5]. Xia C. Q., Jin Z. P. Examination of the diffusion path in a niobium-steel explosion welding weld interface during heat treatment, *Materials Science and Engineering A* 1996. 221: 173–178.
[6]. Blazynski T. Z. Explosive Welding, Forming and Compaction. *Applied Science* 1983, London.
[7]. Findik F. Recent developments in explosive welding. *Materials and Design* 2011. 32.
[8]. Walczak W. Zgrzewanie wybuchowe metali i jego zastosowania. *Wydawnictwa Naukowo-Techniczne* 1989. Warszawa.
[9]. Szachogluchowicz I., Sniezek L., Sulym H., et al. Testing and verification modeling of wave-shape formation under explosion welding to laminate AA 2519-Ti6Al4V. *Procedia Structural Integrity* 2016. 2: 2375–2380.
[10]. Sniezek L., Szachogluchowicz I., Sulym H., et al. Elastodynamic modeling of wave initiation processes during the explosion welding. *Intelligent Technologies in Logistics and Mechatronics Systems* 2015. pp. 245–250.
[11]. Babul W., Ziemba S. Materiały wybuchowe w technologicznych procesach obróbki tworzyw. *Państwowe Wydawnictwo Naukowe* 1972. Warszawa.
[12]. Babul W. Odkształcanie metali wybuchem, *Wydawnictwa Naukowo-Techniczne* 1980. Warszawa.
[13]. Kosturek R., Najwer M., Nieslony P., Wachowski M. Effect of Heat Treatment on Mechanical Properties of Inconel 625/Steel P355NH Bimetal Clad Plate Manufactured by Explosive Welding, *Advances in Manufacturing, Lecture Notes in Mechanical Engineering* 2018. pp. 681–686.
[14]. Findik F., Yilmaz R., Somyurek T. The effects of heat treatment on the microstructure and microhardness of explosive welding. *Sci. Res. Essays* 2011. 6: 4141–4151.
[15]. Prażmowski M., Paul H., Żok F. The effect of heat treatment on the properties of zirconium – carbon steel bimetal produced by explosion welding. *Arch. Metall. Mater.* 2014. 59; 3: 1143–1149.
[16]. Gałka A., Najwer M. Explosive cladding of titanium and aluminium alloys on the example of Ti6Al4V-AA2519, *Arch. Metall. Mater.* 2015. 60: 2985–2991.

ITMS'2018

The 12th International Scientific Conference Intelligent Technologies in Logistics and Mechatronics Systems (ITELMS'2018), 26–27 April 2018, Panevėžys, Lithuania

Environmental Sustainability Through the Geosynthetics Application at of the Subgrade on Weak Foundation Soils

Andrey Zaytsev[a], Andrey Petryaev[b], Laura Černiauskaite[c]

[a]*Russian University of Transport (MIIT), Moscow 127055, Russia*
[b]*Saint-Petersburg State University of Railway Engineering (PGUPS), Saint-Petersburg 190031, Russia*
[c]*Vilnius College of Technologies and Design, Vilnius 01100, Lithuania*

Abstract

Nowadays there is a problem of the environmental sustainability supply at the construction and exploitation of the subgrade (railway infrastructure) on the weak foundation, including the efficiency assessing of subgrade stability assessing methods on the weak silty soils. Especially this problem has a significant importance if the foundation soil is silty clay. There is the evaluation of the performance models for the foundation reinforce by geomatrasses, filled with stone material. The optimal solution in this case is the lightweight embankments with a core of expanded polystyrene (foam). This solution is effective to reduce the total embankment deformations on soft soils and to limit the lateral displacement.

© 2018 A. Zaytsev, A. Petryaev, L. Černiauskaite
Peer-review under responsibility of the Kaunas University of Technology, Panevėžys Faculty of Technologies and Business

Keywords: subgrade; weak foundations, highway embankments, geodrain, geomatresses, geofoam

1. Introduction

The construction of the subgrade on weak soils leads to the increasing pore pressure very often. The effective stresses stay low as the result of this undrained behavior and contractors use special stages for the compaction and relaxation to improve the earth-works. The exceed pore pressure dissipates during of the compaction and the soil shear strength is in-creased up to values which allow the construction resumption. The main objective of this work is to estimate the workability of the project technical decisions for the reconstruction of the approach highway embankment and typical cross-sections for the new-constructed railway embankment.

© 2018 A. Zaytsev, A. Petryaev, L. Černiauskaite. **B427H0038** . Ed*Learning*.it
Peer-review under responsibility of the Kaunas University of Technology, Panevėžys Faculty of Technologies and Business

The geology of the area was studied up to the depth of 30 m: modern overburden soils, lacustrine deposits and speckled (bluish) clay. The lacustrine deposits are presented the clays silts with fluid or fluid-plastic consistency of coarse lens from saturat-ed sands. The silts are filled with the depth from 1 to 24 m.

As it is seen from the reference, the stabilization of the embankments on weak foundation in the highway and railway industries is provided by using berms, which requires much costs and big volumes of draining soils: sand, gravel, ground columns, cement mixing etc. These technologies where described numerous papers include Yakovleva, T. G., Ivanov, D. I. [12], Almeida [1], Arulrajah et al [2], Aslamet al [3], Hayashi et al [4], Petryaev et al, Petriaev [5, 6], Svatovskaya et al [7], Vinogradov et al [8, 9], Zaytsev [10, 11]. Some alternative engineering decisions exist at present time. Different geosynthetics: geogrids, geomatrasses, geodrains, geofoams are used in the major of them.

2. Modelling and monitoring of highway embankment

2.1. The prototypes, models and the characterization of the possible deformations

The distinctive cross-section was chosen for the estimations of the project decision. The characteristic cross-section was chosen to evaluate the design solution. This reconstructed structure consists of the jointed structure with the existing old-term embankment, which had the height of 2.4 m and the slope 1:2.3 for the left slope and 1:5 for the right slope, and a new-constructed part of embankments which is elevated and filled under the existing embankment. The height of the projected new-constructed highway prototype embankment is 6 m and the slope 1:1.5 (Fig. 1, *a*). The prototype embankment was filled of the drain fine sand, this embankment had the height of 6 m too (Fig. 1, *b*). The first models of the embankments were prepared without any reinforcements to estimate the behavior of the embankments on these weak foundations.

Several hypotheses are preliminarily suggested to solve the problem of embankment stability. There are deformations – loss of slope stability; settlements of embankments as the results of the foundation soil uplift and settlement caused by the compression of the embankment and foundation soils. The scale factor N = 75 was chosen for the physical centrifuge modelling.

The weak clayed soil laid at the foundation, the main characteristic of this soil – clayed silt. The laboratory tests were done for the estimation of the main physical and mechanical parameters. The clayed silt of the fluid-plastic consistency was defined as the results of the lab tests. The soil moisture was increased up to 50 % during the preparation of the foundation model.

The foundation soil was compacted layer by layer until it reached the density value which is equal to maximum value of the dry unit weight of the 1.67 g/sm^3 with the moisture 50 %.

The filling of embankment soil (existing and newly constructed in prototype terms) was provided after the foundation preparation. Some probes of the moisture and density of soil were taken for the estimation of the physical characteristics of the model soil.

The laser tape-measures (on three cross-sections) were provided for each model after the finishing of the model preparation and the same measurements were provided after centrifuge modeling.

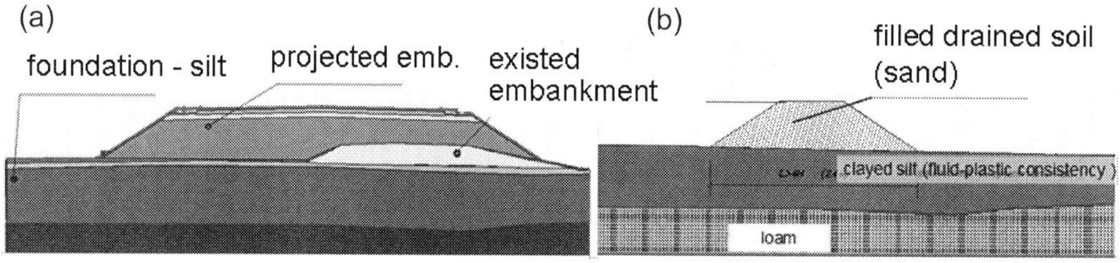

Fig. 1. (a) Scheme of the structure highway ; (b) embankments (scale factor for the modeling N = 75)

The results of physical modelling show, that embankment has stability, but values of displacements exceed the admissible valuation. It was defined that soil moisture of the next experiments must be increased up to 50 %

The settlements of the main top of the model at the prototype scale were equal to 0.45 m.

2.2. The results of modelling and design – construction stage of reinforced highway embankment

The next model was prepared as the embankment model reinforced by vertical sand drains and geomatrasses, which were filled with gravel sand (Fig. 2). The foundation soil was compacted layer by layer until it reached the density value equal to maximum value of the dry unit weight of 1.67 g/sm3 at the moisture of 50 % and then at the foundation the holes were drilled for the sand drains with depth of 10.5 sm. These holes were filled with sand and compaction was produced after moistening.

The model of embankments was prepared after the foundation and elements – models of the geomattress jointed together and with foundation. These geomattress models were filled with coarse sand by size fraction of 3 mm and compacted. Dimensions of the models equal to scale N = 75.

For the physical modelling of the dual-dual high-way car load on the top ground of the subgrade was chosen of the distributed static load 45 kPa. The elements of the static load where made from ten separate pasteboard sections with filling – sand lead mix filling.

The settlements of the main top of the model at the prototype scale were equal to 0.075 m.

The results of the modelling allowed to recommend these decisions for the prototype and to confirm the stabilization of the foundation to the state of admissible flat settlements and to start the construction.

To increase the foundation consolidation, the vertical drains were placed on the section of a new construction site for the highway embankment. The installation of these drains led to increase of the water drainage factor and to the decrease of hydrostatic and pore pressure. For the prototype conditions: the distance between drain centres 0.8 m, depth of installation from 8 to 12 m.

Intensive part of consolidation stopped six months later after the full filling and compaction of embankment soil.

In the triangular scheme of the placing, the distance between the drains was 0.80 m, and between the rows the distance was 0.70 m. Some special technical solutions were used: geomattress were presented to provide a foundation bearing capacity; the geofoam blocks were placed above the existent embankment on the right side to light the embankment.

This structure of geofoam blocks was covered by geotextile. The geomembrane, with the thickness of 1 mm, was laid under the pavement to protect the blocks from oil. The geofoam blocks were placed on the subgrade after cutting the existing pavement on the depth of 1.2 m and these blocks did not add load on the foundation (Fig. 2, *b*).

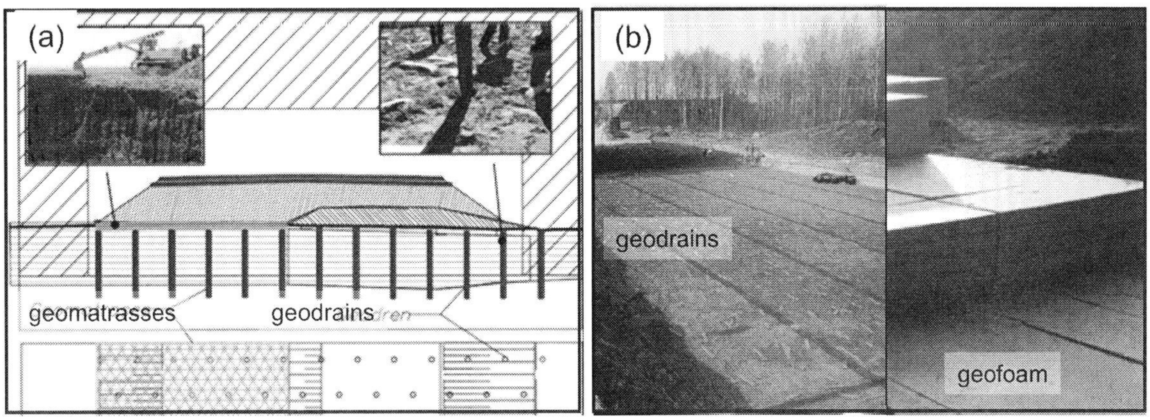

Fig. 2. a) The scheme of the highway emb. reinforced by sand drains and geomatrasses; b) Geodrains and geofoam blocks

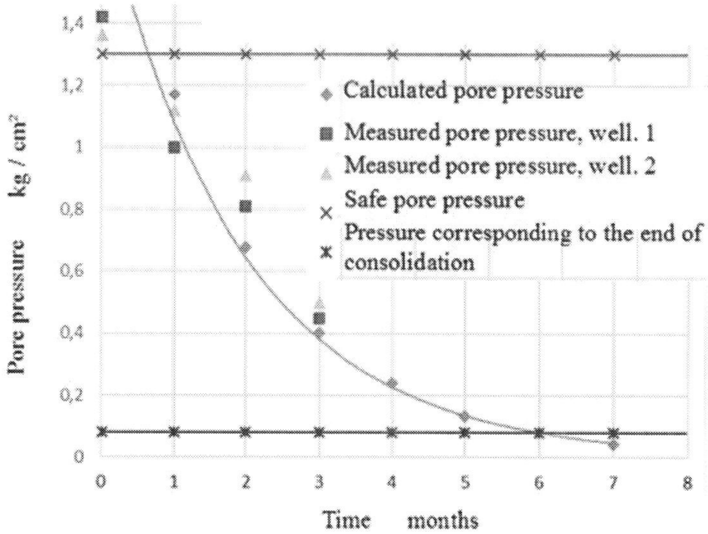

Fig. 3. The monitoring of the pore pressure on the construction site

In connection with the insufficient bearing capacity of the subgrade foundation, when a new embankment is built on weak soils, it is necessary to fill the layers of the embankment layer in stages, using the method of preliminary consolidation.

When filling a new embankment on weak soil foundations it is possible to exceed the permissible values of pore pressure, which can lead to the erosion of the foundation soil of the embankment. To prevent this, it is necessary to monitor the amount of excess pore pressure and the dimensions of the embankment.

Conclusions

Environmental sustainability of the subgrade on weak foundation soils is described through the geosynthetics application. The results of the modelling showed that stability of the embankment and stability of the foundation soil are ensured both for the variant without reinforcing and for the variant with reinforcing.

The analysis of the results of modelling and physical characteristics of soils has shown that when a weak foundation model is constructed, then the clayed soils have a soft-plastic consistency, the stability of the foundation soils is provided, and the appearance of elastic-plastic deformations does not occur.

The analysis of the comparative results for the highway embankment of the model settlements before and after reinforcing on the corresponding (among themselves) sections in the prototype scale reveals that the values of the settlements of the model with the reinforcement of the foundation by sand piles and the geomattress turned out to be considerably less than the model settlements without reinforcement. As a result of the reinforcement of the soil foundation of the reconstructed embankment, the settlements values decreased from 50 to 83 %, compared to the embankment, where the reinforcement of the foundation was not performed, which indicates a certain efficiency and for the prototype-technical solution for increasing the hardness of the weak foundation soils by the sand piles with a cover construction from geomatress filled with stone material.

The control of the road bed construction by monitoring the pore pressure allowed it to be completed without any losses of bearing capacity of the foundation.

At the future investigations this environmental sustainability effect planned to remove at the railway infrastructure, for example Kryptys Kyvishkes–Valchiunai (Fig. 4), according to Lithuanian Standards [13].

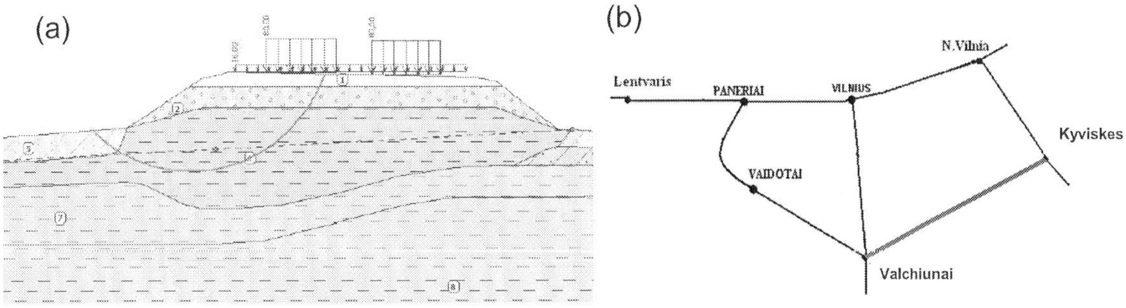

Fig. 4. a) The slope stability analysis for the section; b) Fragment of the scheme with Kyviskes-Valchiunai route

References

[1] Almeida M. S. S., Parry R. H. G. Centrifuge studies of embankment foundations strengthened with granular columns. *Proc. 3rd Geotech. Sem. Soil Imp.*, Singapore; 1985. pp. 153–156.

[2] Arulrajah A., Abdullah A., Bo M. W., Bouazza A. Ground improvement techniques for railway embankments. *Ground Improvement 162(1)* ; 2009. pp. 3–14.

[3] Aslam R., Ellis E. A. Centrifuge modelling of piled embankments. *In S.M. Springman, J. Laue & L. Seward (eds), Phys. Mod. Geotech* London: T&F.; 2010. pp. 1297–1302.

[4] Hayashi H., Nishimoto S., Sawai K. Peat ground treated by deep cement mixing with low improvement ratio. *Proceedings of the International Symposium on Engineering Practice and Performance of Soft Deposits*, IS-Osaka; 2004. pp. 229–234.

[5] Petriaev A. V. Thawing railroad bed and methods of its reinforcing. *Computer Methods and Recent Advances in Geomechanics. In: Proceedings of the 14th International Conference of International Association For Computer Methods and Recent Advances in Geomechanics*. Kyoto, Japan; 2015. 265 p.

[6] Petryaev A. A., Ganchits V. V. The use of lightweight embankments in the construction and reconstruction of the subgrade on weak grounds. Proceedings of international scientific conference Transportation system infrastructure problems. Saint-Petersburg, Russia, September 30th - October 1st.; 2015. pp. 70–73.

[7] Svatovskaya L. B., Baidarashvily M. M., Sakharova A. S., Petryaev A. V. Using of geomembrane is in ecoprotective aims. *Transport construction* N 8; 2012. pp. 26–28.

[8] Vinogradov V. V., Yakovleva T. G., Frolovsky Y. K., Zaitsev, A. A. Centrifugal modelling of the railway embankments with reinforcement by the various reinforced earth constructions. *Proceedings of the International Conference on Physical Modelling in Geotechnics*, St. John's, Newfoundland, Canada, 10–12 July; 2002. pp. 987–991.

[9] Vinogradov V. V., Yakovleva T. G., Frolovsky Y. K., Zaitsev, A. A. Physical modelling of railway embankments on peat foundations. *Proceedings of the 6th International Conference on Physical Modelling in Geotechnics – 6th ICPMGE'06*, Hong Kong, 4–6 August; 2006. pp. 591–595.

[10] Zaytsev, A.A. Modeling of stability of the railway embankments on the weak peat foundations, geotechnical problems on sedimentary soils in seismic region. Proceedings of the third international Central-Asian geotechnical symposium, Tadzhikistan, Dushanbe, 10–12 November; 2005. pp. 171–174.

[11] Zaytsev, A. A. Physical modeling embankment on peat foundation with reinforcing of the wooden piles. *Physical Modelling in Geotechnics. 8th International Conference on Physical Modelling in Geotechnics (ICPMG)*, Perth, Australia, 13–17 January; 2014.p. 877–881

[12] Yakovleva T. G., Ivanov D. I. Modelling of Stiffness and Stability of the Subgrade. Moscow Transport; 1980.

[13] Lietuvos standartas LST EN 1997-1 Eurokodas 7. Geotechninis projeltavimas. 1 dalis. Pagrindines tasukles; 2006 (in Lithuanian).

Author Index